Practical Microwave Circuits

For a complete listing of titles in the
Artech House Microwave Library,
turn to the back of this book.

Practical Microwave Circuits

Stephen A. Maas

ARTECH HOUSE
BOSTON | LONDON
artechhouse.com

Library of Congress Cataloging-in-Publication Data
A catalog record for this book is available from the U.S. Library of Congress.

British Library Cataloguing in Publication Data
A catalogue record for this book is available from the British Library.

Cover design by Adam Renvoize

ISBN 13: 978-1-60807-800-4

685 Canton Street
Norwood, MA 02062

10 9 8 7 6 5 4 3 2 1

To Julie

Contents

Preface

This book is a collection of things that I learned, largely on my own, over forty years in microwave circuit design. While almost all of the material in this book existed somewhere in some form, no single source was comprehensive and accessible. I learned most of it by experience, talking to people and digging through technical papers and books. In some cases, I just had to figure it out on my own.

Forty years ago, relatively few books on high-frequency theory and technology, beyond a few classroom electromagnetics textbooks, existed. There are many more today, but few are oriented toward the microwave-circuit specialist. Most begin with electromagnetic concepts, then move on to some aspects of circuit theory. There are two problems with this approach: first, while no one with any sense would doubt the importance of an understanding of electromagnetics in the design of high-frequency systems and components, the actual design of such components rarely uses electromagnetic concepts directly, if at all, and instead is based on circuit-theoretical concepts. Second, because of the view that high-frequency circuit analysis is fundamentally electromagnetic, useful circuit analysis techniques are not presented. The message delivered to the reader too often seems to be that high-frequency components should be viewed as electromagnetic structures or not at all. This leaves out many ideas that are essential for successful microwave circuit design. Indeed, all circuit theory is, at some level, an expression of electromagnetics; perhaps viewing microwave circuits through circuit theory is not really avoiding electromagnetics at all.

This book develops analytical techniques for understanding and designing high-frequency circuits almost entirely from a circuit point of view. Electromagnetic concepts are not avoided, but they are employed only as necessary to support circuit-theoretical ones or to describe phenomena, such as radiation and surface waves in microstrip, which inherently require an electromagnetic description. At the same time, my intention is to go much further than the conventional wisdom about microwave circuit theory, such as it is. For example, the idea that Mason's rule, applied to network graphs, is the only viable method for analyzing circuits described by S parameters is much too narrow. Methods involving direct manipulation of S matrices are more general and often just as easy—if not easier—to employ. Indeed, general graph analysis, a staple of analog circuit theory, is directly applicable to high-frequency circuits. Even Mason's rule itself is just a carryover from control-system theory. It is not specific to the RF and microwave world.

I would like to thank a couple of people. Most important is my wife, Julie, who never complains about being neglected when I steal away to my office to ruffle through piles of technical papers, peck at a keyboard, and stare at a computer screen for full days or longer. I am also grateful to the management of AWR Corporation for making available the software that I used in the design examples.

Stephen Maas
Long Beach, California

Chapter 1

Transmission Lines

In discussing microwave circuits, the logical place to begin is with transmission lines. Transmission lines are used not only for guiding a wave from one place to another, but as circuit elements and essential parts of components such as couplers and hybrids. Thus, it is essential to understand their characteristics.

1.1 Transmission Lines

1.1.1 Fundamental Relations

A transmission line is a wave-guiding structure consisting of two conductors whose cross sections, while not necessarily identical, are uniform over their lengths. A transmission line has series inductance, L, and resistance, R; additionally, it has shunt capacitance, C, and conductance, G. These quantities are distributed along the line, so they must be defined as such; that is, L is inductance per length, and so on. The line can be modeled as in Figure 1.1, which shows those distributed quantities.

Because of the distributed R, L, C, and G, the voltage and current in the line are not constant with position. From Figure 1.1, it is clear that the variation of the voltage along any increment of the line is

$$dV = -(R + j\omega L)\,I\,dz \tag{1.1}$$

where z is the distance along the line, as shown in Figure 1.1(a). Then,

$$\frac{dV}{dz} = -(R + j\omega L)I \tag{1.2}$$

Through similar reasoning, we obtain, for the current,

$$\frac{dI}{dz} = -(G + j\omega C)V \tag{1.3}$$

Differentiating (1.2) and substituting (1.3) gives

$$\frac{d^2V}{dz^2} = \gamma^2 V \tag{1.4}$$

and differentiating (1.3) and substituting (1.2) similarly gives

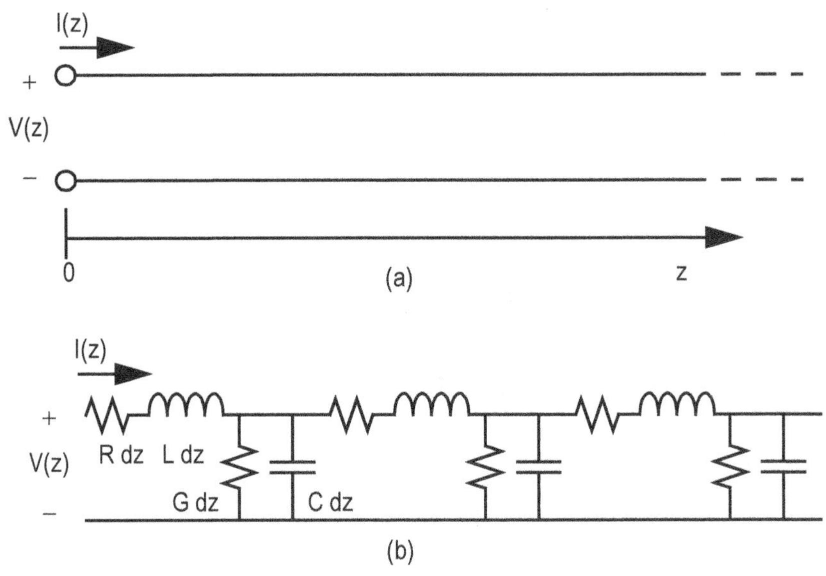

Figure 1.1 A parallel-wire transmission line (a) with the indicated voltage and current along its length. The equivalent circuit (b) consists of distributed *R*, *L*, *C*, and *G*.

$$\frac{d^2I}{dz^2} = \gamma^2 I \tag{1.5}$$

where

$$\gamma = \sqrt{(R + j\omega L)(G + j\omega C)} \tag{1.6}$$

γ is called the *propagation constant*, a complex quantity. It is often expressed as $\gamma = \alpha + j\beta$, where the real part describes attenuation of the wave and the imaginary gives the phase shift along the line. The solutions of (1.4) and (1.5) can be found by inspection[1]:

$$V(z) = V_f \exp(-\gamma z) + V_r \exp(\gamma z) \tag{1.7}$$

$$I(z) = I_f \exp(-\gamma z) - I_r \exp(\gamma z) \tag{1.8}$$

where V_f, I_f, V_r, and I_r are phasors that describe voltage and current waves propagating in the positive (V_f, I_f) or negative (V_r, I_r) z directions. By convention, the reference direction of I_r is the negative-z direction. Therefore, at any point along the line, z_1,

$$\begin{aligned} V(z_1) &= V_f(z_1) + V_r(z_1) \\ I(z_1) &= I_f(z_1) - I_r(z_1) \end{aligned} \tag{1.9}$$

For a wave propagating in one direction, say, the forward wave, we have

$$V(z) = V_f \exp(-\gamma z) = |V_f| \exp(j\phi) \exp(j\omega t - \gamma z) \tag{1.10}$$

where ϕ is an initial phase that we can conveniently assume to be zero. For an ideal, lossless line, where R, $G = 0$, $\gamma = j\omega\sqrt{LC}$,

$$V(z) = |V_f| \exp(j(\omega t - \omega\sqrt{LC} \cdot z)) \tag{1.11}$$

1. That is, inspection of the equations, a good book on differential equations, or the many electromagnetics textbooks that describe transmission lines.

Equation (1.11) describes a wave propagating in the $+z$ direction with a constant phase when $\omega t = \omega\sqrt{LC}\cdot z$. This has two implications: First, that $\beta = \omega\sqrt{LC}$ is the phase shift per length along the line, as the two terms have units of phase. Second, since a constant-phase point requires $t = \sqrt{LC}\cdot z$, the wave, which is the velocity of this constant phase point, must be $v_p = 1/\sqrt{LC}$. We call this quantity the *phase velocity.* Thus, we have

$$\beta = \frac{\omega}{v_p} \tag{1.12}$$

For a line containing no dielectric, v_p is necessarily equal to the free-space velocity of light.

1.1.2 Characteristic Impedance

Solving (1.2) and (1.3) for I, V and dividing them gives

$$\frac{V}{I} = \frac{dI}{dV}\frac{(R + j\omega L)}{(G + j\omega C)} \tag{1.13}$$

Clearly, for a wave propagating in one direction, we can substitute V_f, I_f for V, I and note that $dV/dI = V/I = V_f/I_f$, so

$$\frac{V_f}{I_f} = \sqrt{\frac{R + j\omega L}{G + j\omega C}} \tag{1.14}$$

This quantity is called the *characteristic impedance*, usually designated Z_0. For an ideal, lossless line, or one having negligible R and G, both $R = 0$ and $G = 0$, so

$$Z_0 = \sqrt{\frac{L}{C}} \tag{1.15}$$

This expression can be used to illustrate an interesting fact. We note from (1.12) and the preceding discussion that

$$L = \frac{1}{Cv_p^2} \tag{1.16}$$

Substituting this into (1.15) gives

$$Z_0 = \frac{1}{Cv_p} \tag{1.17}$$

We can similarly derive

$$Z_0 = Lv_p \tag{1.18}$$

We need only know the phase velocity and either the capacitance or inductance of a line to determine its characteristic impedance. If the line contains no dielectric, v_p is known automatically; it is simply the velocity of light, v_c. In a line containing a dielectric, we can determine the capacitances with and without the dielectric. Then, through some simple algebraic manipulations of (1.15) through (1.18), we obtain

$$Z_0 = \frac{1}{v_c}\sqrt{\frac{1}{C_1 C_k}} \tag{1.19}$$

$$v_p = v_c\sqrt{\frac{C_1}{C_k}} \tag{1.20}$$

where C_1 and C_k are the line capacitances, per length, without and with the dielectric in place, respectively. This is an important result, as it allows determination of the characteristic impedance and phase velocity of a low-loss line from only the capacitance; it is not necessary to determine the line's inductance.

1.1.3 Lossy Transmission Lines

Most practical transmission lines have nonnegligible loss, so the ideal line approximation is not valid. The losses of practical lines are low, however (otherwise, they wouldn't be practical), so we can make some appropriate approximations for such lines. We find, fortunately, that the earlier idea of a lossless line actually isn't far from the mark.

For a low-loss transmission line,

$$\begin{aligned} R &\ll \omega L \\ G &\ll \omega C \end{aligned} \tag{1.21}$$

With this approximation,

$$Z_0 = \sqrt{\frac{R + j\omega L}{G + j\omega C}} \approx \sqrt{\frac{L}{C}} \tag{1.22}$$

the same result as in (1.15). The expression for γ is found to be

$$\gamma = \sqrt{(R + j\omega L)(G + j\omega C)} \approx \frac{1}{2}\left(\frac{R}{Z_0} + GZ_0\right) + j\omega\sqrt{LC} \tag{1.23}$$

We obtain essentially what we had for the lossless line, except that γ has picked up a small real part. Remembering that $\gamma = \alpha + j\beta$, we find α and β from comparison to (1.23). The expression for the propagating wave becomes

$$V(z) = V_f \exp(-(\alpha + j\beta)z) + V_r \exp((\alpha + j\beta)z) \tag{1.24}$$

Thus, the effect of a small but nonnegligible R and G is simply to cause an exponentially decreasing amplitude of the wave as it travels down the line.

1.1.4 Conditions at the Ends of Transmission Lines

1.1.4.1 Reflection Coefficient

Suppose we have a wave, happily propagating down a transmission line, and it reaches a termination at the far end. What happens?

Figure 1.2 shows the situation we face. For the forward-propagating wave on the line, we know that $V_f = Z_0 I_f$. For the termination, however, unless $Z_L = Z_0$, the V, I relations on the line at z_1 do not satisfy those of the termination. The only solution is to have a reverse-propagating wave. Then, the conditions are more complex:

$$\begin{aligned} V(z_1) &= V_f(z_1) + V_r(z_1) \\ I(z_1) &= I_f(z_1) - I_r(z_1) \\ \frac{V(z_1)}{I(z_1)} &= Z_L \end{aligned} \tag{1.25}$$

and, as usual,

$$\frac{V_f}{I_f} = \frac{V_r}{I_r} = Z_0 \tag{1.26}$$

Substituting (1.26) into (1.25) eventually gives

$$V_r(z_1) = \frac{Z_L - Z_0}{Z_L + Z_0} V_f(z_1) \tag{1.27}$$

or

$$I_r(z_1) = \frac{Z_L - Z_0}{Z_L + Z_0} I_f(z_1) \tag{1.28}$$

We define the *load reflection coefficient*, $\Gamma(z_1)$, as

$$\Gamma(z_1) = \frac{V_r(z_1)}{V_f(z_1)} = \frac{I_r(z_1)}{I_f(z_1)} = \frac{Z_L - Z_0}{Z_L + Z_0} \tag{1.29}$$

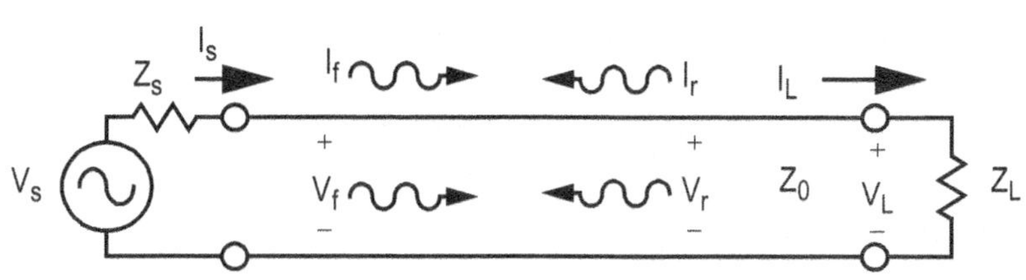

Figure 1.2 Transmission line with voltage and current quantities defined.

Since it is a function of z, the reflection coefficient can be defined for any point on the line. In general,

$$\Gamma(z) = \frac{V_r(z)}{V_f(z)} = \frac{I_r(z)}{I_f(z)} \tag{1.30}$$

and, given the reflection coefficient at some point z_1, it is straightforward to determine the reflection coefficient at some other point, z_2:

$$\begin{aligned} V_f(z_2) &= V_f(z_1)\exp(-j\gamma(z_2 - z_1)) \\ V_r(z_2) &= V_r(z_1)\exp(j\gamma(z_2 - z_1)) \end{aligned} \tag{1.31}$$

Dividing the above expressions gives

$$\Gamma(z_2) = \frac{V_r(z_2)}{V_f(z_2)} = \Gamma(z_1)\exp(j2\gamma(z_2 - z_1)) \tag{1.32}$$

1.1.4.2 Return Loss and VSWR

The return loss, L_r, of a load is simply the magnitude of the reflection coefficient expressed in decibels, usually as a positive quantity:

$$L_r = |10\log(|\Gamma|^2)| \tag{1.33}$$

The reflected wave on a transmission line interferes with the forward-propagating wave in such a way as to create voltage and current maxima and minima along the line. The voltage and current variations are fixed in position and therefore are called a *standing wave*. The voltage standing wave ratio, VSWR, is the ratio of the maximum to the minimum value of the standing wave. The maximum voltage, V_{max}, is

$$V_{max} = |V_f| + |V_r| = |V_f|(1 + |\Gamma|) \tag{1.34}$$

and the minimum, V_{min}, is

$$V_{min} = |V_f| - |V_r| = |V_f|(1 - |\Gamma|) \tag{1.35}$$

The VSWR is the ratio of these:

$$\text{VSWR} = \frac{V_{\max}}{V_{\min}} = \frac{1 + |\Gamma|}{1 - |\Gamma|} \tag{1.36}$$

The current minima and maxima do not occur at the same points as the voltage minima and maxima, but the current standing wave ratio has the same value as the VSWR.

1.1.4.3 Transmission Coefficient

Another occasionally useful concept is that of a *transmission coefficient.* Suppose we have two cascaded transmission lines with differing characteristic impedances, Z_{01} and Z_{02}. We assume that both lines are terminated in their characteristic impedances, as illustrated in Figure 1.3. A wave incident on the discontinuity at the junction is partially transmitted and partially reflected. Incrementally to the left side of the discontinuity, the voltage is

$$V_1 = V_{f1} + V_{r1} = (1 + \Gamma)V_{f1} \tag{1.37}$$

and to the right it is

$$V_2 = V_{f2} \tag{1.38}$$

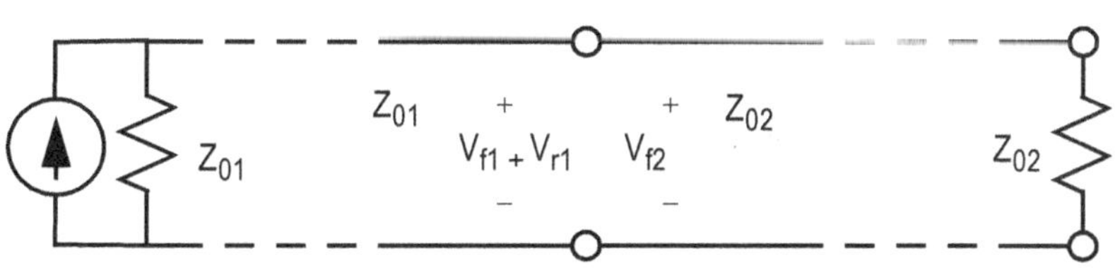

Figure 1.3 A wave incident on the junction between two transmission lines is partially reflected and partially transmitted. The ratio of the transmitted wave, V_{f2}, to the incident one, V_{f1}, is the transmission coefficient.

where the subscripts, 1, 2 refer to the left and right sides of the junction, respectively. Those voltages must be equal; equating (1.37) and (1.38) gives the transmission coefficient, T:

$$T = \frac{V_{f2}}{V_{f1}} = 1 + \Gamma \tag{1.39}$$

From (1.39) and the definition of the reflection coefficient, (1.27), with $Z_L = Z_{02}$,

$$T = \frac{2Z_{02}}{Z_{01} + Z_{02}} \tag{1.40}$$

1.1.4.4 Equivalent Circuits

In microwave circuit analysis, it is helpful to have equivalent circuits of the input and output ends of a transmission line. These are shown in Figure 1.4. At the input, the source excites the forward-propagating wave, $V_f(0)$. In the absence of a reverse wave, the line appears as a a load equal to Z_0. If a reverse wave exists, it is simply reflected by the source impedance and added to $V_f(0)$. Thus,

$$V_f(0) = V_s \frac{Z_0}{Z_s + Z_0} + \Gamma_s V_r(0) \tag{1.41}$$

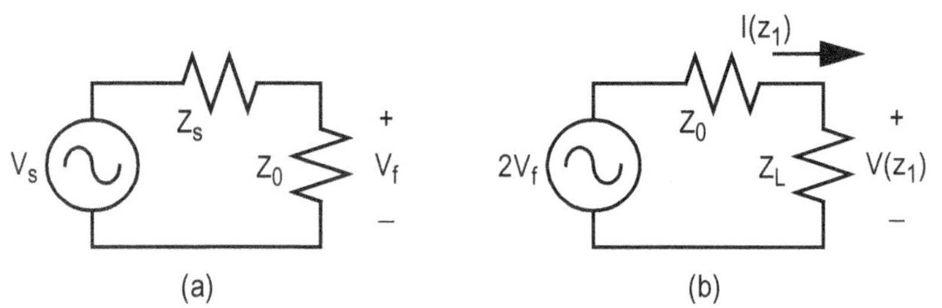

Figure 1.4 Transmission line equivalent circuits. At the input end (a) the initial V_f wave is found from a simple voltage-divider equivalent. To this one must add the component of V_r that is reflected from the source. At the output end (b), the load voltage and current are found from V_f as shown.

where Γ_s is the source reflection coefficient. At the end of the line, the situation is a bit more complicated, as we must find the voltage of the equivalent source. The voltage and current are

$$\begin{aligned} V(z_1) &= V_f(z_1) + V_r(z_1) = V_f(z_1)(1 + \Gamma_L) \\ I(z_1) &= I_f(z_1) - I_r(z_1) = I_f(z_1)(1 - \Gamma_L) = \frac{V_f(z_1)}{Z_0}(1 - \Gamma_L) \end{aligned} \tag{1.42}$$

where Γ_L is the load reflection coefficient. The equivalent source voltage, V_{eq}, is

$$V_{eq} = V(z_1) + Z_0 I(z_1) \tag{1.43}$$

and substituting (1.42) gives $V_{eq} = 2V_f(z_1)$. The reflected wave, as before, is

$$V_r(z_1) = \Gamma_L V_f(z_1) \tag{1.44}$$

Although we have considered only the sinusoidal steady state, the waves on transmission lines can have any functional form. The equivalent circuits of the line ends, in Figure 1.4, are valid in those cases as well and, in fact, may be essential for determining the behavior of nonsinusoidal waves on transmission lines.

1.1.5 Matrix Relationships

To use transmission lines in microwave circuit analysis, it is necessary to have some kind of two-port (or, as we shall see in Section 2.2, multiport) relation for them. In doing this, we must use the hyperbolic-function solutions to (1.2) and (1.3). These have the form

$$\begin{aligned} V &= K_1 \cosh(\gamma z) + K_2 \sinh(\gamma z) \\ I &= K_3 \cosh(\gamma z) + K_4 \sinh(\gamma z) \end{aligned} \tag{1.45}$$

The K_n coefficients can be found by applying the boundary conditions,

$$
\begin{aligned}
V &= V_1\big|_{z=0} & V &= V_2\big|_{z=z_1} \\
I &= I_1\big|_{z=0} & I &= -I_2\big|_{z=z_1}
\end{aligned}
\tag{1.46}
$$

where I_1, I_2 have the sign conventions used in conventional two-port notation. Substituting these quantities into (1.45), with a little help from (1.2) and (1.3), we obtain

$$
\begin{aligned}
V_2 &= V_1 \cosh(\gamma z_1) - I_1 Z_0 \sinh(\gamma z_1) \\
I_2 &= \frac{V_1}{Z_0}\sinh(\gamma z_1) - I_1 \cosh(\gamma z_1)
\end{aligned}
\tag{1.47}
$$

This can be expressed in matrix form as

$$
\begin{bmatrix} V_2 \\ I_2 \end{bmatrix} = \begin{bmatrix} \cosh(\gamma z_1) & Z_0 \sinh(\gamma z_1) \\ \dfrac{1}{Z_0}\sinh(\gamma z_1) & \cosh(\gamma z_1) \end{bmatrix} \begin{bmatrix} V_1 \\ -I_1 \end{bmatrix}
\tag{1.48}
$$

which is an *ABCD* or *transmission* matrix, albeit in the reverse of the usual form. Inverting this gives the more common *ABCD* form:

$$
\begin{bmatrix} V_1 \\ I_1 \end{bmatrix} = \begin{bmatrix} \cosh(\gamma z_1) & Z_0 \sinh(\gamma z_1) \\ \dfrac{1}{Z_0}\sinh(\gamma z_1) & \cosh(\gamma z_1) \end{bmatrix} \begin{bmatrix} V_2 \\ -I_2 \end{bmatrix}
\tag{1.49}
$$

which is identical to (1.48). This should not be a surprise, as reversing the ends of a transmission line results in the same transmission line. In any case, (1.49) is easily converted to an admittance (Y) matrix:

$$
\begin{bmatrix} I_1 \\ I_2 \end{bmatrix} = \frac{1}{Z_0}\begin{bmatrix} \coth(\gamma z_1) & -\operatorname{csch}(\gamma z_1) \\ -\operatorname{csch}(\gamma z_1) & \coth(\gamma z_1) \end{bmatrix} \begin{bmatrix} V_1 \\ V_2 \end{bmatrix}
\tag{1.50}
$$

For a lossless line, $\gamma z_1 = j\beta l$, where l is the line length, and the Y matrix has the more common form,

$$\mathbf{Y} = \frac{j}{Z_0}\begin{bmatrix} -\cot(\beta l) & \csc(\beta l) \\ \csc(\beta l) & -\cot(\beta l) \end{bmatrix} = \frac{j}{Z_0 \sin(\beta l)}\begin{bmatrix} -\cos(\beta l) & 1 \\ 1 & -\cos(\beta l) \end{bmatrix} \qquad (1.51)$$

Other forms, such as impedance matrices, can be found through the use of the standard conversion formulae. See [1.1] for a comprehensive table of those conversions.

1.1.6 Input Impedance and Power Transfer

The input impedance of a terminated transmission line is a useful thing to know. In effect, we are now locating the load at $z = 0$ and Z_{in} is the value of the input impedance at $z = -l$. Using (1.49) and noting that the terminating impedance Z_L is $V_2 / (-I_2)$, we obtain easily

$$Z_{in} = \frac{V_1}{I_1} = Z_0\frac{Z_L \cosh(\gamma l) + Z_0 \sinh(\gamma l)}{Z_0 \cosh(\gamma l) + Z_L \sinh(\gamma l)} \qquad (1.52)$$

or, for a lossless line,

$$Z_{in} = Z_0\frac{Z_L \cos(\beta l) + jZ_0 \sin(\beta l)}{Z_0 \cos(\beta l) + jZ_L \sin(\beta l)} \qquad (1.53)$$

Equation (1.32), with $z_1 = 0$ and $z_2 = -l$, gives the value of the input reflection coefficient:

$$\Gamma_{in} = \Gamma_L \exp(-2j\gamma l) \qquad (1.54)$$

Another useful quantity is the *transducer gain* of the system, defined as the power delivered to the load divided by the power available from the source. We assume, for simplicity, that the load is real, the source impedance is Z_0, and the line is lossless. Then, the power available from the source, P_{av}, and the power delivered to the load, P_L, are

$$P_{av} = \frac{|V_f|^2}{2Z_0}$$
$$P_L = \frac{|V_f + V_r|^2}{2Z_L} \tag{1.55}$$

so the transducer gain, G_t, is

$$G_t = \frac{P_L}{P_{av}} = \frac{|V_f + V_r|^2}{|V_f|^2}\frac{Z_0}{Z_L} = |1 + \Gamma_L|^2 \frac{Z_0}{Z_L} \tag{1.56}$$

This expression can be simplified somewhat by substituting the expression,

$$\frac{Z_L}{Z_0} = \frac{1 + \Gamma_L}{1 - \Gamma_L} \tag{1.57}$$

which is found from (1.29). We obtain

$$G_t = 1 - |\Gamma_L|^2 \tag{1.58}$$

If the transmission line is lossless, $|\Gamma_{in}| = |\Gamma_L|$. Then, the gain can be viewed as a function of the input reflection coefficient. Recognizing this point allows us to use the reflection coefficient as an alternative to gain in synthesizing passive, lossless, microwave circuits; an example is the quarter-wave transformer of Section 1.4.

Although we used the rather restrictive assumption of a real load to obtain this result, it is valid for complex loads as well. This result is not valid, however, when the source impedance does not equal Z_0. Then, the expression is much more complex. We shall examine that case in a later chapter.

1.2 Practical Considerations

Descriptions of the many types of transmission lines are well beyond the scope of this book. Better sources for that information can be found in the references [1.2–1.5]. Instead, we shall focus on only a few types of transmission lines, largely as a way to explain the characteristics of such lines and practical considerations in their use.

1.2.1 Transmission Line Types

1.2.1.1 Parallel-Wire Line

It is ironic that the type of transmission line most frequently used in textbook treatments of transmission-line theory is the least commonly used in practice. Parallel-wire line, as the name implies, consists of a pair of parallel wires separated by insulators or occasionally embedded in dielectric. It is an example of a *balanced* line, one in which neither conductor is at ground potential.

Parallel-wire line is unshielded and therefore tends to radiate unless the line is very straight, well separated from conductive obstructions, and the currents in the line are well balanced.[2] In the absence of radiation, however, its losses are fairly low, as its characteristic impedance is high (usually several hundred ohms) so the currents are relatively low, for a given power level. Its dimensions are generally large compared to other practical transmission lines.

Parallel-wire line is sometimes used to feed large, low-frequency dipole antennas. A 300-ohm "twin lead," a parallel-wire line embedded in polyethylene dielectric, was commonly used for television antenna lead-in cables until the mid-1980s.

1.2.1.2 Coaxial Line

Coaxial line (colloquially *coax*) consists of two circular conductors, one inside the other. The outer conductor is grounded and provides a shield for the inner conductor. The inner conductor is usually supported by dielectric that fills the space between the conductors, but, especially for high-power applications, it may have a helical dielectric support. Figure 1.5 shows a coaxial line.

Coaxial lines can be flexible, semirigid, or rigid. Flexible lines have a braided wire outer conductor, a solid or stranded inner conductor, and a flexible dielectric (usually foamed polyethylene) between them. The outer conductor has a plastic covering. Flexible lines are used to interconnect equipment and as antenna lead-in cables. Semirigid cable has solid outer and inner conductors and soft dielectric insulation. The conductors are composed of a soft metal, usually copper, often silver or gold plated. These cables are designed to be bent to a desired shape but not for repeated flex-

2. Conductive surfaces in the vicinity of the line can unbalance the currents and create multiple modes along it. We discuss balanced and unbalanced structures more fully in Chapter 2.

ing. They are most useful for permanent interconnections between components in a larger system. Rigid coax is not designed to be formed; it is most often used for fixed, high-power applications, such as the feedline to a radio broadcast antenna.

A problem in a coaxial line is its support of spurious modes at high frequencies. Spurious modes are field structures in the line that differ from the simple, dominant TEM mode (Section 1.2.2.1) and have different characteristics. A coaxial line must be sized to prevent such modes; it invariably must decrease in diameter as frequency increases. The reduction in diameter, combined with the natural increase in loss as frequency increases, causes coax loss to increase dramatically at high frequencies. Although "mode-free" coaxial lines and connectors are available for millimeter wavelengths, their losses are disturbingly high.

The characteristic impedance of a coaxial line, when the gap between the conductors is filled with a dielectric, is

$$Z_0 = \frac{138}{\sqrt{\varepsilon_r}}\left(\log\left(\frac{D}{d}\right)\right) \tag{1.59}$$

where ε_r is the effective dielectric constant and the logarithm is base 10. The phase velocity is simply $v_c / \sqrt{\varepsilon_r}$.

1.2.1.3 Planar Transmission Structures

The most commonly used transmission structures in modern RF and microwave engineering are strip transmission lines. These lines consist of one or more flat metal strips, supported by a dielectric, and a ground plane. As such, they are unbalanced lines.

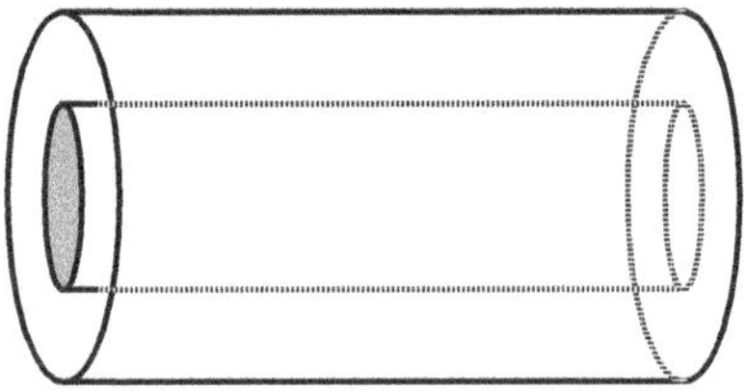

Figure 1.5 A coaxial line consists of cylindrical inner and outer conductors having a common axis.

A wide variety of planar transmission lines are in common use. The most common is microstrip, consisting of a flat metal strip supported by a dielectric slab, with a metallic ground plane on the opposite side. Several types of planar transmission-line structures are shown in Table 1.1. These are examples of *nonhomogeneous* transmission lines. A homogenous line is one in which the fields are entirely contained within a single dielectric medium or air. If the structure includes a dielectric, as in virtually all strip transmission media, and the fields are only partially contained in that dielectric, it is nonhomogeneous. Virtually all structures that are only partially filled with a dielectric are nonhomogeneous. The distinction is important, as the phase velocity in a nonhomogeneous line is not obvious, as it is in a homogeneous one. It is also important to note that nonhomogeneous lines do not support a true TEM mode. We shall examine this further in Section 1.2.2.1.

1.2.2 Properties

1.2.2.1 TEM Modes, Group Velocity, and the Quasi-TEM Approximation

So far we have implicitly assumed that all transmission lines support a transverse electromagnetic (TEM) mode. In a TEM mode, the electric and magnetic fields, like those of a wave in free space, are perpendicular to the direction of propagation; there are no longitudinal field components. That assumption is strictly true only for ideal, lossless, homogeneous transmission lines. In reality, lines have longitudinal field components, because of either loss or a nonhomogeneous dielectric. This causes the phase velocity to vary with frequency; specifically, the phase-shift per meter, β (1.12), is not proportional to ω.

We can define a second quantity, group velocity, v_g, which is in many ways more useful. It is the velocity of the *envelope* of a waveform on the transmission line, and as such is the velocity of information propagation, not simply that of a constant-phase point. While, in some transmission structures, v_p can be greater than the speed of light, v_g is always less than that quantity.

To understand group velocity, consider a wave consisting of two sinusoids closely spaced in frequency, propagating on the line in the forward direction:

$$\begin{aligned} V_f &= \sin((\omega_0 + d\omega)t - (\beta_0 + d\beta)z) \\ &+ \sin((\omega_0 - d\omega)t - (\beta_0 - d\beta)z) \end{aligned} \tag{1.60}$$

Table 1.1 Planar Transmission-Line Structures

Type	Structure	Notes
Microstrip	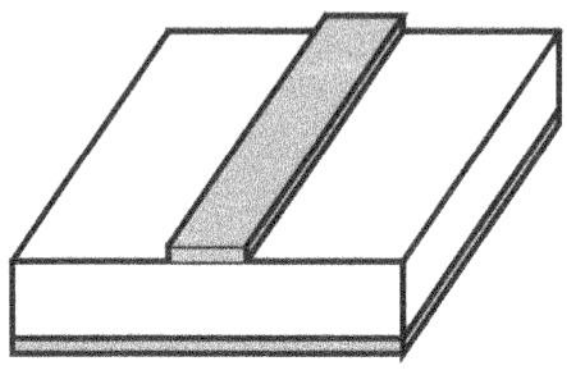	The most frequently used type of transmission line. Moderately dispersive at high frequencies.
Coplanar Waveguide (CPW)	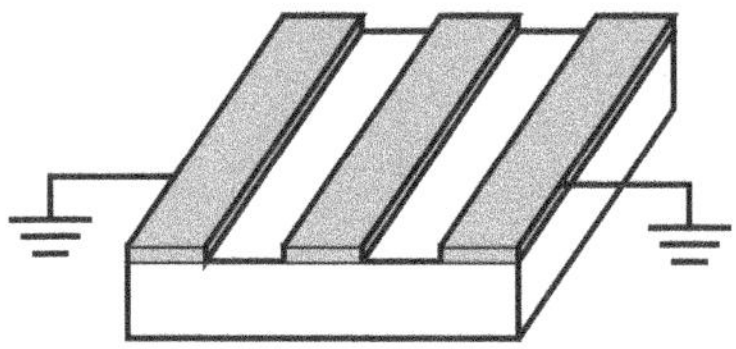	Somewhat lossier and more dispersive than microstrip, but minimizes the parasitic inductance of ground connections.
Stripline	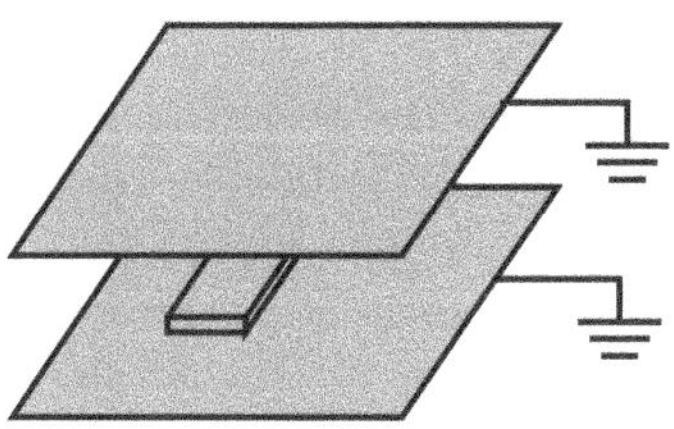	Best for passive components. Low loss, TEM, good transition to coax.
Suspended-Substrate Stripline	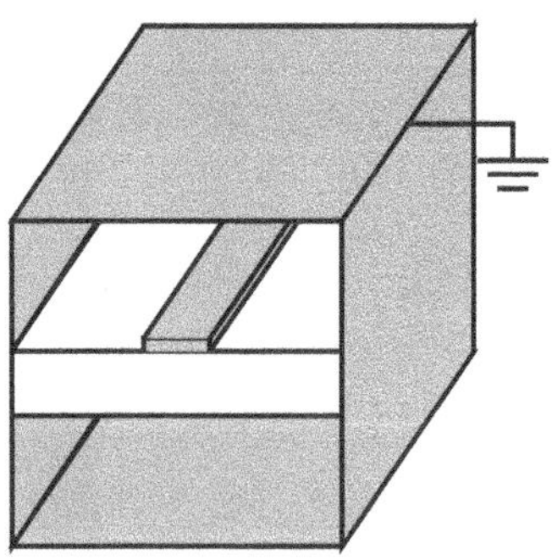	Similar to stripline, but easier to fabricate in many types of circuits. Low loss, low effective dielectric constant, good transition to coax.

where ω_0 is a central value and $d\omega$ is the deviation of each wave's frequency from that value. We have included an increment in β, $d\beta$, in the expectation that it will vary with ω. This can be expressed as

$$V_f = 2\cos(d\omega t - d\beta z)\sin(\omega t - \beta z) \tag{1.61}$$

showing a sinusoid with a multiplicative envelope. The velocity of a point on the envelope occurs when $d\omega t = d\beta z$ so

$$v_g = \frac{d\omega}{d\beta} \tag{1.62}$$

An ideal, homogeneous, lossless transmission line supports a TEM wave. In that case $\omega / \beta = d\omega / d\beta$ and $v_p = v_g$. If losses are small, the discussion in Section 1.1.3 implies that this will still be true, to a good approximation. As losses increase, however, (1.6) indicates that the propagation constant is no longer proportional to frequency, so v_p is not constant with frequency. Then, in general, v_p and v_g are unequal.

Most common types of planar transmission lines are nonhomogeneous. Those exhibit non-TEM propagation even if ideally lossless, as longitudinal field components are necessary to match the field structures at the air-to-dielectric interface. At frequencies where the wavelength is large relative to the cross-sectional dimensions of the structure, however, longitudinal field components are small and the wave is, to a good approximation, TEM. We sometimes call this *quasi-TEM* propagation.

When $v_p \neq v_g$, either because of loss or non-TEM effects, the line is called *dispersive*. In a dispersive line, the frequency components of a nonsinusoidal wave propagate at different velocities, so the information-bearing part of the wave becomes distorted. If that wave is a rectangular pulse, for example, it may become significantly smoothed, or "dispersed," as it travels along the line.

In quasi-TEM lines, dispersion usually is modeled by empirical corrections to the phase velocity. This is accomplished by modifying the effective dielectric constant (which we define precisely in Section 1.2.2.4) as a function of frequency and line dimensions. An example of such an expression, for microstrip, can be found in [1.6].

1.2.2.2 Quasistatic Analysis

The characteristic impedance and phase velocities of only a very few types of transmission lines can be found analytically. Virtually all of these are homogeneous lines or have a simple nonhomogeneous geometry. Examples are coaxial line that is completely or partially filled by a cylindrical dielectric and homogenous stripline. A dynamic, electromagnetic analysis of a strip transmission line such as microstrip is a difficult task, but it can be made much simpler under the quasi-TEM assumption.

In low-loss, quasi-TEM lines, a *quasistatic* analysis is often used. This means that the line characteristics are determined from a dc analysis of static capacitances, and the results are assumed to be valid at RF and microwave frequencies. The assumption is valid when the line's cross-sectional dimensions are small compared to a wavelength.

The analysis involves finding the charge distribution on the strip in the cross-sectional dimension with a potential (typically 1V) applied to it. The capacitance is then determined after integrating the charge. A variety of methods can be used to determine the charge distribution; the method of moments is a favored one [1.7, 1.8]. Capacitance is determined first with the dielectric in place, and second with the dielectric absent. Equations (1.19) and (1.20) are then used to calculate Z_0 and v_p.

1.2.2.3 Loss

In virtually all types of transmission lines, power dissipation in the series resistance, R, is the dominant loss mechanism. Shunt conduction due to dielectric loss is small, in part because dielectric bulk conductivities are invariably negligible. Many dielectrics, however, have molecular resonances that contribute to loss, and phenomena such as water absorption (from the atmosphere or in fabrication) alter their losses and dielectric constants. These effects contribute to line loss, although that contribution is usually small compared to the loss contributed by the metal conductors.

At high frequencies, current does not penetrate the conductor deeply and instead remains close to its surface. This phenomenon is called *skin effect*. The current density varies exponentially with depth into the conductor; the skin depth, δ, the point at which it decreases from its surface value by a factor of exp(−1), is

$$\delta = \frac{1}{\sqrt{\pi f \mu \sigma}} \tag{1.63}$$

where μ is the magnetic permeability of the metal, σ is the bulk conductivity, and f is frequency. As an example, at 6 GHz the skin depth in a good conductor, such as copper or silver, is approximately 1 μm.

At high frequencies, the current remains in a narrow sheet near the conductor's surface. For this reason, we usually view a conductor's resistivity at high frequencies as a *sheet resistance*, the resistance of the sheet of metal that contains the current. If the thickness of a metal is more than a few skin depths, its theoretical sheet resistance is that of the conductor surface to a depth δ. In practice, however, the resistance is considerably greater than that value, as the surface of the conductor is not smooth; it is invariably rough and porous, at a microscopic level, and consists of randomly bonded grains, small crystals of the metal. Additionally, fabrication techniques, such as the etching of strip transmission lines, leave rough edges that are precisely where the current density is greatest. These all make the resistance greater than would be obtained from a naive use of δ and handbook values of metal conductivity. One workable solution to this dilemma is to use an effective conductivity approximately half that of the bulk material. This generally agrees better with measured losses.

In a strip transmission line, where the current density is not uniform, one cannot determine the series resistance per length simply as the sheet resistance multiplied by the line's width and use that quantity in (1.23). The loss is determined by an integration of the squared current distribution times the sheet resistance. This is usually a fairly complex calculation. As a result, empirical expressions have been developed to describe loss in strip transmission lines.

The behavior of loss as a function of frequency and characteristic impedance is a matter of considerable interest. It is important to note that transmission line loss per meter varies approximately as the square root of frequency, as indicated by (1.23) and (1.63). This means, however, that the loss per wavelength *decreases* approximately as the square root of frequency. Thus, the loss in transmission-line components such as couplers, hybrids, and power dividers, which are designed on the basis of a fraction of a wavelength, actually decreases as frequency increases. That decrease is observed over a broad frequency range, but at some point, phenomena such as non-TEM effects come into play, and the dimensions of the surface roughness become a significant fraction of a wavelength. Then, the loss begins to increase with frequency.

Loss varies in a more complex manner with characteristic impedance. For low impedances, the current is relatively great, for a given power, so losses are high and decrease as impedance increases. Conversely, for high impedances, geometries (e.g., strip width in microstrip or center-conductor diameter in coax) are small, so the resistance is large, again implying high loss. Thus, somewhere between these two extremes there should be a minimum. That minimum depends largely on how the problem is normalized; for example, in coax, if the outer conductor's diameter remains constant and impedance is varied by changing the inner conductor's diameter, the optimum impedance, for lowest loss, is around 70 ohms. (This is the reason for the standardization of 70 ohms for coaxial lines in the cable-television industry.) Conversely, in microstrip, that optimum usually occurs at very low impedance values, which are unlikely to be used, so, for all practical purposes, microstrip loss increases with characteristic impedance.

1.2.2.4 Nonhomogeneous Lines

In homogeneous lines, the phase velocity is simply

$$v_p = \frac{v_c}{\sqrt{\varepsilon_r}} \tag{1.64}$$

where ε_r is the relative dielectric constant of the material in which the fields are contained. This is an obvious result, the same as for an unguided wave in a dielectric.

Most strip transmission lines are nonhomogeneous. The phase velocity in such lines is trickier to determine. There are a number of ways to calculate the characteristic impedance and phase velocity in nonhomogeneous lines; one is described in Section 1.2.2.2. It is traditional to describe the phase velocity via the *effective dielectric constant,* ε_{eff}, defined from (1.64) as

$$\varepsilon_{eff} = \left(\frac{v_c}{v_p}\right)^2 \tag{1.65}$$

This quantity is based solely on non-TEM dispersive effects; dispersion due to loss is comparatively small.

Most practical nonhomogeneous lines are not strongly dispersive. As an example, ε_{eff} in a 50-ohm microstrip line on a 0.635-mm-thick alumina substrate varies only 6% between 0 and 12 GHz. There is a small change in the characteristic impedance as well, less than 1 ohm.

Dispersion tends to be worse in transmission media where substantial amounts of field energy exist both inside and outside of the dielectric. In microstrip, most field energy remains inside the dielectric; in contrast, in coplanar waveguide (CPW), much of the field is outside, and thus CPW tends to be more dispersive than microstrip. Conversely, for single lines in suspended-substrate stripline, the dielectric is usually just a thin support, so most of the fields are in air, and its dispersion is minimal. (For coupled lines, however, which we discuss in Chapter 2, the situation is more complicated.) Nonhomogeneous transmission media using high dielectric constants invariably have worse dispersion, as well.

1.3 Application: RC Transmission Line

An RC transmission line consists of only series resistance and shunt capacitance. Many kinds of structures in modern integrated circuits, such as gate-to-channel junctions of field-effect transistors (FETs) and interconnections in silicon integrated circuits, are well modeled by them. In such applications, the series resistance is much greater than the inductive reactance and the shunt conductance is negligible. Clearly, the low-loss assumption is not valid, and we must employ the general line equations, (1.4) through (1.8) and, perhaps, (1.47).

Under these conditions, $\omega L \ll R$ and $G \approx 0$, so the propagation constant (1.6) and characteristic impedance (1.14) become

$$\gamma = \sqrt{j\omega RC} \tag{1.66}$$

$$Z_0 = \sqrt{\frac{R}{j\omega C}} \tag{1.67}$$

The imaginary element inside the radical is easily resolved:

$$\begin{aligned} \sqrt{j} &= \left(\exp\left(j\frac{\pi}{2}\right)\right)^{1/2} = \exp\left(j\frac{\pi}{4}\right) = \frac{\sqrt{2}}{2}(1+j) \\ \sqrt{\frac{1}{j}} &= \left(\exp\left(-j\frac{\pi}{2}\right)\right)^{1/2} = \exp\left(-j\frac{\pi}{4}\right) = \frac{\sqrt{2}}{2}(1-j) \end{aligned} \tag{1.68}$$

Then

$$\gamma = \frac{\sqrt{2}}{2}(1+j)\sqrt{\omega RC} \tag{1.69}$$

$$Z_0 = \frac{\sqrt{2}}{2}(1-j)\sqrt{\frac{R}{\omega C}} \tag{1.70}$$

Note that the real and imaginary parts of the propagation constant are equal, and that the characteristic impedance is reactive. Nevertheless, the general transmission line equations are valid for such a structure, and the ABCD and admittance matrices are given by (1.49) and (1.50), respectively.

The RC transmission line has some interesting properties. The first is that, if the line is excited as in Figure 1.6(a), the real part of the input impedance, Z_{in}, is $R_L / 3$, where R_L is the total line resistance. Secondly, when excited as in Figure 1.6(b), the real part of the transfer impedance $Z_{21} = V_2 / I_1$ is $R_L / 2$. These are valid up to a cutoff frequency of $1 / \pi R_L C_L$. Among other applications, these relations are useful in creating equivalent circuits for field-effect transistors.

1.4 Application: Multisection Quarter-Wave Transformer

The simple, single-section quarter-wave transformer is a staple of undergraduate electromagnetics courses. It consists of a quarter-wavelength line whose characteristic impedance is selected so that, when terminated in real load, Z_L, its input impedance is some desired real value. The input impedance of a terminated line is given by (1.53). With $\beta l = \pi / 2$, we obtain the input impedance, Z_{in},

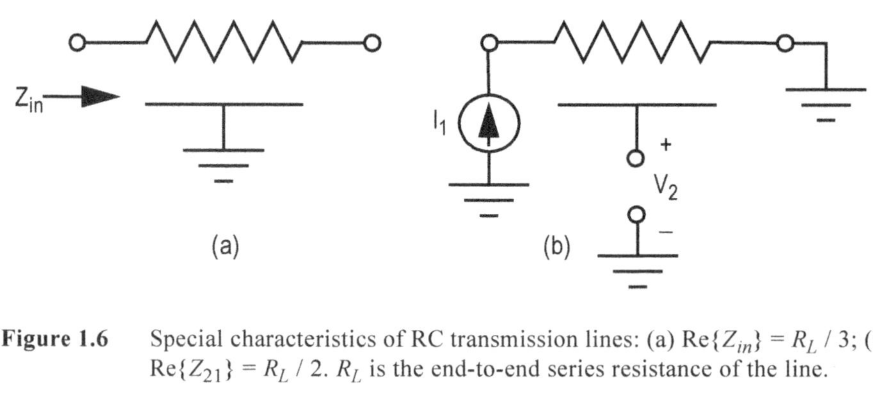

Figure 1.6 Special characteristics of RC transmission lines: (a) $\text{Re}\{Z_{in}\} = R_L / 3$; (b) $\text{Re}\{Z_{21}\} = R_L / 2$. R_L is the end-to-end series resistance of the line.

$$Z_{in} = \frac{Z_L \cos\left(\frac{\pi}{2}\right) + jZ_0 \sin\left(\frac{\pi}{2}\right)}{Z_0 \cos\left(\frac{\pi}{2}\right) + jZ_L \sin\left(\frac{\pi}{2}\right)} = \frac{Z_0^2}{Z_L} \tag{1.71}$$

The bandwidth of a single-section quarter-wave transformer is relatively narrow. Much greater bandwidth can be achieved by cascading a number of sections, as shown in Figure 1.7, each having an optimum characteristic impedance.[3] (We assume that the sections are equal in length.) One of the best ways to design such a structure is to have its input reflection coefficient approximate a Chebyshev, or equal-ripple, characteristic. This is accomplished by equating the input reflection coefficient of the cascade to a scaled version of a Chebyshev polynomial.

First, we consider the reflection coefficient of the cascade. Determining the input reflection coefficient exactly is a difficult problem, as there are multiple reflections and transmissions at every junction. However, if the impedance steps are small, the reflection coefficient at each junction is also small, and multiple, internal reflections can be neglected. The input reflection coefficient, Γ_{in}, is then

$$\Gamma_{in} \approx \Gamma_0 + \Gamma_1 \exp(-2j\theta) + \Gamma_2 \exp(-4j\theta) + \ldots + \Gamma_N \exp(-2jN\theta) \tag{1.72}$$

The Γ_n are the interface reflection coefficients,

$$\Gamma_n = \frac{Z_{n+1} - Z_n}{Z_{n+1} + Z_n} \tag{1.73}$$

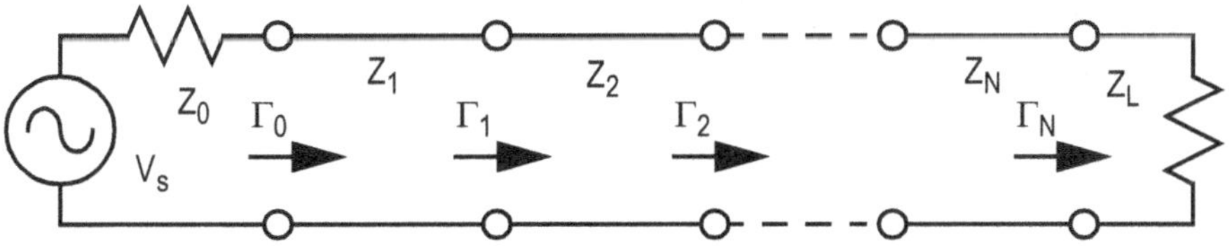

Figure 1.7 Multisection transmission line transformer. Each section is one-quarter wavelength long at the center frequency.

3. The development in this section largely follows that of [1.4].

Equation (1.72) is sometimes called the *small-reflection approximation.* Since the Z_n are (real) characteristic impedances, the Γ_n are real. We assume that the transformer is symmetrical; that is, $\Gamma_0 = \Gamma_N$, $\Gamma_1 = \Gamma_{N-1}$, $\Gamma_2 = \Gamma_{N-2}$, and so on. Then (1.72) can be written

$$\begin{aligned}\Gamma_{\text{in}} \approx{} & 2\exp(-jN\theta)[\Gamma_0\cos(N\theta) + \Gamma_1\cos((n-2)\theta) + \ldots \\ & + \Gamma_n\cos((N-2n)\theta) + \ldots]\end{aligned} \tag{1.74}$$

The last term of (1.74) is $1 / 2\ \Gamma_{N/2}$ when N is even and $\Gamma_{(N-1)/2}$ when N is odd.

Chebyshev polynomials are special functions whose values oscillate between (−1, 1) for arguments in the (−1, 1) range. A Chebyshev polynomial of order n is designated $T_n(x)$; the first few Chebyshev polynomials are

$$\begin{aligned}T_1(x) &= x \\ T_2(x) &= 2x^2 - 1 \\ T_3(x) &= 4x^3 - 3x \\ T_n(x) &= 2xT_{n-1}(x) - T_{n-2}(x)\end{aligned} \tag{1.75}$$

The final expression in (1.75) is a recurrence relation, from which polynomials of any order can be determined. Figure 1.8 shows a plot of the first four orders of Chebyshev polynomials.

Our task is to find a way to scale the Chebyshev polynomials so they are in the form of (1.74). We then equate the coefficients to those of (1.74) to find the reflection coefficients between each pair of sections. First, for x, we substitute the quantity $\cos(\theta) / \cos(\theta_m)$, where θ_m is the electrical length of each section at the band edge. Thus, when $\theta = \theta_m$ and $\theta = \pi - \theta_m$, $x = 1$ and $x = -1$, respectively; when $\theta = \pi / 2$, $x = 0$. Also, we must scale the magnitude of the polynomial to the desired maximum in-band reflection coefficient, which we call Γ_m. With these considerations, it is necessary that

$$\begin{aligned}\Gamma_{\text{in}} &= 2\exp(-jN\theta)[\Gamma_0\cos(N\theta) + \Gamma_1\cos((N-2)\theta) + \ldots \\ &\quad + \Gamma_n\cos((N-2n)\theta) + \ldots] \\ &= \Gamma_c\exp(-jN\theta)T_N\left(\frac{\cos(\theta)}{\cos(\theta_m)}\right)\end{aligned} \tag{1.76}$$

where Γ_c is a constant. To determine it, we note that, at $\theta = 0$,

$$\Gamma_{\text{in}} = \frac{Z_L - Z_0}{Z_L + Z_0} = \Gamma_c T_N\left(\frac{1}{\cos(\theta_m)}\right) \tag{1.77}$$

so

$$\Gamma_c = \frac{Z_L - Z_0}{Z_L + Z_0} \frac{1}{T_N\left(\frac{1}{\cos(\theta_m)}\right)} \tag{1.78}$$

The rightmost side of (1.76) becomes

$$\Gamma_{\text{in}} = \exp(-jN\theta) \frac{Z_L - Z_0}{Z_L + Z_0} \frac{T_N\left(\frac{\cos(\theta)}{\cos(\theta_m)}\right)}{T_N\left(\frac{1}{\cos(\theta_m)}\right)} \tag{1.79}$$

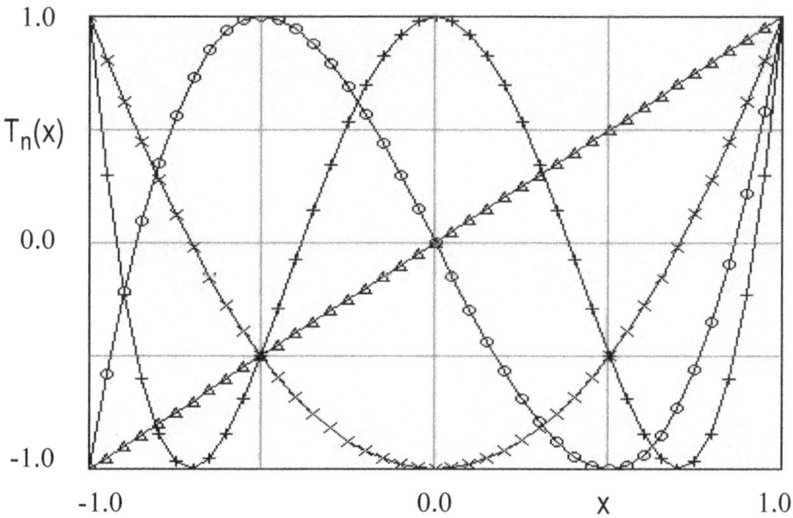

Figure 1.8 A plot of the first four Chebyshev polynomials in the range (−1.0, 1.0). $T_1(x)$: Δ; $T_2(x)$: X; $T_3(x)$: o; $T_4(x)$: +.

When $\Gamma_{in} = \Gamma_m$, $T_N = 1$, so (1.79) gives us the relationship between Γ_m and θ_m; clearly, they cannot be specified independently. From now on, we ignore the exponential term, as we are equating magnitudes:

$$T_N\left(\frac{1}{\cos(\theta_m)}\right) = \frac{Z_L - Z_0}{Z_L + Z_0}\frac{1}{\Gamma_m} \tag{1.80}$$

We now have everything we need to design a transformer. As an example, we consider a two-section transformer, $N = 2$. From (1.76) we equate

$$\Gamma_{in} = 2(\Gamma_0\cos(2\theta) + \Gamma_1) = \Gamma_m T_2\left(\frac{\cos(\theta)}{\cos(\theta_m)}\right) \tag{1.81}$$

Equation (1.75), and a little trigonometry, yields

$$2\Gamma_0\cos(2\theta) + \Gamma_1 = \Gamma_m\left(\frac{\cos(2\theta)}{\cos^2(\theta_m)} + \frac{1}{\cos^2(\theta_m)} - 1\right) \tag{1.82}$$

The reflection coefficients Γ_0 and Γ_1 are found from inspection, and, because of symmetry, $\Gamma_2 = \Gamma_0$.

To complete the example, we apply some numbers to it. Assume that $Z_0 = 50\Omega$, $Z_L = 90\Omega$, and $\Gamma_m = 0.025$. First, solving (1.80), we obtain $\theta_m = 1.158$. This value maps into the lower band edge, and $\pi - \theta_m$ maps into the upper. The fractional bandwidth is

$$\frac{\Delta f}{f_0} = \frac{\Delta\theta}{\pi/2} = \frac{\pi - 2\theta_m}{\pi/2} = 0.525 \tag{1.83}$$

Then, from (1.82)

$$\begin{aligned}\Gamma_0 &= \Gamma_2 = \frac{1}{2}\frac{\Gamma_m}{\cos^2(\theta_m)} = 0.0777\\ \Gamma_1 &= \Gamma_m\left(\frac{1}{\cos^2(\theta_m)} - 1\right) = 0.1304\end{aligned} \tag{1.84}$$

and finally, after inverting (1.29),

$$Z_1 = Z_0\left(\frac{1+\Gamma_0}{1-\Gamma_0}\right) = 58.4\Omega$$
$$Z_2 = Z_1\left(\frac{1+\Gamma_1}{1-\Gamma_1}\right) = 75.9\Omega \qquad (1.85)$$

A plot of the reflection coefficient of the two-section transformer, compared to that of a single-section transformer, is shown in Figure 1.9. It is clear that the use of even a single extra section improves the bandwidth substantially. It also shows that the small-reflection approximation is quite accurate, although in this case not perfect. In the author's experience, the approximation begins to fall apart when the reflection coefficients at the junctions approach 0.2. If that occurs, the reflection coefficients can be decreased by increasing the number of sections. Even when the approximation is questionable, the design approach may still be useful, as the transformer can be adjusted by the use of circuit-analysis software. Conversely, even when the approximation is good, accounting for such effects

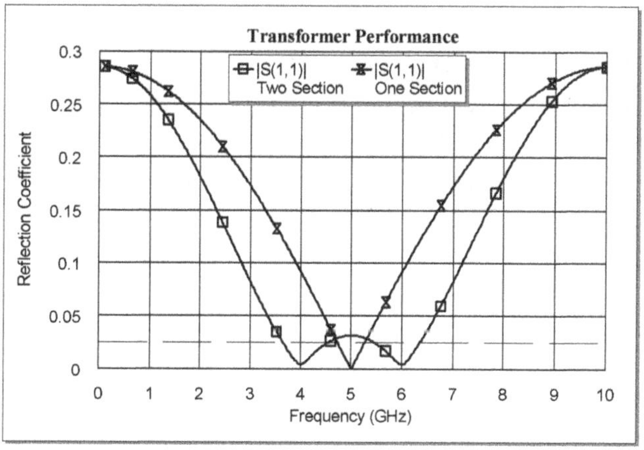

Figure 1.9 The input reflection coefficients of single- and two-section transformers, with $Z_s = 50\Omega$ and $Z_L = 90\Omega$, are compared. The horizontal line marks the design value of Γ_m, 0.025; Γ_m is slightly high, the result of the small-reflection approximation. Even so, it is very close to the desired value, and the two-section transformer has substantially greater bandwidth than a single-section one.

as step discontinuities at the junctions may require some "tweaking" on the computer.

Tables of design data for multisection quarter-wave transformers can be found in [1.9].

References

[1.1] G. Gonzalez, *Microwave Transistor Amplifiers*, Englewood Cliffs, NJ: Prentice-Hall, 1984.

[1.2] B. C. Wadell, *Transmission Line Design Handbook*, Norwood, MA: Artech House, 1991.

[1.3] D. M. Pozar, *Microwave Engineering*, New York: Addison Wesley, 1990.

[1.4] Collin, R. E., *Foundations for Microwave Engineering*, 2nd ed., New York: McGraw-Hill, 1996.

[1.5] S. A. Maas, *The RF and Microwave Circuit Design Cookbook*, Norwood, MA: Artech House, 1998.

[1.6] Kirschning, M., and R. Jansen, "Accurate Model of Effective Dielectric Constant of Microstrip With Validity Up to Millimeter Wave Frequencies," *Electron. Ltrs.*, Vol. 18, 1982, p. 272.

[1.7] Alexopoulos, N., and S. Maas, "Performance of Microstrip Couplers on an Anisotropic Substrate with an Isotropic Superstrate," *IEEE Trans. Microwave Theory Tech.*, Vol. MTT-31, 1983, p. 671.

[1.8] Swanson, D. G., and W. J. R. W. Hoeffer, *Microwave Circuit Modeling Using Electromagnetic Field Simulation*, Norwood, MA: Artech House, 2003.

[1.9] Matthaei, G. L., L. Young, and E. M. T. Jones, *Microwave Filters, Impedance-Matching Networks, and Coupling Structures*, Norwood, MA: Artech House, 1980.

Chapter 2

Coupled Transmission Lines and Modal Analysis

Coupled transmission lines are useful for many types of components, most notably directional couplers, but also for filters, baluns, and control components. The even- and odd-mode analysis used for directional couplers is also useful in a variety of applications and for understanding a number of concepts in microwave circuit theory. We will utilize it often in later chapters.

2.1 Even- and Odd-Mode Analysis

2.1.1 Even and Odd Modes

Our problem is to analyze a pair of identical transmission lines, usually strip transmission lines, that are closely spaced, allowing energy from one to be coupled to the other. Coupled lines are realized in virtually every transmission medium, but they are especially common in microstrip and homogeneous stripline. For this reason, our discussion will use microstrip as an example, but of course the theory is applicable generally.

A pair of coupled microstrip lines is illustrated in Figure 2.1. The spacing between the lines is small, so it seems clear that a wave travelling on one might excite a similar wave on the other. Less clear is the way to analyze the structure. It seems at first to be a difficult problem in electromagnetics.

Considerable simplification can be achieved from a clever trick. We recognize that the lines are a linear system, so we can divide the problem into two parts, analyze them separately, and obtain the solution by sum-

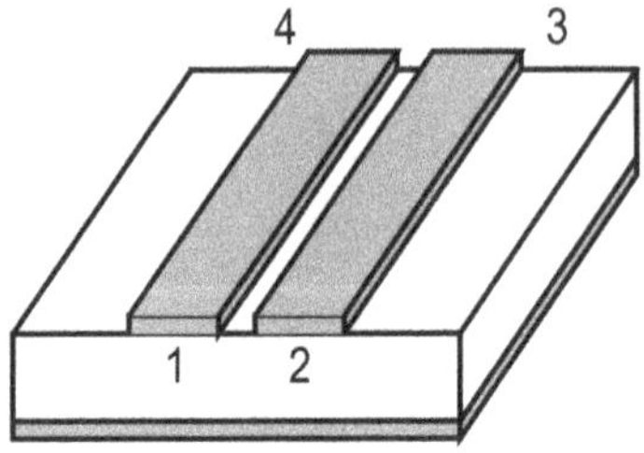

Figure 2.1 Coupled, symmetrical microstrip lines. The lines are equal in width and their spacing is usually much less than the strip width.

ming. Specifically, we treat the problem in Figure 2.1 as shown in Figure 2.2. A single-strip excitation, shown in Figure 2.2(a), can be viewed as the sum of the two cases in Figure 2.2(b) and (c), which we call the *even* and *odd modes*, respectively. Because the system is linear, the sum of the responses must be the response to the sum of the excitations. Since the sum of the excitations of circuits (b) and (c) is simply the excitation in (a), summing the responses of the even- and odd-mode circuits gives us the response of the original circuit.

This process allows us to take advantage of the symmetries in Figure 2.2(b) and (c). Clearly, in the odd-mode case, the electric field must be perpendicular to a vertical surface halfway between the strips. We call this an *electric wall.* We could insert a thin, grounded metal wall halfway between the strips without changing the fields, and we are left with two identical, *single* strips, so we need analyze only one. In the even-mode case, we have the converse, a *magnetic wall,* where the tangential magnetic field is zero, and the even-mode circuit is likewise reduced to a single strip. Thus, the analysis of the coupled strips can be reduced to two analyses of single strips, and all our knowledge of single transmission lines can be applied. The even-mode characteristic impedance, Z_{0e}, and phase velocity, v_{pe}, are those of a single strip when the coupled lines are excited in an even mode; the odd-mode characteristic impedance, Z_{0o}, and phase velocity, v_{po}, are the respective quantities for the odd mode. The only thing left to do is an electromagnetic analysis of the two single-strip equivalent structures. That analysis is beyond the scope of this book, so from now on, we shall assume that we can determine Z_{0e}, v_{pe}, Z_{0o}, v_{po} for any symmetrical pair of coupled lines. For an example of such analysis, for microstrip lines, see [1.7].

2.1.2 Even- and Odd-Mode Characteristics

Figure 2.3 shows even- and odd-mode equivalents of a set of coupled microstrip lines. The shaded areas represent conductors and the dashed line represents a magnetic wall. Just as the electric wall, the vertical conductor in Figure 2.3(a), represents a kind of spatial short circuit, the magnetic wall represents a similar kind of open circuit, as both the electric and magnetic fields do not cross it.

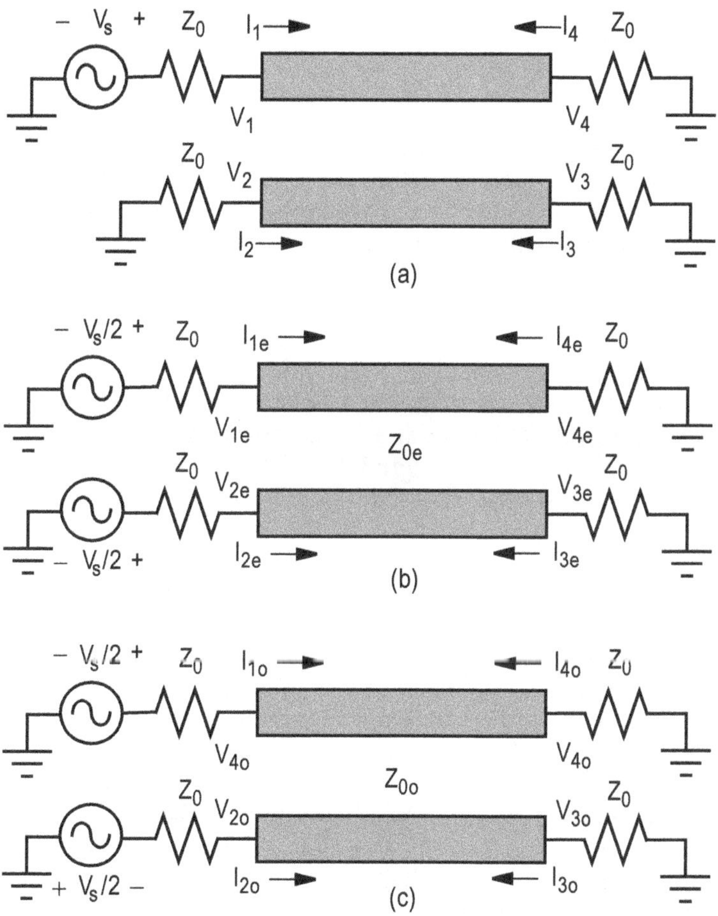

Figure 2.2 The single-strip excitation of a pair of coupled lines (a) can be viewed as the sum of an even mode (b) and odd mode (c).

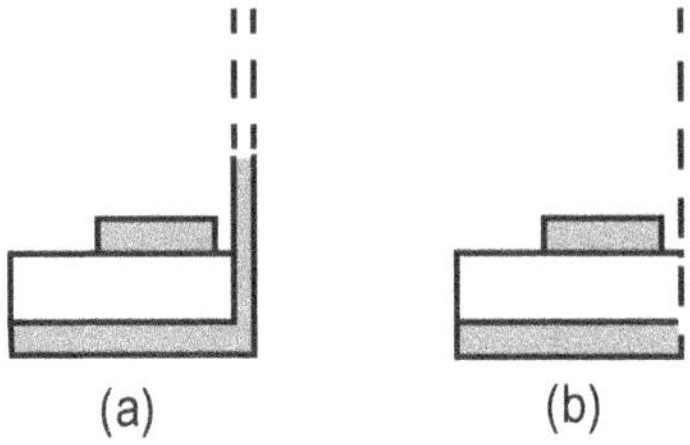

Figure 2.3 Cross-sectional view of a microstrip line, odd mode (a) and even mode (b). The dashed line in (b) signifies a magnetic wall. The capacitance between the line and ground is greater for the odd mode, implying that the characteristic impedance is lower.

We can deduce some properties of the characteristic impedance and phase velocity from these structures. Clearly, the capacitance between the strip and ground conductors is greater in the odd-mode equivalent, so, from (1.19), the characteristic impedance is lower. Because of the magnetic wall in Figure 2.3(b), the fringing fields on one side of the conductor are cut off, so the capacitance is less than in a simple microstrip having the same dimensions. Thus, the even-mode characteristic impedance is greater than that of a simple microstrip.

Nonhomogeneous coupled lines have unequal even- and odd-mode phase velocities and effective dielectric constants. We relate these quantities in a manner identical to single lines (1.65):

$$\varepsilon_{eff,\,e} = \left(\frac{v_c}{v_{pe}}\right)^2 \qquad \varepsilon_{eff,\,o} = \left(\frac{v_c}{v_{po}}\right)^2 \tag{2.1}$$

where the additional *e* or *o* subscript refers to the even or odd mode, respectively. In Figure 2.3(a) the vertical wall of the odd mode causes much of the electric field to be in the air above the strip conductor; in (b), it is more concentrated in the dielectric. This causes the effective dielectric constant in the even-mode case to be greater than in the odd mode, implying that the even-mode phase velocity is lower than the odd mode. Of course, in homogeneous transmission structures, where the area containing the electric field is completely filled with dielectric, the effective dielectric constants of the two modes are equal and thus the phase velocities are also equal. In microstrip, however, the inequality of the two mode velocities usually has a significant (and usually deleterious) effect on the performance of any component realized in that medium.

2.1.3 Coupled-Line Analysis

We now have all the tools we need to analyze coupled lines. Suppose we have a pair of symmetrical coupled lines, terminated in impedance Z_0 as shown in Figure 2.2(a). For example, suppose we wish to find the current at port 2, which will give us the power at that port as well. Voltages and currents at the other ports can be found similarly.

The current at port 2 is the sum of the even- and odd-mode currents at the ends of the lines. From Figures 2.2 and 2.3,

$$I_2 = \frac{V_s/2}{Z_0 + Z_{\text{in},e}} - \frac{V_s/2}{Z_0 + Z_{\text{in},o}} \tag{2.2}$$

where $Z_{\text{in},e}$ and $Z_{\text{in},o}$ are the even- and odd-mode input impedances, respectively. From (1.53), with Z_{0e} as the characteristic impedance and Z_0 as the load,

$$Z_{\text{in},e} = Z_{0e}\frac{Z_0\cos(\theta_e) + jZ_{0e}\sin(\theta_e)}{Z_{0e}\cos(\theta_e) + jZ_0\sin(\theta_e)} \tag{2.3}$$

We have ignored line loss, as all practical couplers use low-loss lines. Unless the lines are homogeneous, the even- and odd-mode phase velocities are unequal, so we use $\theta_e = \beta_e l$ as the even-mode electrical length. Similarly, for the odd mode,

$$Z_{\text{in},o} = Z_{0o}\frac{Z_0\cos(\theta_o) + jZ_{0o}\sin(\theta_o)}{Z_{0o}\cos(\theta_o) + jZ_0\sin(\theta_o)} \tag{2.4}$$

where $\theta_o = \beta_o l$. The power dissipated in the termination at port 2 is

$$P_2 = \frac{1}{2}|I_2|^2 Z_0 \tag{2.5}$$

and the available power from the source is

$$P_{av} = \frac{|V_s|^2}{8Z_0} \tag{2.6}$$

so the power coupling to port 2, C, is

$$C = \frac{P_2}{P_{av}} = 4Z_0^2\frac{|I_2|^2}{|V_s|^2} \tag{2.7}$$

Other important quantities can be defined in a manner similar to (2.7). The isolation, I_c, is the loss from the input to the isolated port:

$$I_c = \frac{P_3}{P_{av}} = 4Z_0^2\frac{|I_3|^2}{|V_s|^2} \tag{2.8}$$

and the directivity, D, is

$$D = \frac{P_2}{P_3} = \left|\frac{I_2}{I_3}\right|^2 \tag{2.9}$$

where P_n is the power at port n. In an ideal coupler, the isolation and directivity are perfect; imperfect isolation occurs when port reflections exist or the even- and odd-mode phase velocities are unequal. When the coupler is matched, $P_1 = P_{av}$. These quantities are normally expressed in decibels and usually, for simplicity, as positive values.

2.1.4 Application: Coupled-Line Directional Coupler

Properly designed, a quarter-wavelength pair of symmetrical coupled lines can realize a directional coupler. In such a device, power incident on port 1 is coupled to port 2 of Figure 2.2(a); the wave on the coupled line propagates in a direction opposite to the one on the main line, and as such it is sometimes called a *backward-wave coupler*. The remaining power is delivered to port 4. Port 3 is isolated; ideally, no power is delivered to it, but in practical couplers, that isolation is never perfect.

Our analysis is for an ideal coupler, one which has no losses and the even- and odd-mode phase velocities are equal, so $\theta_e = \theta_o = \theta$. We first consider the conditions for an input match. From the even- and odd-mode circuits, the input impedance is

$$Z_{in} = \frac{V_{1e} + V_{1o}}{I_{1e} + I_{1o}} \tag{2.10}$$

where the quantities are defined in Figure 2.2. V_{1e}, V_{1o}, I_{1e}, and I_{1o} are

$$\begin{aligned} V_{1e} &= \frac{V_s}{2}\frac{Z_{in,\,e}}{(Z_{in,\,e} + Z_0)} \\ V_{1o} &= \frac{V_s}{2}\frac{Z_{in,\,o}}{(Z_{in,\,o} + Z_0)} \\ I_{1e} &= \frac{V_s}{2}\frac{1}{(Z_{in,\,e} + Z_0)} \\ I_{1o} &= \frac{V_s}{2}\frac{1}{(Z_{in,\,o} + Z_0)} \end{aligned} \tag{2.11}$$

where $Z_{in,e}$ and $Z_{in,o}$ are given by (2.3) and (2.4), respectively. Substituting these and (2.11) into (2.10) gives

$$Z_{in} = Z_0 + \frac{2(Z_{in,\,o}Z_{in,\,e} - Z_0^2)}{Z_{in,\,e} + Z_{in,\,o} + Z_0} \tag{2.12}$$

Clearly, if $Z_{in,e}Z_{in,o} = Z_0^2$, $Z_{in} = Z_0$, as desired. We note that, when

$$Z_0 = \sqrt{Z_{0e}Z_{0o}} \tag{2.13}$$

equations (2.3) and (2.4) become

$$Z_{in,\,e} = Z_{0e}\frac{\sqrt{Z_{0o}}\cos(\theta) + j\sqrt{Z_{0e}}\sin(\theta)}{\sqrt{Z_{0e}}\cos(\theta) + j\sqrt{Z_{0o}}\sin(\theta)} \tag{2.14}$$

$$Z_{\text{in},\,o} = Z_{0o}\frac{\sqrt{Z_{0e}}\cos(\theta) + j\sqrt{Z_{0o}}\sin(\theta)}{\sqrt{Z_{0o}}\cos(\theta) + j\sqrt{Z_{0e}}\sin(\theta)} \tag{2.15}$$

Multiplying (2.14) by (2.15) gives

$$Z_{\text{in},\,e}Z_{\text{in},\,o} = Z_{0e}Z_{0o} = Z_0^2 \tag{2.16}$$

so (2.13) is the condition for matched ports.

To find an expression for the coupling, in terms of voltage, we need to find V_2 / V_s. Clearly,

$$V_2 = V_{2e} + V_{2o} = \frac{V_s}{2}\left(\frac{Z_{\text{in},\,e}}{Z_{\text{in},\,e} + Z_0} - \frac{Z_{\text{in},\,o}}{Z_{\text{in},\,o} + Z_0}\right) \tag{2.17}$$

Substituting our previous expressions and making use of (2.13), we obtain

$$\frac{V_2}{V_1} = \frac{jc\sin(\theta)}{\sqrt{1-c^2}\cos(\theta) + j\sin(\theta)} \tag{2.18}$$

where

$$c = \frac{Z_{0e} - Z_{0o}}{Z_{0e} + Z_{0o}} \tag{2.19}$$

c is the voltage coupling ratio at center frequency ($\theta = \pi/2$), the square root of the power coupling ratio, C, in (2.7).

From a similar process, we obtain the voltage at port 4:

$$\frac{V_4}{V_1} = \frac{\sqrt{1-c^2}}{\sqrt{1-c^2}\cos(\theta) + j\sin(\theta)} \tag{2.20}$$

and, finally, we can use that process to show that the voltage at port 3 is zero.

Equations (2.13) and (2.19) can be manipulated to give

$$Z_{0e} = Z_0\sqrt{\frac{1 + c}{1 - c}} \tag{2.21}$$

$$Z_{0o} = Z_0\sqrt{\frac{1 - c}{1 + c}} \tag{2.22}$$

These are somewhat more convenient for coupler design.

It is important to remember our assumption that even- and odd-mode phase velocities are equal. In nonhomogeneous couplers, that is not the case, and some degree of error in input match and coupling results. The magnitude of that error depends on the magnitude of the phase-velocity mismatch, so efforts are frequently made to equalize phase velocities in practical couplers. See [2.1] and [2.2] for examples of some of those methods. One solution is simply, when applicable, to use a homogeneous coupler. For this reason, dielectrically loaded stripline is a favored medium for directional couplers.

To design a single-section coupler, we determine Z_{0e} and Z_{0o} from (2.21) and (2.22), then make the lines one-quarter wavelength long ($\theta = \pi / 2$). If the even- and odd-mode phase velocities are unequal, the length is chosen to minimize the error for both modes. For example, for an ideal 3-dB coupler realized in a homogeneous medium, (2.21) and (2.22) give the even- and odd-mode characteristic impedances, 120.7Ω and 20.7Ω, respectively. Figure 2.4 shows the coupling as a function of frequency.

Equations (2.13), (2.18), and (2.20) show that the coupler has some remarkable properties:

1. The ports are perfectly matched at all frequencies;
2. The coupler has infinite directivity at all frequencies;
3. The phase difference between the coupled port and output ("through") port is $\pi / 2$ at all frequencies.

As we shall see in later chapters, these properties make a coupler with 3-dB coupling remarkably valuable for many applications, especially in high-frequency amplifiers.

As with multisection transformers, it is possible to design multisection couplers having remarkably broad bandwidths. See [1.3] or [1.9] for more information and a design procedure.

2.1.5 Effect of Unequal Modal Phase Velocities

When the even- and odd-mode phase velocities are unequal, the performance of the coupler is degraded. The isolation and directivity are most strongly affected, most significantly when the coupling is weak.

Figure 2.5 shows an example of the coupling and isolation of a 10-dB coupler as a function of the percentage difference between the even- and odd-mode dielectric constants. It shows that the effect on coupling is minor, probably not noticeable in a practical coupler, but the reduction in the isolation is profound.

For other values of coupling, the isolation does not change much, only a few decibels from 3- to 20-dB coupling. Thus, the directivity is most strongly affected; for a 20-dB coupler with a 20% difference in the effective dielectric constants, the directivity is only 2.6 dB. For the 10-dB coupler, it is 13 dB and for 3 dB, it is 25 dB.

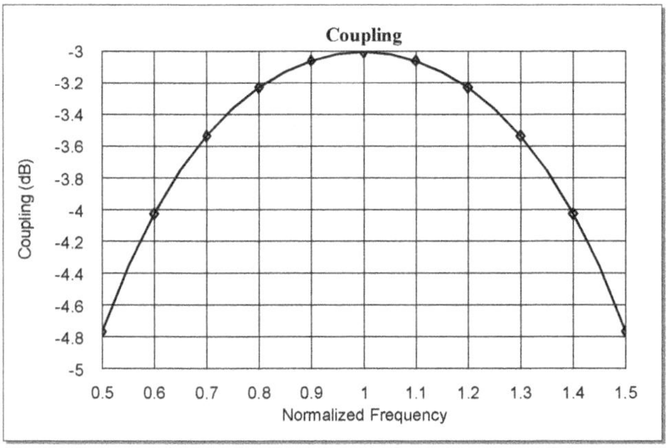

Figure 2.4 Coupling of a 3-dB homogeneous directional coupler vs. frequency. The 0.5-dB bandwidth is approximately 60%. The bandwidth can be increased by overcoupling at the center frequency; increasing the coupling to −2.5 dB at the center frequency increases the 0.5-dB bandwidth to 90%.

2.2 General, Multiple Coupled Lines

Multiple coupled lines have many uses, most importantly as directional couplers and baluns. In this section we describe a general method for the analysis of such lines. The theory is sufficiently general to include nonhomogeneous lines as well as homogeneous ones. It does not require that the strips have equal widths, spacings, or even cross sections. The strips can also be on different layers of a multilayer substrate. The only requirement is that the strips be uniform in the longitudinal (z) direction.

The theory is similar to single-line theory, although the expressions are in matrix, not scalar, form. As such, it could be viewed as a generalization of the transmission-line equations. We begin by examining the concepts of the line capacitance, inductance, resistance, and conductance for the multiple-line case.

2.2.1 R, L, G, and C Matrices

Much coupled-line theory is based on a quasistatic analysis (Section 1.2.2.2); the line is described by capacitance (C) and inductance (L) matrices. If losses are included, series resistance (R) and shunt conductance (G) matrices must also be included.

In a set of N parallel lines, the line charges, Q_n, are given by

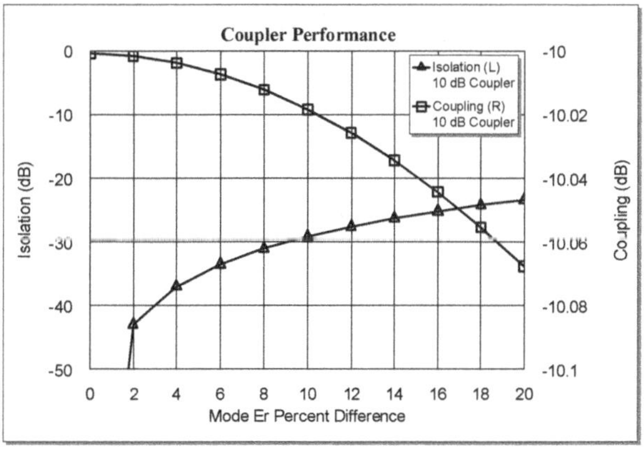

Figure 2.5 Coupling and isolation of a 10-dB coupler as a function of the percentage difference between the even- and odd-mode effective dielectric constants.

$$\begin{bmatrix} Q_1 \\ Q_2 \\ \dots \\ Q_N \end{bmatrix} = \begin{bmatrix} C_{11} & C_{12} & \dots & C_{1N} \\ C_{21} & C_{22} & \dots & C_{2N} \\ \dots & \dots & \dots & \dots \\ C_{N1} & C_{N2} & \dots & C_{NN} \end{bmatrix} \begin{bmatrix} V_1 \\ V_2 \\ \dots \\ V_N \end{bmatrix} \tag{2.23}$$

where Q_n is the charge on the nth strip, V_n is the voltage on the nth strip, and

$$C_{mn} = \left.\frac{Q_m}{V_n}\right|_{V_p = 0,\, p \neq n} \tag{2.24}$$

This formulation accounts for the capacitive couplings between all pairs of lines. In the older literature on couplers, however, it is common to consider only the coupling to adjacent lines; in that case, the capacitance matrix, as well as the following L and G matrices, is a tridiagonal one. The inductance matrix is analogous:

$$\begin{bmatrix} \Phi_1 \\ \Phi_2 \\ \dots \\ \Phi_N \end{bmatrix} = \begin{bmatrix} L_{11} & L_{12} & \dots & L_{1N} \\ L_{21} & L_{22} & \dots & L_{2N} \\ \dots & \dots & \dots & \dots \\ L_{N1} & L_{N2} & \dots & L_{NN} \end{bmatrix} \begin{bmatrix} I_1 \\ I_2 \\ \dots \\ I_N \end{bmatrix} \tag{2.25}$$

where Φ_n is the magnetic flux surrounding the nth strip and I_n is the current in the nth strip. Again,

$$L_{mn} = \left.\frac{\Phi_m}{I_n}\right|_{I_p = 0,\, p \neq n} \tag{2.26}$$

The conductance matrix is given by

$$\begin{bmatrix} I_1 \\ I_2 \\ \dots \\ I_N \end{bmatrix} = \begin{bmatrix} G_{11} & G_{12} & \dots & G_{1N} \\ G_{21} & G_{22} & \dots & G_{2N} \\ \dots & \dots & \dots & \dots \\ G_{N1} & G_{N2} & \dots & G_{NN} \end{bmatrix} \begin{bmatrix} V_1 \\ V_2 \\ \dots \\ V_N \end{bmatrix} \tag{2.27}$$

with

$$G_{mn} = \left.\frac{I_m}{V_n}\right|_{V_p = 0,\, p \neq n} \tag{2.28}$$

Finally, the resistance matrix is

$$\begin{bmatrix} V_1 \\ V_2 \\ \dots \\ V_N \end{bmatrix} = \begin{bmatrix} R_{11} & R_{12} & \dots & R_{1N} \\ R_{21} & R_{22} & \dots & R_{2N} \\ \dots & \dots & \dots & \dots \\ R_{N1} & R_{N2} & \dots & R_{NN} \end{bmatrix} \begin{bmatrix} I_1 \\ I_2 \\ \dots \\ I_N \end{bmatrix} \tag{2.29}$$

and

$$R_{mn} = \frac{V_m}{I_n} \tag{2.30}$$

While off-diagonal terms in the G matrix are physically plausible, the matrix is usually treated as a diagonal. In ordinary situations, however, off-diagonal terms in the R matrix are not physically meaningful. Although the formulation allows for off-diagonal elements in both, the G and R matrices are invariably assumed to be diagonal.

2.2.2 Transmission Line Equations

The equations describing multiple coupled lines are a generalization of (1.1) to (1.5) [2.3]. The transmission-line equations, in matrix form, become

$$\frac{d^2\mathbf{V}}{dz^2} = \mathbf{ZYV} \tag{2.31}$$

where boldface indicates a matrix or vector, as appropriate, and

$$\frac{d^2\mathbf{I}}{dz^2} = \mathbf{YZI} \tag{2.32}$$

where

$$\mathbf{Z} = \mathbf{R} + j\omega\mathbf{L} \tag{2.33}$$

$$\mathbf{Y} = \mathbf{G} + j\omega\mathbf{C} \tag{2.34}$$

The voltage and current vectors are represented as orthogonal modes and any set of line voltages is expressed as a linear combination of those modes. A set of N lines (not including the ground plane) can support N modes for both voltage and current, each of which has its own propagation constant. The nth voltage mode is expressed as

$$\mathbf{V}_n(z) = \mathbf{V}_{nf}\exp(-\gamma_m z) + \mathbf{V}_{nr}\exp(\gamma_n z) \qquad n = 1\ldots N \tag{2.35}$$

Substituting (2.35) into (2.31) gives

$$(\gamma_n^2\mathbf{1} - \mathbf{ZY})(\mathbf{V}_{nf}\exp(-\gamma_n z) + \mathbf{V}_{nr}\exp(\gamma_n z)) = 0 \tag{2.36}$$

where **1** is the identity matrix. This equation has a nontrivial solution only when

$$\det(\gamma_n^2\mathbf{1} - \mathbf{ZY}) = 0 \tag{2.37}$$

The solutions to (2.37) yield the propagation constants for each of the N modes. Furthermore, (2.36) can be identified as a classical eigenvalue problem, where the eigenvectors are the voltage modes that can exist on the lines. Those eigenvectors are orthogonal, and we shall assume that they are normalized. As such, we express the vector of voltage $\mathbf{V}(z)$ along the N lines as

$$\mathbf{V}(z) = \mathbf{V}_f(z) + \mathbf{V}_r(z) = \mathbf{S_V}(\mathbf{G}_f(z) + \mathbf{G}_r(z)) \tag{2.38}$$

$\mathbf{G}_f$ and $\mathbf{G}_r$ are column vectors whose elements are the complex magnitudes of the respective modes, and, as usual, the f and r subscripts refer, respectively, to forward- and reverse-propagating waves. These vary exponentially along the lines as in (2.35). $\mathbf{S_V}$ is a matrix whose columns are the modal eigenvectors of voltage; thus, the set of line voltages is a linear combination of the N modes. Similarly, for the currents, we obtain

$$\mathbf{I}(z) = \mathbf{I}_f(z) - \mathbf{I}_r(z) = \mathbf{S_I}(\mathbf{G}_f(z) - \mathbf{G}_r(z)) \tag{2.39}$$

where $\mathbf{S_I}$ is a matrix whose columns are the current-mode eigenvectors. As with single transmission lines, we can define a characteristic impedance, although in this case it is a matrix:

$$\mathbf{V}_f(z) = \mathbf{Z}_0\mathbf{I}_f(z) \tag{2.40}$$

where

$$\mathbf{Z}_0 = \mathbf{S_V}\mathbf{S_I}^{-1} \tag{2.41}$$

By now we can see that the two symmetrical coupled lines are simply a special case of multiple lines. For the two lossless coupled lines, the voltage modes are the even and odd modes,

$$\mathbf{S_V} = \begin{bmatrix} \frac{1}{2} & \frac{1}{2} \\ -\frac{1}{2} & \frac{1}{2} \end{bmatrix} \tag{2.42}$$

and the propagation constants are

$$\begin{aligned} \gamma_1 &= j\frac{\omega}{v_{po}} \\ \gamma_2 &= j\frac{\omega}{v_{pe}} \end{aligned} \tag{2.43}$$

2.2.3 Matrices

In the two-line case, the even- and odd-mode view was useful and tractable. When we have multiple lines, however, which may not be symmetrical, dealing analytically with a large number of modes is simply not practical. The best option is to define an admittance matrix for the set of lines, and convert that matrix, as needed, to other kinds of parameters. We shall see, in later chapters, that the admittance representation is also ideal for use in circuit simulators.

The situation is shown in Figure 2.6, where we consider a set of N lines of length l. We want to find a Y matrix for these lines, which consists of the $N \times N$ submatrices $\mathbf{Y}_{ij}$, where N is the number of coupled strips:

$$\begin{bmatrix} \mathbf{I}(0) \\ -\mathbf{I}(l) \end{bmatrix} = \begin{bmatrix} \mathbf{Y}_{11} & \mathbf{Y}_{12} \\ \mathbf{Y}_{21} & \mathbf{Y}_{22} \end{bmatrix} \begin{bmatrix} \mathbf{V}(0) \\ \mathbf{V}(l) \end{bmatrix} \tag{2.44}$$

From symmetry, $\mathbf{Y}_{11} = \mathbf{Y}_{22}$ and $\mathbf{Y}_{21} = \mathbf{Y}_{12}$. The submatrices are defined in a manner analogous to scalar quantities. The negative sign before $\mathbf{I}(l)$ is a consequence of the reference direction of $\mathbf{I}(z)$ in Figure 2.6.

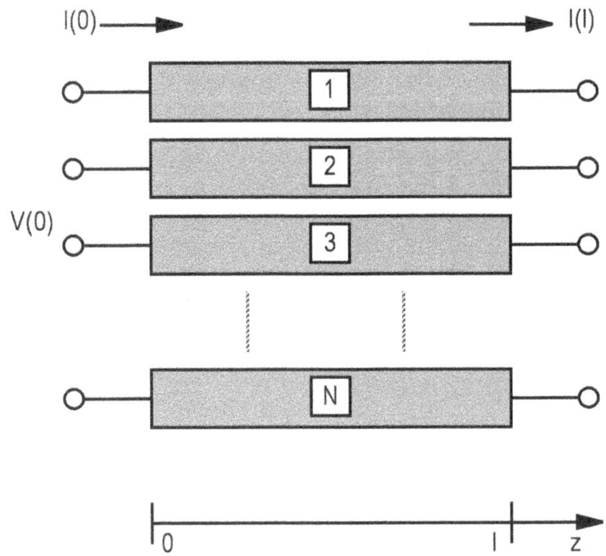

Figure 2.6 The set of N coupled lines of length l, for which we derive the Y matrix.

From the previous section, we note that

$$\mathbf{G}_f(l) = \Gamma_l \mathbf{G}_f(0) \tag{2.45}$$

and

$$\mathbf{G}_r(0) = \Gamma_l \mathbf{G}_r(l) \tag{2.46}$$

where

$$\Gamma_l = \mathrm{diag}(\exp(-\gamma_n l)) \qquad n = 1 \ldots N \tag{2.47}$$

From (2.38) the voltage at $z = 0$ can be expressed as

$$\mathbf{V}(0) = \mathbf{S_V}(\mathbf{G}_f(0) + \mathbf{G}_r(0)) \tag{2.48}$$

For an admittance matrix, we set $\mathbf{V}(l) = 0$ so

$$\begin{aligned} \mathbf{V}(0) &= \mathbf{S_V}(\mathbf{G}_f(0) - \Gamma_{2l}\mathbf{G}_f(0)) \\ &= \mathbf{S_V}(1 - \Gamma_{2l})\mathbf{G}_f(0) \end{aligned} \tag{2.49}$$

where

$$\Gamma_{2l} = \mathrm{diag}(\exp(-2\gamma_n l)) \tag{2.50}$$

Solving for $\mathbf{G}_f(0)$ gives

$$\mathbf{G}_f(0) = (1 - \Gamma_{2l})^{-1}\mathbf{S}_\mathbf{V}^{-1}\mathbf{V}(0) \tag{2.51}$$

Similarly, from (2.39) we obtain

$$\mathbf{I}(0) = \mathbf{S_I}(1 + \Gamma_{2l})\mathbf{G}_f(0) \tag{2.52}$$

Substituting (2.52) into (2.51) gives

$$\mathbf{I}(0) = \mathbf{S_I}(1 + \Gamma_{2l})(1 - \Gamma_{2l})^{-1}\mathbf{S}_\mathbf{V}^{-1}\mathbf{V}(0) \tag{2.53}$$

so

$$\mathbf{Y}_{11} = \mathbf{S}_\mathbf{I}(1 + \mathbf{\Gamma}_{2l})(1 - \mathbf{\Gamma}_{2l})^{-1}\mathbf{S}_\mathbf{V}^{-1} \tag{2.54}$$

which, from symmetry, equals $\mathbf{Y}_{22}$. To find $\mathbf{Y}_{21}$, we set $\mathbf{V}(l) = 0$ and note that

$$-\mathbf{I}(l) = -\mathbf{S}_\mathbf{I}(\mathbf{G}_f(l) - \mathbf{G}_r(l)) \tag{2.55}$$

when $\mathbf{V}(l) = 0$, $\mathbf{G}_r(l) = -\mathbf{G}_f(l)$ so

$$-\mathbf{I}(l) = -\mathbf{S}_\mathbf{I}2\mathbf{G}_f(l) = -\mathbf{S}_\mathbf{I}2\mathbf{\Gamma}_l\mathbf{G}_f(0) \tag{2.56}$$

Substituting our earlier expression for $\mathbf{G}_f(0)$, we obtain

$$-\mathbf{I}(l) = -\mathbf{S}_\mathbf{I}2\mathbf{\Gamma}_l(1 - \mathbf{\Gamma}_{2l})^{-1}\mathbf{S}_\mathbf{V}^{-1}\mathbf{V}(0) \tag{2.57}$$

and, by inspection,

$$\mathbf{Y}_{21} = -\mathbf{S}_\mathbf{I}2\mathbf{\Gamma}_l(1 - \mathbf{\Gamma}_{2l})^{-1}\mathbf{S}_\mathbf{V}^{-1} \tag{2.58}$$

which, again from symmetry, equals $\mathbf{Y}_{12}$.

We now have a general method for obtaining a circuit matrix representing an arbitrary set of coupled transmission lines. Those lines, and any dielectric supporting them, need not be symmetrical or uniform in cross section; they need only have equal lengths, and the cross-sectional dimensions must be independent of z. Of course, we have used the quasistatic assumption, so the lines are treated as a quasi-TEM transmission medium, and non-TEM dispersion is not included. The formulation does, however, include allowance for dispersion due to loss.

The quasi-TEM analysis is not as limited as one might expect, for two reasons: first, it is sometimes possible to apply empirical dispersion corrections to the quasi-TEM results, and second, as a practical matter, it is usually possible to design a coupler to have minimal non-TEM dispersion. This is done by making the substrate sufficiently thin, the dielectric constant low, and the line dimensions small relative to a wavelength. Indeed, non-TEM dispersion is not a problem unique to couplers; if it is significant, the entire circuit has probably been realized in an inappropriate manner, with a too-thick substrate or a similar problem.

2.2.4 Application: Lange Coupler

A multistrip equivalent of a two-strip coupler, realized in microstrip, was described by J. Lange in 1969 [2.4]. The Lange coupler solved many problems in the use of 3-dB directional couplers (technically, *hybrid junctions*; see Section 3.2.5) for several types of circuits, especially amplifiers, but introduced the problem of determining its even- and odd-mode characteristics. Solving that problem has been a source of gratifying labor for electromagneticists for many years [2.5–2.9].

The Lange coupler, shown in Figure 2.7, elegantly solves a problem in two-strip planar couplers: the inability to achieve sufficient coupling between two edge-coupled strips to realize a 3-dB coupler. (Achieving 3-dB coupling in other media, especially broadside-coupled stripline, is often fairly easy.) In a Lange coupler, the four strips are interconnected in such a way as to create a two-strip equivalent, but the additional adjacent pairs of edges increase the coupling substantially. It also places both outputs on the same side of the structure; for example, if port 1 is excited, ports 2 and 4 are the outputs. This is a significant practical advantage in realizing quadrature-coupled pairs of amplifiers (Chapter 8).

The even- and odd-mode properties can be determined if the coupler is first viewed as a set of four identical parallel-coupled lines. The lines are numbered as in Figure 2.6. In the even mode, all the voltages V_n are set to +1, so the even-mode charge becomes $Q_1 + Q_3 = Q_2 + Q_4$. Furthermore, we note some obvious symmetries in the C matrix, $C_{33} = C_{22}$, $C_{34} = C_{12}$, and $C_{13} = C_{31}$; then, from (2.23),

$$C_e = C_{11} + C_{22} + 2C_{13} + 2C_{12} + C_{23} + C_{14} \tag{2.59}$$

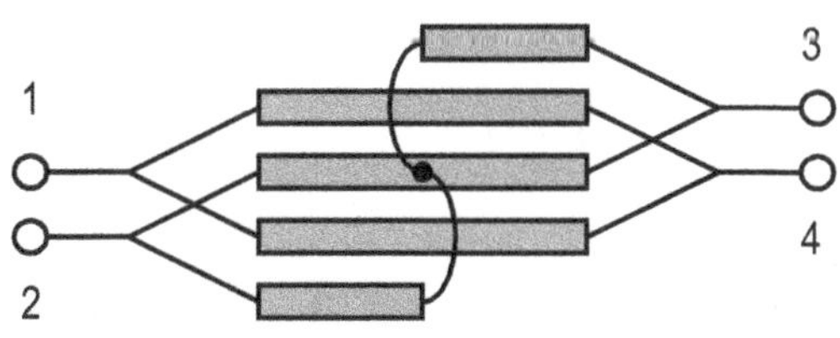

Figure 2.7 A Lange coupler is equivalent to a two-strip directional coupler. The strips are split, rearranged, and connected as shown to increase coupling. the arrangement also places both outputs on the same side of the structure.

In the odd mode, alternate line voltages are set to +1 and –1; $Q_1 + Q_3$ now gives the odd-mode charge, and the resulting odd-mode capacitance is

$$C_o = C_{11} + C_{22} + 2C_{13} - 2C_{12} - C_{23} - C_{14} \tag{2.60}$$

Note that the off-diagonal elements of the C matrix are negative, so (2.60) is greater than (2.59), as expected.

Finding the even- and odd-mode inductances is a little tricky. The voltages on the strips are known, not the currents, so we must invert (2.25) to obtain $\mathbf{I} = \mathbf{L}^{-1}\,\boldsymbol{\Phi}$. Since $V_n = j\omega\Phi_n$, Φ_n is proportional to V_n and we can treat Φ_n as the excitation. We set Φ_n to +1 or –1, as for the capacitances, and obtain

$$L_e = \frac{1}{L_{i11} + L_{i22} + 2L_{i13} + 2L_{i12} + L_{i23} + L_{i14}} \tag{2.61}$$

and

$$L_o = \frac{1}{L_{i11} + L_{i22} + 2L_{i13} - 2L_{i12} - L_{i23} - L_{i14}} \tag{2.62}$$

where L_{ijk} is the (j, k) element of the inverted inductance matrix. Finally, the even- and odd-mode characteristic impedances are

$$Z_{0e} = \sqrt{\frac{L_e}{C_e}} \qquad Z_{0o} = \sqrt{\frac{L_o}{C_o}} \tag{2.63}$$

and the phase velocities are

$$v_{pe} = \frac{1}{\sqrt{L_e C_e}} \qquad v_{po} = \frac{1}{\sqrt{L_o C_o}} \tag{2.64}$$

As with nonhomogeneous transmission lines, these are often expressed as effective dielectric constants:

$$\varepsilon_{eff,\,e} = \left(\frac{v_c}{v_{pe}}\right)^2 \qquad \varepsilon_{eff,\,o} = \left(\frac{v_c}{v_{po}}\right)^2 \tag{2.65}$$

The Lange coupler, of course, is not just a simple, four-strip structure. The many interconnections, especially the wire jumpers at the center, have parasitic inductance and capacitance that often cannot be neglected, especially at high frequencies. One option in such cases is to treat the coupler as a cascade of two eighth-wavelength structures. We have shown that those parasitics have no significant effect in ordinary monolithic integrated circuits at least up to 40 GHz [2.10].

Another way to approach this problem would be to create a Y matrix for the set of four lines, as described in Section 2.2.3. This would be an 8×8 matrix, which would not be practical to generate or manipulate analytically, although when formulated numerically, a computer could make short work of it. We would need to reduce it, using ordinary circuit-matrix manipulations, to a 4×4 matrix. By reducing it to a two-strip equivalent, we are able to use concepts that are simple and intuitive.

One problem remains: how do we determine the C and L matrices of the set of four lines? As with the determination of single-line properties, this is a problem in electromagnetics. It has been addressed in a number of ways; a particularly elegant one is embodied in a program called *LINPAR* [2.11], based on theory from [2.12]. The LINPAR user's manual also describes the analytical method in substantial detail. That software, in turn, is built into the coupled-line models of Microwave Office, a popular circuit simulator [2.13].

2.3 Balun Design

A *balun* is a transition between balanced and unbalanced transmission media. An unbalanced transmission line uses a ground surface as one conductor; that ground surface is a metal plane, as in microstrip, a shield, as in coaxial line, or some similar structure that is very different from the ungrounded conductor. A balanced line consists of two conductors of identical cross section surrounded by free space. A ground surface is not needed for a balanced line, but if one imagines a ground to exist, the balanced line can be viewed as a pair of coupled, symmetrical, unbalanced lines carrying only an odd mode. The characteristic impedance of the balanced line is twice the odd-mode impedance of the two conductors viewed as coupled, unbalanced lines. Its even-mode characteristic impedance is infinite.

In many cases, an ideal balun can be viewed as a device that is excited in an unbalanced mode and converts it solely to an odd mode. Viewing the operation of a balun in terms of its modal characteristics is the most useful and fruitful way to describe and to understand a balun, and we shall depend on it heavily in the following sections.

Many classical microwave balun designs date from the MIT Radiation Laboratory of the 1940s. The application, in those days, usually was an unbalanced coaxial line feeding a balanced, dipole antenna, which often was a feed for a large reflector antenna. In such an application, the balanced output has no adjacent ground surface and the concept of an even mode has little meaning. Therefore, the balun was described as an interconnection of ordinary transmission lines. In modern applications, though, the balun is more likely to be used in a circuit, often a microstrip or suspended-substrate medium, in which a ground plane is present and very close to the strips. In that case, the possibility of an even mode exists, and it cannot be ignored in the analysis. Therefore, the old, antenna-balun analyses are not very useful for planar baluns, and the full analyses of Sections 2.1 and 2.2 must be used.

2.3.1 Balun Properties

Although a balun is strictly a two-port component, it is useful for design purposes to view it as a three-port component having one input and two outputs. Realized as an electromagnetic structure, it is ideally passive, lossless, and reciprocal. Electronic baluns are sometimes realized, but we do not consider them here.

Figure 2.8 shows a balun. The grounded loads, Z_L, may be surprising, as ground has no meaning (beyond a simple interconnection node) in an ideal balanced transmission medium. We include a ground to account for the nonideal case, in which a weak even mode may exist. If no even mode is present at the output, the output voltages V_2 and V_3 are equal and opposite in phase. The presence of an even mode unbalances the output.

A balun has a number of properties:

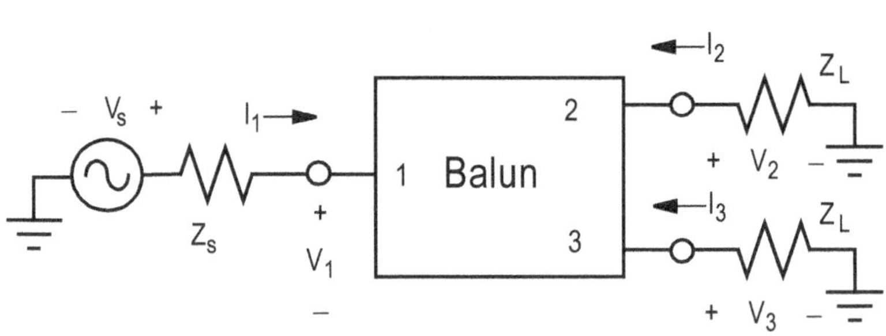

Figure 2.8 A balun is an ideally lossless transition from a balanced to an unbalanced transmission medium. As such, the input must be matched and $V_2 = -V_3$.

1. Above all, a balun is not a power divider. While a balun occasionally can be used to provide power division, the properties of a balun differ from those of a power divider. As we shall see in Section 3.2.4, baluns make poor power dividers.

2. A balun also is not a hybrid junction, although a 180-degree hybrid can sometimes be used in place of a balun. We discuss hybrids in Section 3.2.5.

3. The source impedance, Z_S, need not equal the load impedances, Z_L. The load impedances, however, must be equal, and the input is matched.

4. One of the most important differences between a balun and a hybrid or power divider is in the output match. As we shall see in Section 3.2.3, it is impossible to match simultaneously all three ports of a passive, lossless, reciprocal three-port junction, so the loads Z_L are not matched to their respective ports. If the ground is removed, however, the resulting single output port, consisting of the ungrounded terminals of ports 2 and 3 in Figure 2.8, can be matched, as the structure is simply a two-port component.

5. The even- and odd-mode output characteristics are important to note. If a balun's output is excited in an odd mode, the balun appears matched; that is, if we excite the balun with $I_2 = -I_3$, we find that $V_2 / I_2 = V_3 / I_3$. Most baluns exhibit an output open circuit for even-mode excitation; a few types, however, exhibit a short circuit. Conversely, in a hybrid or power divider, both the even- and odd-mode output impedances are always Z_0. This distinction is an important consideration in the selection of a balun or a hybrid in a circuit application, as some types of circuits make use of the component's even- or odd-mode output-impedance properties. This matter is covered further in Section 8.3.

In a balun configured as in Figure 2.8, the output balance tells us a lot about how effectively we are suppressing the even mode. The amplitude balance is defined as $|20 \log(V_3 / V_2)|$ and the phase balance is the difference in phase of V_3 and V_2, not including the necessary 180 degrees. In a balanced component such as a mixer, frequency multiplier, or amplifier, the existence of an even mode creates a path by which a signal at one port can be coupled to another port, so imperfect even-mode rejection in the balun causes port isolation to suffer. Isolation is an extremely important property of those components.

2.3.2 Application: Parallel-Strip Balun

Figure 2.9 shows a parallel-strip balun. Such baluns are used in a wide variety of microwave and RF components. As with the directional coupler, the balun can be analyzed in a straightforward manner by the use of even- and odd-mode analysis.

In Figure 2.9 we have left out the source impedances to facilitate the analysis. The load impedances are identical, Z_0, and we want the output voltages V_{L1} and V_{L2} to be equal and to differ in phase by 180 degrees. As in the previous section, we use grounded ports to assess the performance of the balun.

Using the port numbering of Figure 2.2, we note that

$$\begin{aligned} V_{L1} &= V_4 = V_{4e} + V_{4o} \\ V_{L2} &= V_3 = V_{3e} + V_{3o} \end{aligned} \tag{2.66}$$

and

$$\begin{aligned} V_{1e} &= V_{2e} = \frac{V_s}{2} \\ I_{1e} &= I_{2e} = \frac{V_s/2}{Z_{\text{in},e}} \end{aligned} \tag{2.67}$$

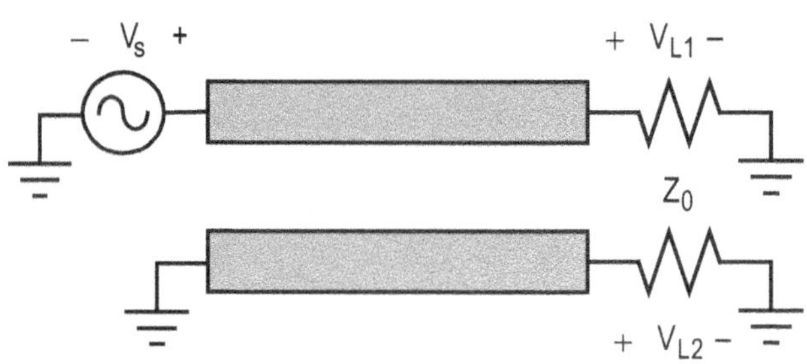

Figure 2.9 A parallel-strip balun consists of a pair of strips that can be edge or broadside coupled.

$$V_{1o} = -V_{2o} = \frac{V_s}{2}$$
$$I_{1o} = -I_{2o} = \frac{V_s/2}{Z_{\text{in},o}} \tag{2.68}$$

From (1.48), and assuming that the structure is lossless and homogeneous,

$$\begin{bmatrix} V_{4e} \\ I_{4e} \end{bmatrix} = \begin{bmatrix} \cos(\theta) & -jZ_{0e}\sin(\theta) \\ \dfrac{j}{Z_{0e}}\sin(\theta) & -\cos(\theta) \end{bmatrix} \begin{bmatrix} V_s/2 \\ \dfrac{V_s/2}{Z_{\text{in},e}} \end{bmatrix} \tag{2.69}$$

so

$$V_{4e} = \frac{V_s}{2}\left(\cos(\theta) - j\frac{Z_{0e}}{Z_{\text{in},e}}\sin(\theta)\right) \tag{2.70}$$

and, similarly, we obtain

$$V_{3e} = \frac{V_s}{2}\left(\cos(\theta) - j\frac{Z_{0e}}{Z_{\text{in},e}}\sin(\theta)\right) \tag{2.71}$$

$$V_{4o} = \frac{V_s}{2}\left(\cos(\theta) - j\frac{Z_{0o}}{Z_{\text{in},o}}\sin(\theta)\right) \tag{2.72}$$

$$V_{3o} = \frac{V_s}{2}\left(-\cos(\theta) + j\frac{Z_{0o}}{Z_{\text{in},o}}\sin(\theta)\right) \tag{2.73}$$

We want $V_{L1} = -V_{L2}$. From this, (2.66), and substituting (2.3) and (2.4) for $Z_{\text{in},e}$ and $Z_{\text{in},o}$, we obtain

$$j\frac{Z_{0e}\cos(\theta) + jZ_0\sin(\theta)}{Z_0\cos(\theta) + jZ_{0e}\sin(\theta)} = \cos(\theta) \tag{2.74}$$

This should be valid at our presumed band center, so we set $\theta = \pi / 2$ and obtain the disturbing result,

$$j\frac{Z_0}{Z_{0e}} = 0 \tag{2.75}$$

This result implies that we must achieve $Z_{0e} \to \infty$ for an ideal balun. While the result is not quite what we might wish, it makes sense intuitively, as the balun selects the odd mode simply by becoming an open circuit for the even mode.

In practice, there is a trade-off between Z_{0e} and output balance. Empirically we find that $Z_{0e} \sim 10Z_{0o}$ is usually adequate to achieve acceptable balance. In microstrip and other dielectric-loaded media, achieving such a high even-mode impedance is virtually impossible, but in others, one can sometimes achieve an adequately high Z_{0e}. A common realization, for example, consists of broadside-coupled lines on a suspended substrate. The substrate thickness is made very thin to minimize the line widths, for a specified Z_{0o}, and air gaps above and below the substrate are as large as possible to maximize Z_{0e}. Another technique, described in more detail in Section 4.3.2, is used in the RF range. In this realization, the balun consists of a twisted pair of wires wrapped around a toroidal ferrite core. The core increases the even-mode inductance, thus increasing the even-mode characteristic impedance (2.63), while having minimal effect on the odd mode. A third option is simply to avoid exciting a strong even mode in the first place. In a transition from a coaxial line to a set of broadside-coupled lines on a suspended substrate, for example, the fields in the coax approximately match those of the odd mode but not the even. As such, they excite the odd mode preferentially.

The rest of the design procedure is straightforward. When $Z_{0e} \to \infty$, the structure behaves as a balanced transmission line whose characteristic impedance is 2 Z_{0o} and is terminated in 2 Z_0. Ideally, the source impedance $Z_s = 2\,Z_0$ and $Z_{0o} = Z_0$; then, the balun is simply a segment of matched transmission line, which has infinite bandwidth. For finite Z_{0e}, however, the situation is somewhat more complex. Usually, Z_{0o} must be modified to account for the finite Z_{0e}, and for reasons discussed below, the length of the balun usually is kept below one-quarter wavelength. Thus, it behaves more like an inductive series line whose impedance can be adjusted to assist in matching a capacitive load.

A practical problem in this type of balun is a resonance occurring at the frequency where the balun is one-quarter wavelength long. This "glitch" decreases in magnitude with increasing Z_{0e}. At any realistic value of Z_{0e}, however, it is troublesome, so the balun length must often be kept below a quarter wavelength at the highest frequency of operation. The lowest frequency depends on the type of circuit in which it is used, but often it is established by the need for a high even-mode output impedance.

2.3.3 Application: Marchand Balun

The Marchand balun has an interesting history. When first conceived, in the early 1940s [2.14], it was realized in coaxial line and intended as an antenna balun. The balun was used only occasionally until approximately 1990, when designers realized that it could be used effectively as a planar balun in microstrip and other strip media. Since then, it has become widely used in both hybrid and monolithic circuits. More modern realizations of the balun differ in concept and structure, but because of their close relationship to Marchand's original design, they are still known as Marchand baluns.

The original coaxial concept of a Marchand balun is shown in Figure 2.10(a) and its equivalent circuit is shown in Figure 2.10(b). It consists of four quarter-wavelength transmission lines whose values can be selected to achieve a Chebyshev characteristic. It is assumed that the lines are homogeneous. As it is an antenna balun, located far from any ground plane, it is assumed that no even mode exists. The balun's sections can then be treated analytically as simple transmission-line segments. Cloete [2.15] shows that a 10:1 bandwidth can be achieved with 15-dB maximum input return loss. Although [2.15] does not include an analysis, it does include graphs of design data. Figure 2.11 shows a graph of impedance values for the design of a balun with a 10:1 bandwidth ratio, where the impedance values are given as a function of maximum reflection coefficient. The balun's design is based on the input reflection coefficient only, as the balance, under the stated assumptions, is theoretically perfect at all frequencies.

The balun of Figure 2.10 is practical for antenna use, but in circuit applications some form of planar realization is needed. In that case, the strip conductors realize Z_1 and Z_2, while Z_3 either does not exist or is realized by the housing ground surfaces. Depending on the application, a quarter-wavelength output line segment, Z_4, may or may not be used. For practical reasons, usually $Z_1 = Z_2$, although greater bandwidth often can be achieved with a lower value of Z_2. The most important difference between a planar realization and the coaxial one is the existence of an even mode. The presence of that mode has a profound effect on the operation of the balun.

Figure 2.12 shows one such balun realized in suspended-substrate stripline with broadside-coupled lines. For and ideal balun, the even- and odd-mode phase velocities on the lines are equal, $Z_{0e} \to \infty$, and $Z_{0o} = Z_0 / \sqrt{2}$. In a real balun, these requirements cannot be met. For best bandwidth, the even-mode impedance must be made as high as possible. The odd-mode impedance and load impedance are then selected to optimize the input return loss over the necessary bandwidth. The resulting load impedance usually is somewhat greater than Z_0. If it is practical to do so, it can be adjusted by a quarter-wave output section (analogous to Z_4 in Figure 2.10), usually at some cost to bandwidth.

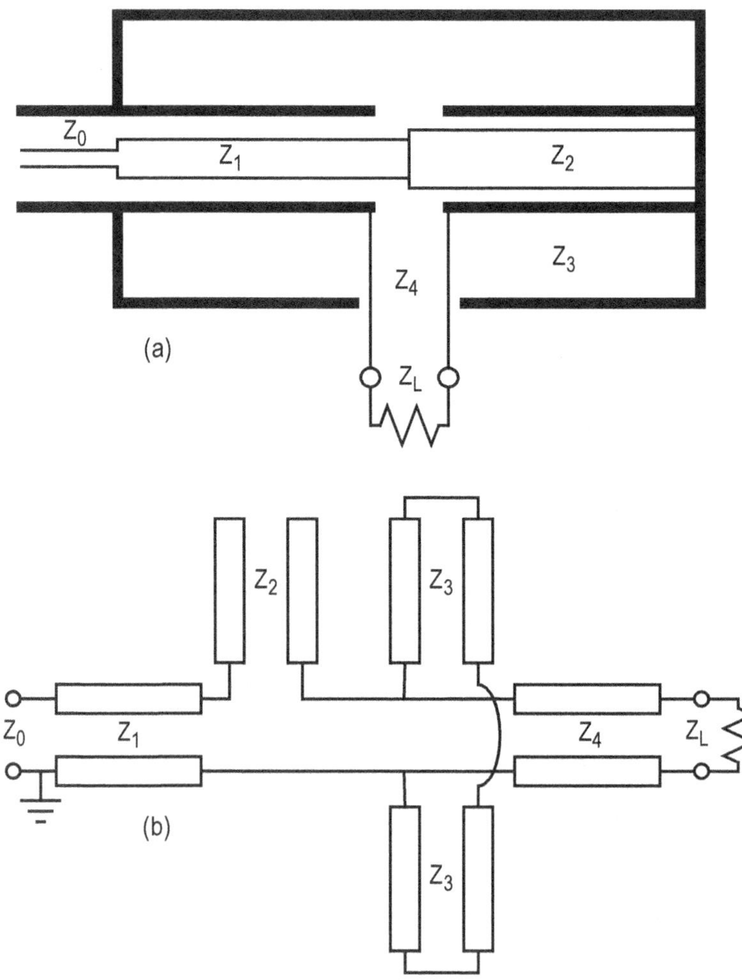

Figure 2.10 A coaxial Marchand balun (a); the equivalent circuit (b). All transmission-line segments are one-quarter wavelength long.

To maximize the even-mode impedance, the strips are made as narrow as possible and the air gap above and below the substrate is made as great as possible, within practical limits. To minimize the strip width for a given odd-mode impedance, the substrate must be thin; thicknesses of 250 μm or even 125 μm are not unusual. As an example, coupled lines on a 125-μm suspended composite substrate with $\varepsilon_r = 2.3$ and an air gap of 3.5 mm have

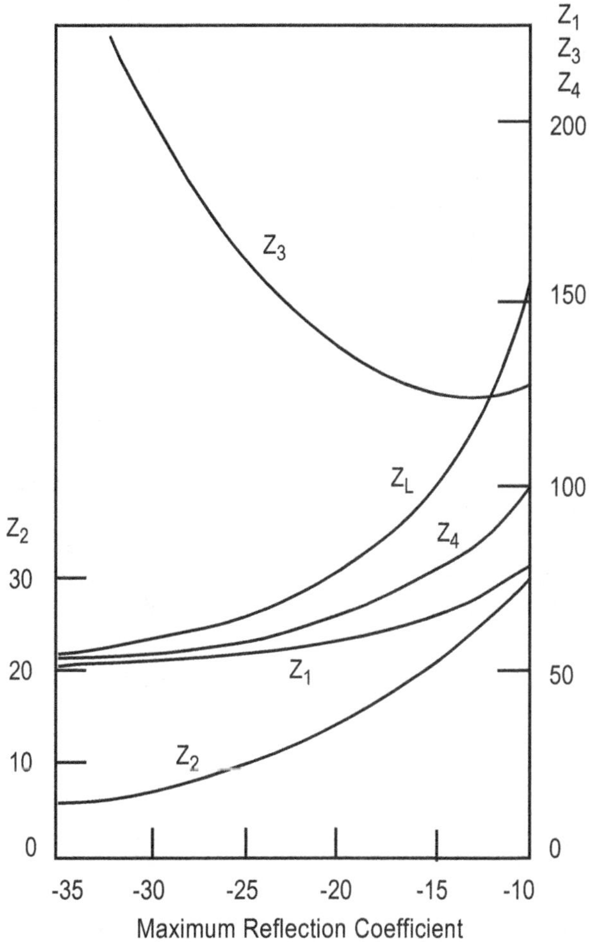

Figure 2.11 Design data for the 10:1 bandwidth Marchand balun in Figure 2.10. This is valid only when the transmission-line sections are coaxial. Reproduced by permission from [2.15], © 1980, Horizon House, Inc.

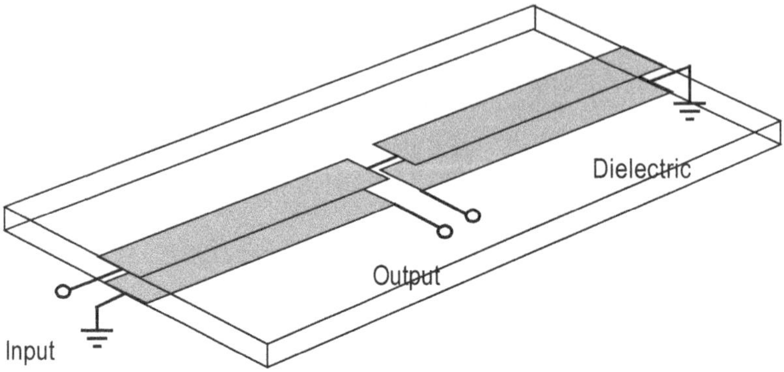

Figure 2.12 Suspended-substrate balun with broadside-coupled strips. The housing top and bottom, which are not shown in the figure, create the ground surfaces, and the lower strips are grounded to the housing walls.

$Z_{0e} \approx 350\Omega$ when $Z_{0o} \approx 30\Omega$. These values are entirely practical for realizing a planar Marchand balun.

In the balun of Figure 2.12, the even- and odd-mode effective dielectric constants are not equal. The even-mode constant is dominated by the air gap, so it is close to unity, while the odd-mode is dominated by the dielectric, so it is only slightly lower than the dielectric constant. The resulting differences in modal phase velocity reduce the bandwidth of the balun and upset its balance. To minimize the difference, a low-dielectric-constant substrate must be used. These are often soft, composite materials, such as polytetrafluoroethylene with glass-fiber reinforcement. "Hard" substrates having high dielectric constants, such as alumina, are not well suited to this kind of balun.

The problem of unequal modal phase velocities can be ameliorated somewhat by placing the coupled lines on the same side of the substrate. On a suspended substrate, the odd-mode effective dielectric constant is decreased slightly, while the even mode is largely unaffected. On a microstrip substrate, the even-mode effective dielectric constant is increased significantly, compared to a suspended substrate, although not to the point where it equals the odd-mode. In microstrip, however, single coupled lines invariably have inadequate coupling, so, as with the Lange coupler, multiple strips must be used. Often (especially in high-dielectric-constant substrates) an odd number of strips gives optimum performance.

Figure 2.13 shows such a balun, realized in microstrip on a gallium arsenide (GaAs) substrate 100 μm thick. The outer strip width is 6 μm and

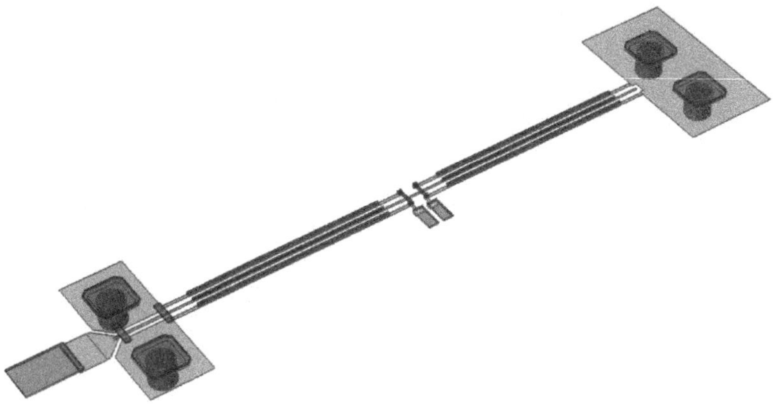

Figure 2.13 A planar Marchand balun can be realized with three strips of unequal width. This type of balun is especially useful for monolithic circuits. The figure shows a three-dimensional layout rendering.

the inner one 4 μm; the strip spacing is 11 μm and the overall length of each coupled section is 484 μm. Figure 2.14 shows a plot of the balun's performance from 30 to 50 GHz. This type of balun is too complex to be treated analytically, but circuit analysis software, based on the formulation of Section 2.2, can be applied. Electromagnetic simulation is used to analyze the via holes and pads, and the strips are characterized by a version of LINPAR embedded in Microwave Office [2.11-2.13]. In this case, the strip thicknesses (~3 μm) are not small relative to the spacing, so it is imperative that the capacitance matrix include the strips' sidewall components. Unlike the rather idealized baluns of Figures 2.10 and 2.12, this analysis includes all important parasitics:

- The inductance of the vias and capacitance of the pads;
- The transition from the input microstrip to the balun;
- The short microstrip connection between the center strips at the center of the balun;
- The air bridges at the output.

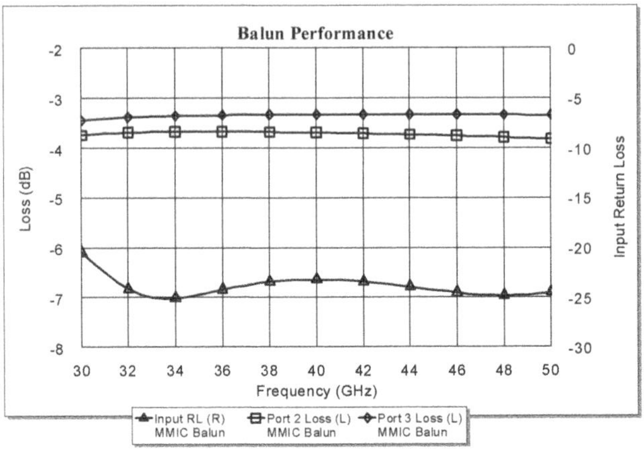

Figure 2.14 Performance of the balun of Figure 2.13. Parasitics and unequal modal phase velocities cause the balance to be imperfect, 0.4 dB worst case.

While this balun's balance is not perfect, it is also not bad: the worst-case amplitude imbalance is 0.4 dB and less than 0.3 dB across most of the band. Although not shown in the figure, the worst-case phase imbalance is only one degree. This type of balun has been used extensively in practical, monolithic mixers [2.16]

2.3.4 Application: Half-Wave Balun

A half-wavelength transmission line can be used as a simple, narrowband balun. A set of half-wavelength lines, with quarter-wave coupling, can realize a balun with much greater bandwidth. The reason for the good performance of the multisection balun can be made clear through the use of even- and odd-mode analysis.

Figure 2.15 shows the balun. The lines generally do not have the same characteristic impedances, and the bandwidth of the balun is a function of the number of sections.

We apply even- and odd-mode analysis by treating the lower left corner of the structure in Figure 2.15 as an open-circuited input, thereby making it a symmetrical four-port. We then generate even- and odd-mode equiva-

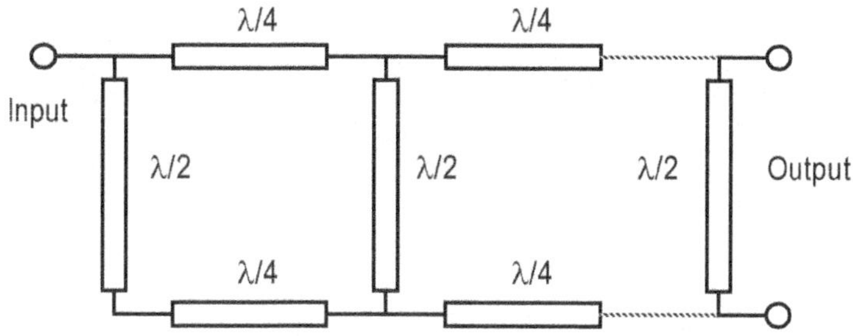

Figure 2.15 The multisection half-wavelength balun consists of half-wavelength branch lines coupled by quarter-wavelength series lines. The characteristic impedances are, in general, unequal.

lents, as shown in Figure 2.16. No terminating impedances are used at the input ports, so the lower port in Figure 2.16 remains an open circuit. In the even mode, symmetry dictates that the centers of the half-wavelength lines are virtual open circuits. These open circuits split the balun into two identical parts, so only one part must be analyzed. The structure is clearly a type of band-reject filter, showing that the balun effectively rejects the even mode. In the odd mode, the centers become virtual grounds. Again, the grounds separate the structure into two identical ones, in this case realizing a bandpass filter.

The even-mode circuit also shows that the balun presents a short circuit to even-mode excitation at the output. This is an important property, as most baluns present an open circuit to even-mode excitation. In many types of balanced nonlinear circuits, the short-circuit termination provides better performance than the open, so this characteristic is valuable in many types of circuits.

It is tempting to design the balun according to classical filter theory. This may not be possible, however, because it involves two separate designs, which probably will result in different line impedances for the two kinds of filters. The design must also account for the singly terminated characteristic of the filters. Even so, empirical experience shows that it is possible to design such a balun to achieve broad bandwidth with good even-mode rejection using a circuit simulator and numerical optimization.

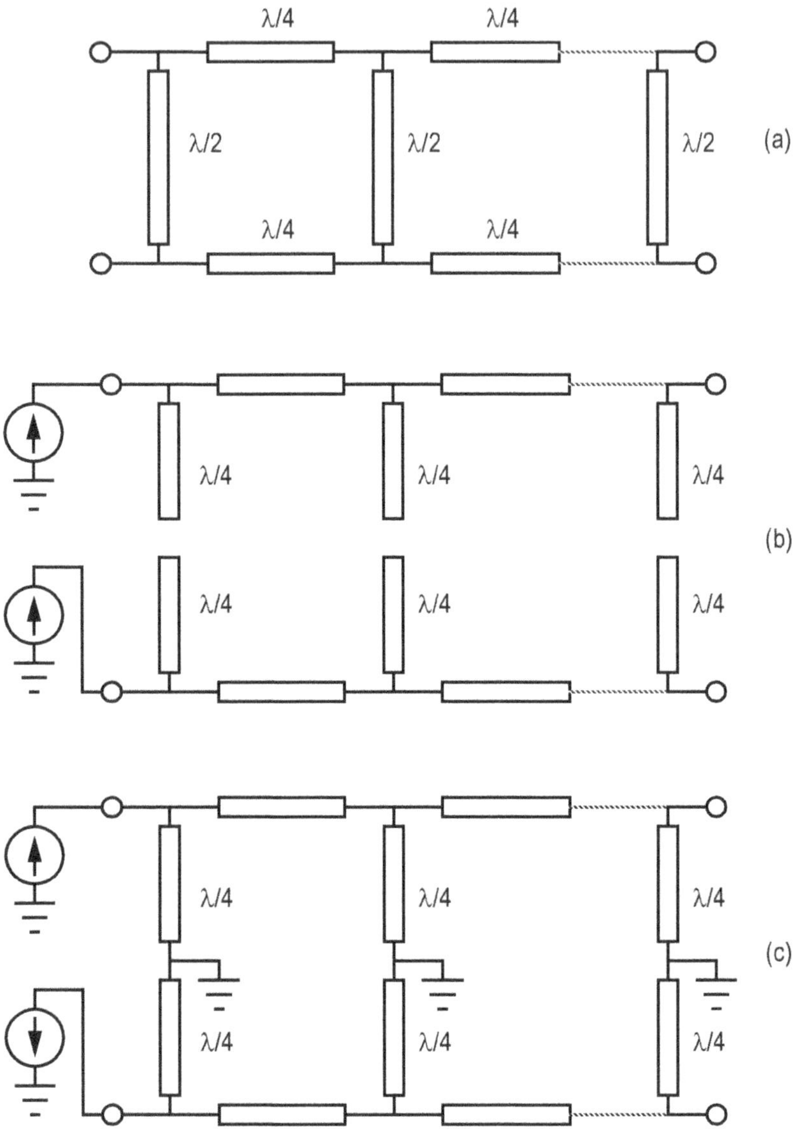

Figure 2.16 Half-wavelength balun with an additional input (a); even-mode equivalent circuit (b); odd-mode equivalent circuit (c).

References

[2.1] Mongia, R., I. Bahl, and P. Bhartia, *RF and Microwave Coupled-Line Circuits*, Norwood, MA: Artech House, 1999.

[2.2] D. Paolino, "MIC Overlay Coupler Design Using Spectral Domain Techniques," *IEEE Trans. Microwave Theory Tech.*, Vol. MTT-26, 1978, p. 646.

[2.3] Scanlan, J. O., "Theory of Microwave Coupled-Line Networks," *Proc. IEEE*, Vol. 68, 1980, p. 209.

[2.4] Lange, J., "Interdigital Stripline Quadrature Hybrid," *IEEE Trans. Microwave Theory Tech.*, Vol. MTT-17, 1969, p. 1363.

[2.5] Tajima, Y., and S. Kamihashi, "Multiconductor Couplers," *IEEE Trans. Microwave Theory Tech.*, Vol. MTT-26, 1978, p. 795.

[2.6] Osmani, R. M., "Synthesis of Lange Couplers," *IEEE Trans. Microwave Theory Tech.*, Vol. MTT-29, 1981, p. 168.

[2.7] Ou, W. P., "Design Equations for an Interdigitated Directional Coupler," *IEEE Trans. Microwave Theory Tech.*, Vol. MTT-25, 1975, p. 253.

[2.8] Presser, A., "Interdigitated Microstrip Coupler Design," *IEEE Trans. Microwave Theory Tech.*, Vol. MTT-26, 1978, p. 801.

[2.9] Paolino, D., "Design More Accurate Interdigitated Couplers," *Microwaves*, May, 1976, p. 34.

[2.10] Maas, S. A., "Accurate Design of Lange Couplers on GaAs," *Microwave J.*, Vol. 39, Aug. 1996, p. 90.

[2.11] Djordjevic, A. R., et al., *LINPAR for Windows*, Norwood, MA: Artech House, 1999.

[2.12] Wei, C., R. F. Harrington, J. R. Mautz, and T. K. Sarkar, "Multiconductor Transmission Lines in Multilayered Media," *IEEE Trans. Microwave Theory Tech.*, Vol. MTT-32, 1984, p. 439.

[2.13] AWR Corp., El Segundo, California, USA.

[2.14] Marchand, N., "Transmission Line Conversion Transformers," *Electronics*, Vol. 17, 1944, p. 142.

[2.15] Cloete, J. H., "Exact Design of the Marchand Balun," *Microwave J.*, Vol. 23, May, 1980, p. 99.

[2.16] Maas, S. A., and K. W. Chang, "A Broadband, Planar, Doubly Balanced Monolithic Ka-Band Diode Mixer," *IEEE Trans. Microwave Theory Tech.*, Vol. MTT-41, 1993, p. 2330.

Chapter 3

Scattering Parameters

Scattering parameters, or S parameters, are a form of multiport parameter based on a wave formulation instead of port voltages and currents. They came into common use in the 1960s and 1970s as accurate, automated microwave network analyzers were created. Since voltages and currents in high-frequency circuits were not directly measurable, but travelling waves were, it seemed logical to employ circuit characterizations that were based on travelling waves.

Today one could argue that the need for S parameters has disappeared. With modern techniques, voltages and currents in microwave circuits can be measured directly, and, in any case, it is possible to define equivalences between S parameters and other multiport characterizations. Indeed, S parameters imported into a circuit simulator are immediately converted to Y parameters. S parameters do, however, have certain advantages in simplifying an analysis and providing an intuitive sense of the behavior of distributed circuits. A significant advantage is that circuits having no Z or Y matrix, such as an ideal transformer, almost always have an S-parameter characterization. They are also, whether for historical or technical reasons, a standard in the microwave world. They are a language that we all must speak.

A comment on the notation in this chapter: In the following sections, boldface type is used to designate vectors and matrices. In vectors and matrices, a superscript asterisk represents a conjugate transpose, while in scalars, it represents a simple complex conjugate. While the use of the same notation to indicate two different operations may be a little confusing, it is, for better or worse, a de facto standard.

3.1 Circuit Description in Terms of Wave Quantities

3.1.1 Voltage Waves and Power Waves

In Chapter 1, we viewed travelling waves as voltages or currents. A voltage wave does not exist without a current wave, nor a current wave without a voltage wave, so those waves also represent power. It turns out to be more convenient, in circuit analysis, to express those waves as the square root of power, rather than as voltage or current.[1] Thus, we define two waves at each port of a multiport network, an incident wave (travelling toward the port) and reflected wave (away from the port), each of which has units of the square root of power. One can view these waves as purely abstract and define them even when no distributed network exists. Alternatively, one can imagine a transmission line connected to each port with the incident and reflected waves propagating along those lines.

In either case, however, we designate the incident wave at the nth port as the a_n wave and the reflected one as the b_n wave. Those waves are defined as[2]

$$a_n = \frac{1}{2}\left(\frac{1}{\sqrt{Z_{0n}}}V_n + \sqrt{Z_{0n}}I_n\right) \tag{3.1}$$

$$b_n = \frac{1}{2}\left(\frac{1}{\sqrt{Z_{0n}}}V_n - \sqrt{Z_{0n}}I_n\right) \tag{3.2}$$

where V_n, I_n are the voltage and current at the nth port and Z_{0n} is the *port normalizing impedance*. That impedance is an abstract quantity and can be set to any convenient value. It need not be the same at all ports. It can be complex, but here we limit our concern only to real impedances. If the component is designed to have a certain port impedance, such as the 50Ω

1. The question of "voltage waves" versus "power waves" received a lot of attention in the mid-20th century, with a consensus eventually forming around power waves. One reason for the preference for power waves was the odd behavior of voltage waves when the normalizing impedances of the circuit were complex. For the real normalizing impedances that we consider exclusively, these problems do not appear.

2. These quantities can be derived in a formal manner. Repeating that derivation is more than we wish to deal with, and, in the end, doesn't add much more than simply defining them. The more academic among us can find this derivation in [3.1].

standard used throughout the RF and microwave industries, it is logical to use that for Z_{0n}. Adding (3.1) and (3.2), we can easily see that

$$V_n = (a_n + b_n)\sqrt{Z_{0n}} \tag{3.3}$$

and subtracting them gives

$$I_n = \frac{a_n - b_n}{\sqrt{Z_{0n}}} \tag{3.4}$$

If we imagine that a transmission line of characteristic impedance Z_{0n} is connected to port n, from (3.3) and Chapter 1 we find that

$$V_n = (a_n + b_n)\sqrt{Z_{0n}} = V_{fn} + V_{rn} \tag{3.5}$$

where V_{fn} and V_{rn} are the forward and reflected waves, respectively, at port n. Since a_n and V_{fn} are incident waves, and b_n, V_{rn} are reflected, we can equate them and see that

$$\begin{aligned} V_{fn} &= a_n\sqrt{Z_{0n}} \\ V_{rn} &= b_n\sqrt{Z_{0n}} \end{aligned} \tag{3.6}$$

Similarly,

$$I_n = \frac{a_n - b_n}{\sqrt{Z_{0n}}} = I_{fn} - I_{rn} \tag{3.7}$$

where I_{fn} and I_{rn} are likewise the incident and reflected current waves at port n. Then,

$$\begin{aligned} I_{fn} &= \frac{a_n}{\sqrt{Z_{0n}}} \\ I_{rn} &= \frac{b_n}{\sqrt{Z_{0n}}} \end{aligned} \tag{3.8}$$

so the a_n and b_n waves are easily related to the voltage and current ones. This is an important point, as it implies that the reflection coefficient in terms of voltage or current waves is the same as that in terms of the a and b waves.

3.1.2 The Scattering Matrix

The vectors of a and b waves at the ports of an N-port network are related as

$$\begin{bmatrix} b_1 \\ b_2 \\ \cdots \\ b_N \end{bmatrix} = \begin{bmatrix} S_{11} & S_{12} & \cdots & S_{1N} \\ S_{21} & S_{22} & \cdots & S_{2N} \\ \cdots & \cdots & \cdots & \cdots \\ S_{N1} & S_{N2} & \cdots & S_{NN} \end{bmatrix} \begin{bmatrix} a_1 \\ a_2 \\ \cdots \\ a_N \end{bmatrix} \tag{3.9}$$

or

$$\mathbf{b} = \mathbf{Sa} \tag{3.10}$$

where $\mathbf{S}$ is called the *scattering matrix* or simply the *S matrix*. One can view the S matrix as a kind of generalized reflection coefficient; where

$$S_{nm} = \left.\frac{b_n}{a_m}\right|_{a_p = 0,\, p \neq m} \tag{3.11}$$

The condition $a_p = 0,\ p \neq m$, means that there is no incident wave on any port other than port m. That can be guaranteed only if the port is terminated in its normalizing impedance, Z_{0p}, and, of course, has no excitation. Thus, S_{nm} in (3.11) simply describes the reflected wave from port n as a response to the incident wave on port m when all other ports have no excitation and are terminated in their normalizing impedances. The terms along the main diagonal of the S matrix are therefore the port reflection coefficients of the terminated multiport. The off-diagonal elements can be interpreted in terms of transducer gain.

The available power from a source having impedance Z_{0m}, exciting port m, is

$$P_{av,m} = \frac{|V_{fm}|^2}{2Z_{0m}} = \frac{|a_m|^2}{2} \tag{3.12}$$

where (3.6) has been used.[3] Similarly, the power delivered to the load at port n is

$$P_{d,n} = \frac{|V_{rn}|^2}{2Z_{0n}} = \frac{|b_n|^2}{2} \tag{3.13}$$

so the transducer gain between ports m and n, G_{tnm}, is

$$G_{tnm} = \frac{P_{d,n}}{P_{av,m}} = \left|\frac{b_n}{a_m}\right|^2 = |S_{nm}|^2 \tag{3.14}$$

The squared magnitude of an off-diagonal element in the (n, m) position is the transducer gain between ports m and n, with all ports, including m and n, terminated in their respective normalizing impedances.

Conversion between the S matrix and other multiport matrices is straightforward. Consider an impedance (Z) matrix,

$$\mathbf{V} = \mathbf{ZI} \tag{3.15}$$

We can write this

$$\mathbf{Z}_0^{-1/2}\mathbf{V} = \mathbf{Z}_0^{-1/2}\mathbf{Z}\mathbf{Z}_0^{-1/2} \cdot \mathbf{Z}_0^{1/2}\mathbf{I} \tag{3.16}$$

where $\mathbf{Z}_0$ is a diagonal matrix whose elements are Z_{0n}. From (3.3) and (3.4), (3.16) becomes

$$\mathbf{a} + \mathbf{b} = \mathbf{Z_n}(\mathbf{a} - \mathbf{b}) \tag{3.17}$$

where $\mathbf{Z_n}$ is the normalized Z matrix,

3. Here, and throughout this book, voltages and currents are phasor quantities. In other treatments [3.1, 3.2], they are treated as RMS quantities and the 1/2 disappears. Usually we deal with ratios of the a and b quantities, so the 1/2 factor cancels and the distinction doesn't matter, but occasionally it can be important.

$$\mathbf{Z_n} = \mathbf{Z}_0^{-1/2}\mathbf{Z}\mathbf{Z}_0^{-1/2} \tag{3.18}$$

Equation (3.17) can be solved easily to obtain

$$\mathbf{b} = (\mathbf{Z_n} + \mathbf{1})^{-1}(\mathbf{Z_n} - \mathbf{1})\mathbf{a} \tag{3.19}$$

and from inspection,

$$\mathbf{S} = (\mathbf{Z_n} + \mathbf{1})^{-1}(\mathbf{Z_n} - \mathbf{1}) \tag{3.20}$$

where **1** is the identity matrix. Equation (3.20) is a multiport generalization of our previous expression for the reflection coefficient of a simple impedance, which can be written

$$\Gamma = \frac{Z/Z_0 - 1}{Z/Z_0 + 1} \tag{3.21}$$

Converting from S to Z is likewise straightforward. One need merely solve (3.20) for $\mathbf{Z_n}$; the result is

$$\mathbf{Z_n} = (1 - \mathbf{S})^{-1}(1 + \mathbf{S}) \tag{3.22}$$

and $\mathbf{Z_n}$ can be returned to unnormalized form by pre- and postmultiplying by $\mathbf{Z}_0^{1/2}$.

Expressions relating the S matrix to other types of multiport parameters, such as Y parameters, can be derived similarly. For Y parameters, we obtain

$$\mathbf{S} = (\mathbf{1} + \mathbf{Y_n})^{-1}(\mathbf{1} - \mathbf{Y_n}) \tag{3.23}$$

where $\mathbf{Y_n}$ is the normalized Y matrix,

$$\mathbf{Y_n} = \mathbf{Z_n}^{-1} \tag{3.24}$$

$\mathbf{Y_n}$, given **S**, comes directly from our previous results:

$$\mathbf{Y_n} = \mathbf{Z_n^{-1}} = ((1 - \mathbf{S})^{-1}(1 + \mathbf{S}))^{-1} = (1 + \mathbf{S})^{-1}(1 - \mathbf{S}) \tag{3.25}$$

It is interesting and often helpful to recognize that the matrix terms in (3.20) and (3.23) commute.

3.1.3 S-Parameter Renormalization

Renormalization is the process of changing the values of a network's normalizing impedances. This is sometimes necessary in dealing with RF and microwave circuits. The derivation is lengthy but straightforward; imagine premultiplying (3.18) by $\hat{Z}_0^{-1/2} Z_0^{1/2}$ and postmultiplying by $Z_0^{1/2} \hat{Z}_0^{-1/2}$, where $\hat{Z}_0^{-1/2}$ is the new normalization matrix, and carrying through the rest of the matrix algebra. We spare the reader this tedious exercise and give only the result:

$$\hat{\mathbf{S}} = \mathbf{A}^{-1}(\mathbf{S} - \mathbf{R})(\mathbf{1} - \mathbf{RS})^{-1}\mathbf{A} \tag{3.26}$$

where $\hat{\mathbf{S}}$ is the renormalized matrix and

$$\mathbf{A} = \operatorname{diag}\left(\sqrt{\frac{\hat{Z}_0}{Z_{0n}}} \frac{1}{\hat{Z}_{0n} + Z_{0n}}\right) \tag{3.27}$$

$$\mathbf{R} = \operatorname{diag}\left(\frac{\hat{Z}_{0n} - Z_{0n}}{\hat{Z}_{0n} + Z_{0n}}\right) \tag{3.28}$$

Z_{0n} is the old normalizing impedance for the nth port, and $\hat{Z}_{0n}$ is the new normalizing impedance.

3.1.4 Circuit Interconnections

S parameters can provide a simple, clean method for finding a matrix representing interconnected multiports, particularly in cases where the use of matrices based on voltage and current is messy and complicated. Here we describe an algebraic approach; in Chapter 5, we will examine methods using network graphs. Consider, for example, the two interconnected circuits in Figure 3.1. We wish to find the S parameters of the combination, between ports 1 and 2.

Network m is described by S matrix S_m, which can be expressed as follows:

$$\begin{bmatrix} b_{1m} \\ b_{2m} \\ \hline b_{3m} \\ b_{4m} \\ b_{5m} \end{bmatrix} = \left[\begin{array}{c|c} \mathbf{S}_{11m} & \mathbf{S}_{12m} \\ \hline \mathbf{S}_{21m} & \mathbf{S}_{22m} \end{array}\right] \begin{bmatrix} a_{1n} \\ a_{2n} \\ \hline a_{3n} \\ a_{4n} \\ a_{5n} \end{bmatrix} \tag{3.29}$$

We have partitioned the matrix to represent it as four submatrices. Circuit n is simply

$$\begin{bmatrix} b_{1n} \\ b_{2n} \\ b_{3n} \end{bmatrix} = \mathbf{S}_n \begin{bmatrix} a_{1n} \\ a_{2n} \\ a_{3n} \end{bmatrix} \tag{3.30}$$

From the interconnection, and with the assumption that the mated ports have the same normalizing impedances,

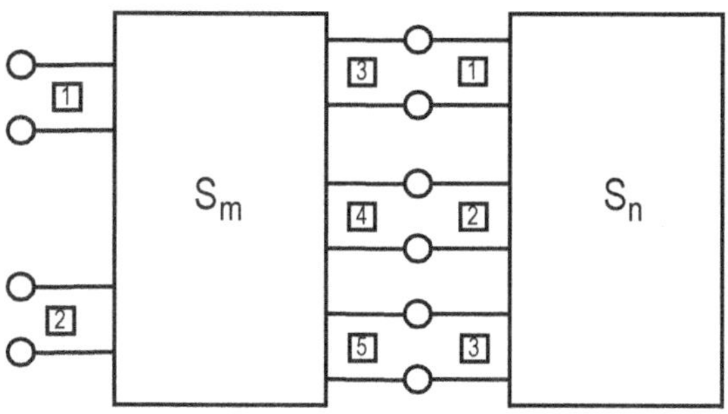

Figure 3.1 Two interconnected multiports. Ports 3, 4, and 5 of network m are connected to ports 1, 2, and 3 of network n. We assume that the connected ports have the same normalizing impedances.

$$\begin{bmatrix} a_{3m} \\ a_{4m} \\ a_{5m} \end{bmatrix} = \begin{bmatrix} b_{1n} \\ b_{2n} \\ b_{3n} \end{bmatrix} \qquad \begin{bmatrix} b_{3m} \\ b_{4m} \\ b_{5m} \end{bmatrix} = \begin{bmatrix} a_{1n} \\ a_{2n} \\ a_{3n} \end{bmatrix} \tag{3.31}$$

When the components are connected, we can express (3.30) as

$$\begin{bmatrix} a_{3m} \\ a_{4m} \\ a_{5m} \end{bmatrix} = \mathbf{S}_n \begin{bmatrix} b_{3m} \\ b_{4m} \\ b_{5m} \end{bmatrix} \tag{3.32}$$

Expanding (3.29) gives

$$\begin{bmatrix} b_{1m} \\ b_{2m} \end{bmatrix} = \mathbf{S}_{11m} \begin{bmatrix} a_{1m} \\ a_{2m} \end{bmatrix} + \mathbf{S}_{12m} \begin{bmatrix} a_{3m} \\ a_{4m} \\ a_{5m} \end{bmatrix} \tag{3.33}$$

$$\begin{bmatrix} b_{3m} \\ b_{4m} \\ b_{5m} \end{bmatrix} = \mathbf{S}_{21m} \begin{bmatrix} a_{1m} \\ a_{2m} \end{bmatrix} + \mathbf{S}_{22m} \begin{bmatrix} a_{3m} \\ a_{4m} \\ a_{5m} \end{bmatrix} \tag{3.34}$$

Substituting (3.34) into (3.32) results in

$$\begin{bmatrix} a_{3m} \\ a_{4m} \\ a_{5m} \end{bmatrix} = \mathbf{S}_n \mathbf{S}_{21m} \begin{bmatrix} a_{1m} \\ a_{2m} \end{bmatrix} + \mathbf{S}_n \mathbf{S}_{22m} \begin{bmatrix} a_{3m} \\ a_{4m} \\ a_{5m} \end{bmatrix} \tag{3.35}$$

Moving the rightmost term to the left side gives

$$(\mathbf{1} - \mathbf{S}_n\mathbf{S}_{22m})\begin{bmatrix} a_{3m} \\ a_{4m} \\ a_{5m} \end{bmatrix} = \mathbf{S}_n\mathbf{S}_{21m}\begin{bmatrix} a_{1m} \\ a_{2m} \end{bmatrix} \tag{3.36}$$

and moving the matrix coefficient to the right gives

$$\begin{bmatrix} a_{3m} \\ a_{4m} \\ a_{5m} \end{bmatrix} = (\mathbf{1} - \mathbf{S}_n\mathbf{S}_{22m})^{-1}\mathbf{S}_n\mathbf{S}_{21m}\begin{bmatrix} a_{1m} \\ a_{2m} \end{bmatrix} \tag{3.37}$$

Substituting (3.37) into (3.33) and collecting terms gives the S matrix of the combination,

$$\begin{bmatrix} b_{1m} \\ b_{2m} \end{bmatrix} = (\mathbf{S}_{11m} + \mathbf{S}_{12m}(\mathbf{1} - \mathbf{S}_n\mathbf{S}_{22m})^{-1}\mathbf{S}_n\mathbf{S}_{21m})\begin{bmatrix} a_{1m} \\ a_{2m} \end{bmatrix} \tag{3.38}$$

We have gone through this in perhaps more algebraic detail than necessary, but with a specific purpose: to show that simple algebraic manipulations can be applied to S matrices in largely the same way that they are applied to scalar quantities in linear equations. Although this should be obvious, from a mathematical view, we have consistently observed reluctance by engineers to apply that kind of algebra to S matrices, as if it is "off limits" to S matrices for some reason. This development may dispel the widely held fear of applying ordinary matrix manipulations to S matrices.

Despite the complicated appearance of (3.38), one can easily show that the submatrices are conformable. As long as the original partition is correct, that will always be the case with this kind of manipulation.

As a final note, we reemphasize that the mated ports must have the same normalizing impedances; given the widespread use of a 50Ω standard impedance in the RF and microwave world, that usually will be the case. If not, one of the matrices can be renormalized (Section 3.1.3) to satisfy this requirement.

3.2 Properties of the Scattering Matrix

3.2.1 General Properties

As scattering parameters are fundamentally based on power concepts, it is useful to examine their properties in terms of power. We begin by generating an S-parameter expression for the power dissipated in a network, then apply special cases to obtain some remarkable results.

The power in a multiport network is

$$\begin{aligned} P &= (1/2)\mathrm{Re}\{\mathbf{V}^*\mathbf{I}\} \\ &= (1/2)\mathrm{Re}\{(\mathbf{Z}_0^{1/2}(\mathbf{a}+\mathbf{b}))^*\mathbf{Z}_0^{-1/2}(\mathbf{a}-\mathbf{b})\} \end{aligned} \tag{3.39}$$

where the superscript asterisk indicates the conjugate transpose of a vector or matrix, and we have used vector generalizations of (3.3) and (3.4). This expression includes both dissipated (real) and reactive (imaginary) power. In the following, we shall make repeated use of a valuable relation, which states that the conjugate transpose of a product equals the product of the conjugate transposes in reverse order. Expressed mathematically, for any conformable matrices **M** and **N**,

$$(\mathbf{MN})^* = \mathbf{N}^*\mathbf{M}^* \tag{3.40}$$

Applying this to (3.39), we obtain

$$P = (1/2)\mathrm{Re}\{(\mathbf{a}^*+\mathbf{b}^*)(\mathbf{Z}_0^{1/2})^*\mathbf{Z}_0^{-1/2}(\mathbf{a}-\mathbf{b})\} \tag{3.41}$$

$\mathbf{Z}_0$ is, of course, real, so

$$\begin{aligned} P &= (1/2)\mathrm{Re}\{(\mathbf{a}^*+\mathbf{b}^*)(\mathbf{a}-\mathbf{b})\} \\ &= (1/2)\mathrm{Re}\{(\mathbf{a}^*\mathbf{a}-\mathbf{b}^*\mathbf{b})+(\mathbf{b}^*\mathbf{a}-\mathbf{a}^*\mathbf{b})\} \\ &= (1/2)\mathrm{Re}\{(\mathbf{a}^*\mathbf{a}-\mathbf{b}^*\mathbf{b})+(\mathbf{b}^*\mathbf{a}-(\mathbf{b}^*\mathbf{a})^*)\} \end{aligned} \tag{3.42}$$

The rightmost term in (3.42) is clearly imaginary, so it does not contribute to the real power, and the leftmost term is clearly real. Thus,

$$P = (1/2)(\mathbf{a}^*\mathbf{a}-\mathbf{b}^*\mathbf{b}) \tag{3.43}$$

Substituting (3.10) in the rightmost product gives

$$P = (1/2)(\mathbf{a}^*\mathbf{a} - \mathbf{a}^*\mathbf{S}^*\mathbf{S}\mathbf{a}) = (1/2)\,\mathbf{a}^*(\mathbf{1} - \mathbf{S}^*\mathbf{S})\mathbf{a} \quad (3.44)$$

which we finally express as

$$P = (1/2)\,\mathbf{a}^*\mathbf{Q}\mathbf{a} \quad (3.45)$$

where

$$\mathbf{Q} = \mathbf{1} - \mathbf{S}^*\mathbf{S} \quad (3.46)$$

Taking the conjugate transpose of $\mathbf{Q}$ and applying (3.40) to the $\mathbf{S}^*\mathbf{S}$ term shows that

$$\mathbf{Q}^* = (\mathbf{1} - \mathbf{S}^*\mathbf{S})^* = \mathbf{1} - \mathbf{S}^*\mathbf{S} = \mathbf{Q} \quad (3.47)$$

The conjugate transpose of $\mathbf{Q}$ is equal to itself; such matrices are called *Hermitian* or *self-adjoint*. Furthermore, if the network is passive, the power dissipation must be zero or positive, so $\mathbf{Q}$ is positive semidefinite. Then, the principal cofactors of $\mathbf{Q}$ must be greater than or equal to zero:

$$c_{jj} = 1 - \sum_i S_{ij}^* S_{ij} = 1 - \sum_i |S_{ij}|^2 \geq 0 \quad (3.48)$$

We conclude that

$$|S_{ij}| \leq 1 \quad (3.49)$$

All S parameters of a passive network must be equal to or less than unity. This really should come as no surprise, in view of the interpretation of the S parameters given in Section 3.1.2.

Finally, if the network is lossless, power dissipation is zero for any vector $\mathbf{a}$. From (3.45), this can happen only if $\mathbf{Q}$ is zero. Then,

$$\mathbf{S}^*\mathbf{S} = \mathbf{S}\mathbf{S}^* = \mathbf{1} \quad (3.50)$$

and the conjugate transpose of **S** is its inverse. The S matrix of a lossless network is therefore *unitary*. From this property, we can derive some useful characteristics of passive multiport components.

3.2.2 Two-Ports

For a lossless passive two-port, (3.50) implies

$$\begin{bmatrix} S_{11}^* & S_{21}^* \\ S_{12}^* & S_{22}^* \end{bmatrix} \begin{bmatrix} S_{11} & S_{12} \\ S_{21} & S_{22} \end{bmatrix} = \begin{bmatrix} 1 & 0 \\ 0 & 1 \end{bmatrix} \tag{3.51}$$

Multiplying the matrices gives, among other terms,

$$\begin{aligned} |S_{11}|^2 + |S_{21}|^2 &= 1 \\ |S_{22}|^2 + |S_{12}|^2 &= 1 \end{aligned} \tag{3.52}$$

These describe a kind of power-conservation law for lossless two ports, stating that all incident power must be either reflected from the port or transferred to the other port. If we reverse the order of multiplication in (3.51), which is legitimate, as the product of a matrix and its inverse always commutes, we obtain additionally

$$|S_{22}|^2 + |S_{21}|^2 = 1 \tag{3.53}$$

From (3.52) and (3.53) it is apparent that

$$|S_{11}| = |S_{22}| \tag{3.54}$$

and

$$|S_{21}| = |S_{12}| \tag{3.55}$$

Equation (3.55) at first seems unsurprising, as any ordinary passive, reciprocal network would have $S_{21} = S_{12}$. However, we have not made any assumptions of reciprocity in the above derivation. Indeed, we shall see that, in the case of lossless, passive three-ports, any assumption of reciprocity would not be in our best interest.

3.2.3 Three-Ports

The unitary property of the S matrix has some remarkable implications for lossless, passive three-ports. For such components, (3.50) becomes

$$\begin{bmatrix} S_{11} & S_{12} & S_{13} \\ S_{21} & S_{22} & S_{23} \\ S_{31} & S_{32} & S_{33} \end{bmatrix} \begin{bmatrix} S_{11}^* & S_{21}^* & S_{31}^* \\ S_{12}^* & S_{22}^* & S_{32}^* \\ S_{13}^* & S_{23}^* & S_{33}^* \end{bmatrix} = \begin{bmatrix} 1 & 0 & 0 \\ 0 & 1 & 0 \\ 0 & 0 & 1 \end{bmatrix} \tag{3.56}$$

It's not unusual that we would like to have all three ports matched, $S_{11} = S_{22} = S_{33} = 0$. As we shall see, however, that's not something we always can do. Setting all the $S_{ii} = 0$, we obtain, from (3.56),

$$|S_{12}|^2 + |S_{13}|^2 = 1 \tag{3.57}$$

$$|S_{21}|^2 + |S_{23}|^2 = 1 \tag{3.58}$$

$$|S_{31}|^2 + |S_{32}|^2 = 1 \tag{3.59}$$

and

$$\begin{aligned} S_{13}S_{23}^* &= 0 \\ S_{12}S_{32}^* &= 0 \\ S_{21}S_{31}^* &= 0 \end{aligned} \tag{3.60}$$

(Three more equations result, but they turn out to be equivalent to the above three.) A little consideration shows that there are only two solutions to (3.57) – (3.59) and (3.60):

$$\begin{aligned} S_{13} &= S_{32} = S_{21} = 1 \\ S_{31} &= S_{23} = S_{12} = 0 \end{aligned} \tag{3.61}$$

or

$$
\begin{aligned}
S_{31} &= S_{23} = S_{12} = 1 \\
S_{13} &= S_{32} = S_{21} = 0
\end{aligned}
\tag{3.62}
$$

These equations describe a *circulator*, a lossless three-port in which all the ports are matched. According to (3.61), a signal incident on port 1 emerges from port 2, a signal incident on port 2 emerges from port 3, and a signal incident on port 3 emerges from port 1. Equation (3.62) simply describes a circulator with the opposite direction of circulation; that is, a signal incident on port 3 emerges from port 2, and so on. A circulator is clearly a nonreciprocal component. Figure 3.2 shows the symbol for a circulator, which does a good job of visually describing its operation.

An interesting equivalent circuit for a circulator can be developed. We convert its S parameters to Z parameters and ground port 3. This gives us

$$
\begin{bmatrix} V_1 \\ V_2 \end{bmatrix} = \begin{bmatrix} 0 & -Z_0 \\ Z_0 & I \end{bmatrix} \begin{bmatrix} I_1 \\ I_2 \end{bmatrix}
\tag{3.63}
$$

which is the Z matrix of a gyrator (Section 6.2.2). This implies that the circulator can be realized by a gyrator in the manner shown in Figure 3.3, as long as all ports have the same normalizing impedances. While this is not a terribly practical approach for realizing circulators, as gyrators are notoriously difficult to create, it is very helpful for modeling and circuit analysis.

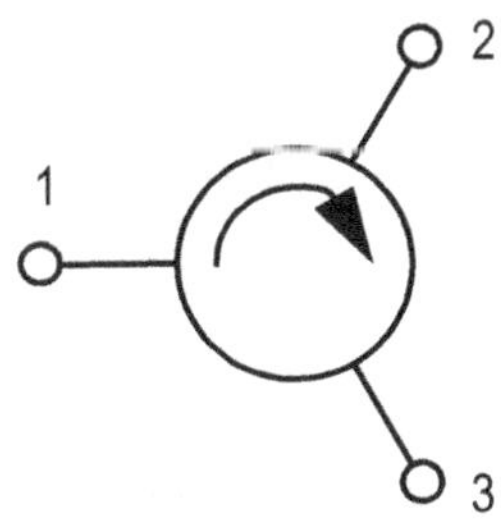

Figure 3.2 A circulator, with its ports numbered and an arrow to show the direction of circulation. The direction of circulation shown here is consistent with (3.61). A circulator consistent with (3.62) would simply circulate in the opposite direction.

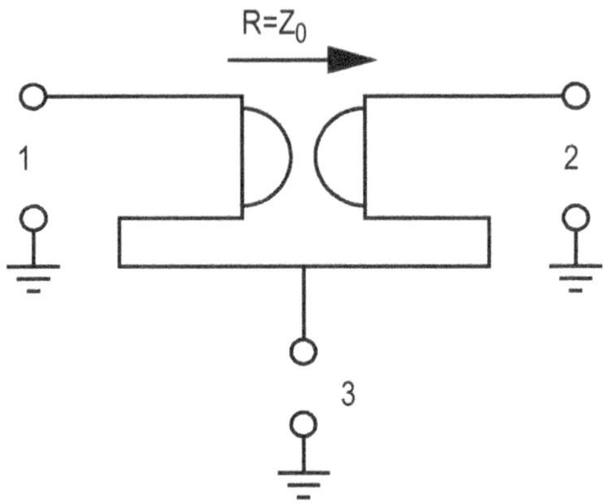

Figure 3.3 The circulator in Figure 3.2 can be realized as a gyrator. The gyrational resistance is equal to the port normalizing impedance, Z_0.

If we assume the three-port to be reciprocal, it quickly becomes clear that no solution is possible. For example, subtracting (3.58) from (3.57) gives

$$|S_{13}|^2 - |S_{23}|^2 = 0 \tag{3.64}$$

where we have assumed, because of reciprocity, $S_{12} = S_{21}$. Then,

$$|S_{13}| = |S_{23}| \tag{3.65}$$

From (3.60), however, at least one of these parameters must be zero, so from (3.65), both must be zero. Continuing in this manner, we discover that all off-diagonal terms must be zero. Given also our original assumption that the on-diagonal terms are zero, we are left with a zero matrix. Not only is such a component not terribly useful, but it violates the unitary property (3.50) with which we began. Thus, we conclude the following:

1. A matched, passive, lossless three-port must be a circulator;
2. It is impossible to match simultaneously all three ports of a lossless, passive, reciprocal three-port.

The second of the above conclusions is especially important. It means, for example, that it is impossible to match all three ports of a balun. This point was mentioned briefly in Section 2.3.1 and will be examined further in the next section. In a balun, it is best to design for an input match with specified load impedances and to select a structure having even- and odd-mode output impedances that are appropriate to the kind of circuit in which it is used.

It is also important to recognize that many apparently lossless, matched, passive three-ports are not really lossless. A coupler with a termination on its isolated port is not lossless, even though the termination absorbs no power in ordinary operation. Similarly, a Wilkinson power divider [1.3], one of many types of three-port power dividers, is not lossless, as it includes a resistor.

3.2.4 Application: Baluns

At this point it is worthwhile to revisit the balun, as it is a lossless, passive, three-port. In Section 2.3.1 we noted that it is impossible to match all three ports of a balun, and by now the reason should be thoroughly clear. However, it is worthwhile to apply the unitary property to a balun's S matrix and see if there is more to be discovered.

Suppose we dispense with the requirement that the balun's output ports must be matched, but keep everything else ideal. The resulting S matrix of the balun in Figure 2.8 can be written

$$\mathbf{S} = \begin{bmatrix} 0 & S & -S \\ S & \Gamma & 0 \\ -S & 0 & \Gamma \end{bmatrix} \tag{3.66}$$

where we have simplified the notation a bit, with S representing the transfer quantities and Γ the output reflection coefficient. Taking $\mathbf{SS}^* = \mathbf{1}$ results in the following requirements:

$$\begin{aligned} 2|S|^2 &= 1 \\ |S|^2 + |\Gamma|^2 &= 1 \\ S\Gamma^* = S^*\Gamma &= 0 \end{aligned} \tag{3.67}$$

These are clearly incompatible, so a balun with a matched input and perfect isolation between the output ports cannot exist. The problem is with the assumption of perfect isolation, $S_{23} = S_{32} = 0$; if we drop that requirement, the S matrix becomes

$$\mathbf{S} = \begin{bmatrix} 0 & S & -S \\ S & \Gamma & I \\ -S & I & \Gamma \end{bmatrix} \quad (3.68)$$

where $I = S_{23} = S_{32}$ represents the coupling between the output ports. Applying the unitary property gives

$$\begin{aligned} 2|S|^2 &= 1 \\ |S|^2 + |\Gamma|^2 + |I|^2 &= 1 \\ S(\Gamma - I)^* &= 0 \\ -|S|^2 + 2\mathrm{Re}\{\Gamma I^*\} &= 0 \end{aligned} \quad (3.69)$$

These are easily satisfied. The first term gives $|S|^2 = 0.5$, the obvious 3-dB power division. The remaining three indicate that $\Gamma = I$ and that $|\Gamma|^2 = |I|^2 = 0.25$. Thus, a lossless balun with a matched input requires that the output return loss and output-port isolation be 6 dB. You really don't want to use one of these as a conventional power divider.

The story goes further than this, of course. Baluns are used in balanced circuits, such as mixers and frequency multipliers, where the even- and odd-mode properties are more important than output-port match and isolation. We examine these matters further in Chapter 8.

3.2.5 Four-Ports

We now consider the possibility of creating a matched, lossless four-port; that is, one having $S_{11} = S_{22} = S_{33} = S_{44} = 0$. Again, we apply the unitary property of a lossless junction (3.50) but with 4×4 S matrices. The results, as one might expect, are a bit more complicated. We obtain

$$|S_{12}|^2 + |S_{13}|^2 + |S_{14}|^2 = 1 \quad (3.70)$$

$$|S_{12}|^2 + |S_{23}|^2 + |S_{24}|^2 = 1 \quad (3.71)$$

$$|S_{13}|^2 + |S_{23}|^2 + |S_{34}|^2 = 1 \tag{3.72}$$

$$|S_{14}|^2 + |S_{24}|^2 + |S_{34}|^2 = 1 \tag{3.73}$$

$$S_{13}S_{23}^* = -S_{14}S_{24}^* \tag{3.74}$$

$$S_{23}S_{12}^* = -S_{34}S_{14}^* \tag{3.75}$$

$$S_{13}S_{12}^* = -S_{24}S_{34}^* \tag{3.76}$$

Adding (3.70) and (3.71) and subtracting (3.72) and (3.73) gives

$$|S_{12}|^2 = |S_{34}|^2 \tag{3.77}$$

Similar manipulations give

$$|S_{13}|^2 = |S_{24}|^2 \tag{3.78}$$

and

$$|S_{23}|^2 = |S_{14}|^2 \tag{3.79}$$

Multiplying (3.77) through (3.79) gives

$$|S_{12}S_{13}S_{23}|^2 = |S_{14}S_{24}S_{34}|^2 \tag{3.80}$$

and multiplying (3.74) through (3.76) gives

$$|S_{12}S_{13}S_{23}|^2 = -|S_{14}S_{24}S_{34}|2 \tag{3.81}$$

Equations (3.80) and (3.81) can be satisfied only if one or more of the terms on each side is zero. Not all can be zero, or course, or the whole matrix becomes zero.

Standing back a little, looking at the previous results, and trying not to be totally confused, we conclude that only one possibility (or its equivalents) provides a nontrivial result. If we take $S_{12} = S_{34} = 0$ and assume that the circuit is reciprocal, the S matrix takes the form

$$\mathbf{S} = \begin{bmatrix} 0 & 0 & S_{13} & S_{14} \\ 0 & 0 & S_{23} & S_{24} \\ S_{13} & S_{23} & 0 & 0 \\ S_{14} & S_{24} & 0 & 0 \end{bmatrix} = \begin{bmatrix} \mathbf{0} & \mathbf{N} \\ \mathbf{N}^{\mathrm{T}} & \mathbf{0} \end{bmatrix} \tag{3.82}$$

with

$$\begin{aligned} |S_{13}|^2 + |S_{14}|^2 &= 1 \\ |S_{23}|^2 + |S_{24}|^2 &= 1 \end{aligned} \tag{3.83}$$

This, if you like impressive terms, is called a *biconjugate four-port*. **N** is a 2×2 submatrix, which, like **S**, is unitary; the superscript T indicates the transpose. Two possibilities for (3.82), with $|S_{13}| = |S_{14}|$ and $|S_{23}| = |S_{24}|$, are the following:

$$\mathbf{S} = \frac{1}{\sqrt{2}} \begin{bmatrix} 0 & 0 & 1 & 1 \\ 0 & 0 & 1 & -1 \\ 1 & 1 & 0 & 0 \\ 1 & -1 & 0 & 0 \end{bmatrix} \tag{3.84}$$

and

$$\mathbf{S} = \frac{1}{\sqrt{2}} \begin{bmatrix} 0 & 0 & -j & 1 \\ 0 & 0 & 1 & -j \\ -j & 1 & 0 & 0 \\ 1 & -j & 0 & 0 \end{bmatrix} \tag{3.85}$$

Equations (3.84) and (3.85) describe two kinds of microwave hybrid junctions, or, less formally, *hybrids*. The first is a 180-degree hybrid; the second is a 90-degree or *quadrature* hybrid, which should be recognizable as the 3-dB directional coupler from Chapter 2. From the above equations and the discussion of couplers in that chapter, it should be clear that the couplings implied by the S matrices are not unique; that is, we have arbitrarily selected $|S_{13}|^2 = |S_{14}|^2 = 1/2$, but other values are possible within the constraints posed by (3.83). That is not surprising; we have already noted that four-port directional couplers having arbitrary coupling values are possible. The same is true of the 180-degree circuit described by (3.84).[4]

Figure 3.4 illustrates schematically the operation of 90- and 180-degree hybrids. The figure illustrates the transmission and isolation properties and is a common way to represent them. Hybrids can also be realized by transformers. Figure 3.5 shows the most common realization, often used in low-frequency mixers. It requires that the impedance of port 3 be half that of the other two ports; while this seems like a disadvantage, it is not a significant impediment for use in mixers. Other transformer realizations having equal port impedances are possible but are more complex.

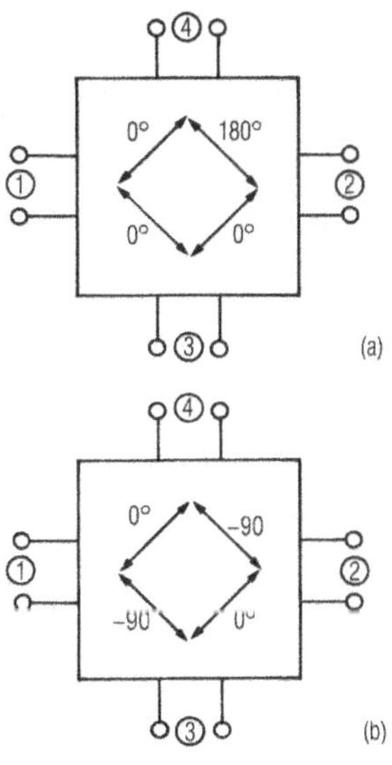

Figure 3.4 180- (a) and 90-degree (b) hybrids, showing the coupling and phase shift between ports. In both cases, ports 1 and 2 and ports 3 and 4 are mutually isolated pairs.

4. The term *hybrid* is usually reserved for circuits having equal (3-dB) coupling.

3.3 S-Parameter Analysis of Two-Ports

The S matrix can provide considerable useful information about the performance of terminated two-ports. This advantage is greatest in the case of small-signal amplifiers, where the solid-state device is described by a set of S parameters. The design of such amplifiers consists of determining the source and load reflection coefficients that should be presented to the device. We then create ideally lossless matching circuits that realize the desired source and load reflection coefficients.

3.3.1 Gain and Reflection Coefficients

We address the situation illustrated in Figure 3.6. The two-port (imagine a high-frequency field-effect or bipolar transistor, if you need something concrete) is driven from a source impedance Z_s having reflection coefficient Γ_s and is terminated at its output in an impedance Z_L having reflection coefficient Γ_L. The port normalizing impedances are Z_{01} and Z_{02} at ports 1 and 2, the input and output, respectively. In most practical amplifier designs, the normalizing impedances of both ports are invariably the same, almost always 50Ω, but this will not be required in the derivations that follow. As an aid to visualizing the wave variables, we imagine the existence of ideal transmission lines, whose characteristic impedances are their respective normalizing impedances, connected from the source to the input

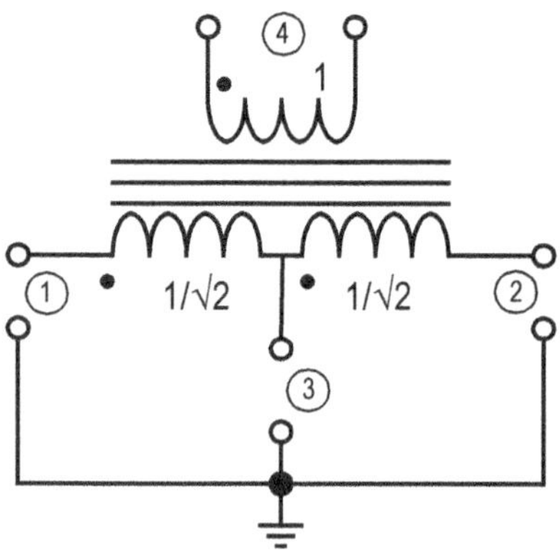

Figure 3.5 A 180-degree hybrid can be realized as a transformer. This is one of many possible realizations. The port numbers are circled and the transformer windings are labeled with their turns ratios. The impedance of port 3 is half that of the other ports, whose impedances are equal.

and from the output to the load. These lines are an integer number of wavelengths long, so they do not affect the results.

A second point is that we shall use the concepts of reflected and incident waves a bit more loosely than before. A *b* wave is simply one propagating away from its port; it may exist because of reflections or because of a source at that port. Similarly, an *a* wave is one incident on its port. A *b* wave sometimes may be the same as an *a* wave, as the *b* wave from one port may be the incident wave at another port.

The wave launched from the source is a *b* wave, which we call b_s. It is the wave on the input transmission line generated by the source. The wave incident on the input of the two-port is

$$a_1 = b_s + \Gamma_s b_1 \tag{3.86}$$

and the wave incident on the output port, port 2, is

$$a_2 = \Gamma_L b_2 \tag{3.87}$$

We now have what we need to determine the input and output reflection coefficients and the gain of the terminated two-port.

3.3.1.1 Gain

We view the gain as the product of three terms, which we call the source, two-port, and load gains. The source gain, G_S, is the power transferred to the input line divided by the available power of the source. The load gain, G_L, is the power delivered to the load divided by the available power of the

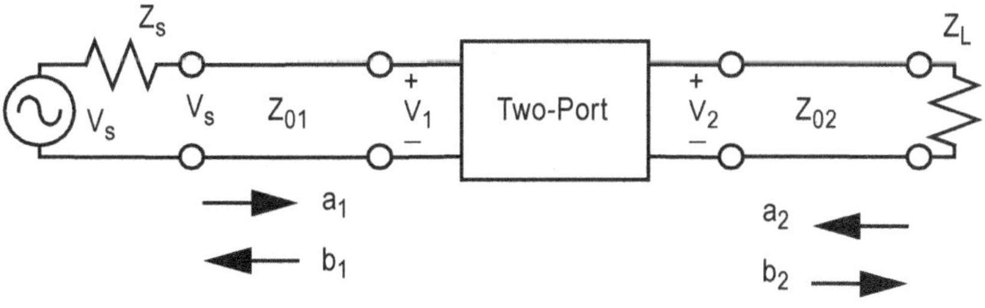

Figure 3.6 A terminated two-port. Transmission lines at the input and output are included as an aid in visualizing the incident and reflected waves. One may think of them as an integer number of wavelengths long, or simply zero length.

wave on the output line, and the two-port gain, G_2, is the power delivered to the output line divided by the power available on the input line. Thus, the transducer gain of the entire structure, G_t, is

$$G_t = \frac{P_d}{P_{av}} = G_s G_2 G_L \tag{3.88}$$

where

$$G_s = \frac{|b_s|^2/2}{P_{av}} \tag{3.89}$$

$$G_2 = \left|\frac{b_2}{b_s}\right|^2 \tag{3.90}$$

and

$$G_L = \frac{P_d}{|b_2|^2/2} \tag{3.91}$$

where P_{av} is the available power from the source and P_d is the power dissipated in the load.

The available power on the output line is simply the power of the b wave. The power dissipated in the load is that available power minus the reflected power:

$$P_d = \frac{|b_2|^2}{2} - \frac{|a_2|^2}{2} \tag{3.92}$$

The load gain is the power dissipated in the load divided by the available power on the line:

$$G_L = \frac{\frac{|b_2|^2}{2} - \frac{|a_2|^2}{2}}{\frac{|b_2|^2}{2}} = 1 - |\Gamma_L|^2 \tag{3.93}$$

The source gain is a little trickier. We know that the available power is

$$P_{av} = \frac{|V_s|^2}{8\mathrm{Re}\{Z_s\}} \tag{3.94}$$

From our equivalent circuit of the source, Figure 1.4,

$$V_f = V_s \frac{Z_{01}}{Z_s + Z_{01}} \tag{3.95}$$

and from (3.6),

$$b_s = \frac{V_f}{\sqrt{Z_{01}}} = V_s \frac{\sqrt{Z_{01}}}{Z_s + Z_{01}} \tag{3.96}$$

Solving (3.96) for V_S and substituting into (3.94) gives the available power of the source in terms of b_S:

$$P_{av} = \frac{|b_s|^2 |Z_s + Z_{01}|^2}{8Z_{01}\mathrm{Re}\{Z_s\}} \tag{3.97}$$

From the definition of the reflection coefficient, we can show that

$$1 - |\Gamma_s|^2 = \frac{4Z_{01}\mathrm{Re}\{Z_s\}}{|Z_s + Z_{01}|^2} \tag{3.98}$$

so

$$P_{av} = \frac{|b_s|^2}{2} \frac{1}{(1 - |\Gamma_s|^2)} \tag{3.99}$$

That is,

$$G_s = \frac{|b_s|^2 / 2}{P_{av}} = 1 - |\Gamma_s|^2 \tag{3.100}$$

This result should not be a surprise. As we showed in (3.93), the loss for a wave incident on a source having reflection coefficient Γ is $1 - |\Gamma|^2$. From reciprocity, the loss in the other direction must be the same.

We now need to determine $G_2 = |b_2 / b_s|^2$. We begin by substituting our early expressions for a_1 and a_2, (3.86) and (3.87):

$$\begin{bmatrix} b_1 \\ b_2 \end{bmatrix} = \begin{bmatrix} S_{11} & S_{12} \\ S_{21} & S_{22} \end{bmatrix} \begin{bmatrix} b_s + \Gamma_s b_1 \\ \Gamma_L b_2 \end{bmatrix} \tag{3.101}$$

Solving for b_2 / b_s requires some tedious algebra but is otherwise straightforward. We obtain

$$\frac{b_2}{b_s} = \frac{S_{21}}{(1 - S_{11}\Gamma_s)(1 - S_{22}\Gamma_L) - S_{12}S_{21}\Gamma_s\Gamma_L} \tag{3.102}$$

Finally, substituting all this into (3.88) gives an expression for the transducer gain:

$$G_t = \frac{|S_{21}|^2(1 - |\Gamma_s|^2)(1 - |\Gamma_L|^2)}{|(1 - S_{11}\Gamma_s)(1 - S_{22}\Gamma_L) - S_{12}S_{21}\Gamma_s\Gamma_L|^2} \tag{3.103}$$

3.3.1.2 Input and Output Reflection Coefficients

We now determine the input reflection coefficient of the terminated two-port. This quantity is Γ_{in},

$$\Gamma_{in} = \frac{b_1}{a_1} \tag{3.104}$$

when the output is terminated in Γ_L, so

$$a_2 = \Gamma_L b_2 \tag{3.105}$$

Γ_{in} is a property of the two-port and load; it is independent of the source, so we need not consider the source reflection coefficient. We substitute, as before,

$$\begin{bmatrix} b_1 \\ b_2 \end{bmatrix} = \begin{bmatrix} S_{11} & S_{12} \\ S_{21} & S_{22} \end{bmatrix} \begin{bmatrix} a_1 \\ \Gamma_L b_2 \end{bmatrix} \tag{3.106}$$

and solve for b_1 / a_1. This results in the expression,

$$\Gamma_{in} = \frac{b_1}{a_1} = S_{11} + \frac{S_{12}S_{21}\Gamma_L}{1 - S_{22}\Gamma_L} \tag{3.107}$$

The output reflection coefficient with the input terminated, Γ_{out}, can be found by symmetry; we just turn the two-port around and obtain

$$\Gamma_{out} = S_{22} + \frac{S_{12}S_{21}\Gamma_s}{1 - S_{11}\Gamma_s} \tag{3.108}$$

3.3.1.3 Determining S Parameters from Nodal Analysis

Computers analyze circuits by nodal methods, which we shall examine in detail in Chapter 5. To do so, they must use numerical techniques, in which specific voltages and currents are found when specific excitations are used. To determine S parameters, we need to determine them from some particular solution V_1, V_2, I_1 and I_2 resulting from excitation V_s.

Suppose the two-port is excited by V_s as in Figure 3.6 and is terminated in $Z_s = Z_{01}$ and $Z_L = Z_{02}$. We wish to find the S parameters of the circuit normalized to those impedances, given the resulting port voltages V_1 and V_2. Since the circuit is linear, V_s can have any value, and V_2 and V_1 can be determined. From (3.3),

$$a_1 + b_1 = a_1(1 + S_{11}) = \frac{V_1}{\sqrt{Z_{01}}} \tag{3.109}$$

From (3.1),

$$
\begin{aligned}
a_1 &= \frac{1}{2}\left(\frac{V_1}{\sqrt{Z_{01}}} + I_1\sqrt{Z_{01}}\right) = \frac{1}{2}\left(\frac{V_1}{\sqrt{Z_{01}}} + \sqrt{Z_{01}}\frac{(V_s - V_1)}{Z_{01}}\right) \\
&= \frac{1}{2}\frac{V_s}{\sqrt{Z_{01}}}
\end{aligned}
\tag{3.110}
$$

Solving (3.109) for a_1 and equating to (3.110) gives

$$
S_{11} = 2\frac{V_1}{V_s} - 1 \tag{3.111}
$$

To find S_{21}, we recognize that $a_2 = 0$ and from (3.3),

$$
b_2 = \frac{V_2}{\sqrt{Z_{02}}} \tag{3.112}
$$

Dividing (3.112) by (3.110) gives

$$
S_{21} = \frac{2V_2}{V_s}\sqrt{\frac{Z_{01}}{Z_{02}}} \tag{3.113}
$$

The other two S parameters are found by exciting the output with V_s and using the previous results. Then, from symmetry,

$$
\begin{aligned}
S_{22} &= 2\frac{V_2}{V_s} - 1 \\
S_{12} &= \frac{2V_1}{V_s}\sqrt{\frac{Z_{02}}{Z_{01}}}
\end{aligned}
\tag{3.114}
$$

As a final note, the voltage gain can be found easily from (3.111) and (3.113). The result is

$$\frac{V_2}{V_1} = \frac{S_{21}}{(1+S_{11})}\sqrt{\frac{Z_{02}}{Z_{01}}}$$
$$\frac{V_2}{V_s} = \frac{S_{21}}{2}\sqrt{\frac{Z_{02}}{Z_{01}}} \tag{3.115}$$

3.3.2 Two-Port Gain Definitions

So far, we have considered only one definition of gain using power units. That is transducer gain, G_t, defined as

$$G_t = \frac{P_d}{P_{av}} \tag{3.116}$$

where P_d is the power delivered to the load and P_{av} is the available power from the source. Transducer gain is a logical "default" quantity for a gain definition, as it is the gain that is obtained from a straightforward laboratory measurement: attach the (matched) source to the (matched) power sensor to obtain available power, then attach the power sensor to the output of the excited two-port to obtain the power dissipated in the load. Divide those, and you have transducer gain.

Two other possibilities are called *power gain* and *available gain*. The power gain, G_p, is defined as

$$G_p = \frac{P_d}{P_{in}} \tag{3.117}$$

where P_{in} is the input power, the power dissipated in port 1 of the two-port. Available gain, G_a, is

$$G_a = \frac{P_{av2}}{P_{av}} \tag{3.118}$$

where P_{av2} is the available power from the output port of the two-port.

These secondary definitions are not used commonly to specify components; invariably, when we loosely speak of "gain" we mean transducer gain. They are more often useful as part of the design for active components, especially low-noise and power amplifiers. We address those design techniques more extensively in Chapter 7.

We have already developed an expression for transducer gain of a two-port, (3.103). It is possible to derive similar expressions for power and available gain. As the procedure is almost identical to the derivation of G_t, we present only the results:

$$G_p = \frac{|S_{21}|^2(1 - |\Gamma_L|^2)}{|1 - S_{22}\Gamma_L|^2 - |S_{11} - (S_{11}S_{22} - S_{12}S_{21})\Gamma_L|^2} \tag{3.119}$$

This can also be written

$$G_p = \frac{|S_{21}|^2(1 - |\Gamma_L|^2)}{(1 - |\Gamma_{\text{in}}|^2)|1 - S_{22}\Gamma_L|^2} \tag{3.120}$$

where Γ_{in} is given by (3.107). For G_a we have

$$G_a = \frac{|S_{21}|^2(1 - |\Gamma_s|^2)}{|1 - S_{11}\Gamma_s|^2 - |S_{22} - (S_{11}S_{22} - S_{12}S_{21})\Gamma_s|^2} \tag{3.121}$$

which also can be simplified to

$$G_a = \frac{|S_{21}|^2(1 - |\Gamma_s|^2)}{(1 - |\Gamma_{\text{out}}|^2)|1 - S_{11}\Gamma_s|^2} \tag{3.122}$$

and Γ_{out} is given by (3.108).

3.4 Stability

3.4.1 Two-Port Stability

It is well known that a network function is stable if and only if all its poles are in the left half of the *s* plane, where *s* is the LaPlace transform variable. Applying this rule to a microwave circuit presents several problems, however. The first is that it requires a complete circuit; once the circuit is reduced to a two-port, described by a set of S, Y, or similar parameters, it is no longer possible to assess its stability correctly. Figure 3.7 shows an admittedly contrived circuit that oscillates if an impulse is applied to its input. The ordinary two-port stability tests we describe below, based on the circuit's S parameters, indicate that it is just fine, but an analysis of the complete circuit readily shows that it is unstable. A second problem is that a circuit often consists of little more than a solid-state device and a pair of matching circuits. The device is not a circuit and can be included in a full stability analysis (i.e., beyond the "external" stability considerations we examine in the next section) only if it is represented by an equivalent circuit.[5] While an equivalent circuit can be generated for microwave devices, it is a substantial task, one that is often not needed for other aspects of the design and for which adequate information may not be available. Moreover, there is no guarantee that the equivalent circuit has captured all phenomena that could lead to instability. Indeed, in a low-noise amplifier (LNA) design, the device characterization often consists only of S parameters.

3.4.2 Port Terminations and External Stability

Fortunately, the situation is not as hopeless as the previous discussion implies. It is often adequate to assume that the device itself is stable; after all, if it were unstable, it would not be manufactured and sold in the first place. The greater problem is the selection of source and load reflection coefficients that do not introduce instability in operation. It is important to emphasize that this criterion is different from the usual idea of stability in

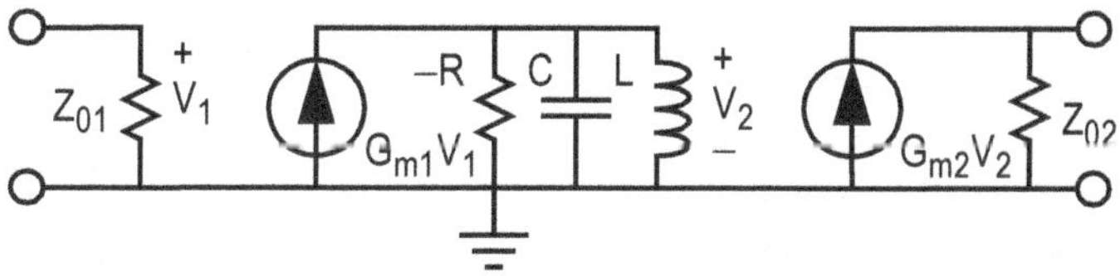

Figure 3.7 Example of a circuit that passes conventional two-port stability tests but is clearly unstable, having a negative resistance embedded in it.

5. This statement is not entirely true. Methods exist for estimating poles of a two-port network that is described by a Y, S, Z, or similar parameter set over a wide range of frequencies. These are generally impractical for stability analysis in microwave design, and, in any case, their theory is well beyond the scope of this book.

analog circuits, as it does not prove the absence of right-plane poles in the entire circuit. As such, it is sometimes called *external stability.* It is quite valuable, however, in assessing certain practical aspects of stability in an RF or microwave circuit.

The rule for external stability is simple; if all passive values of Γ_L cause $|\Gamma_{in}| < 1$ and all passive values of Γ_s cause $|\Gamma_{out}| < 1$, the circuit is externally stable. Otherwise, the input or output impedance has a negative real part and oscillation can occur at the frequency where $|\Gamma_{in,\,out}| > 1$. With some fairly ugly but otherwise straightforward algebra applied to (3.107) and (3.108), we can show that the terminated two-port can have no $|\Gamma_{in,\,out}| > 1$ for any passive $\Gamma_{L,s}$ if the following two conditions are met:

$$K = \frac{1 - |S_{11}|^2 - |S_{22}|^2 + |\Delta|^2}{2|S_{21}S_{12}|} > 1 \tag{3.123}$$

and

$$|\Delta| = |S_{11}S_{22} - S_{21}S_{22}| < 1 \tag{3.124}$$

where Δ is the determinant of the S matrix. Rollet's criterion, used somewhat less frequently, is another external stability criterion, formulated in terms of two-port voltage and current matrices:

$$\begin{gathered} \mathrm{Re}\{\kappa_{11}\} > 0 \\ \mathrm{Re}\{\kappa_{22}\} > 0 \\ \frac{|\kappa_{12}\kappa_{21}|}{2\mathrm{Re}\{\kappa_{11}\}(\mathrm{Re}\{\kappa_{22}\} - \mathrm{Re}\{\kappa_{12}\kappa_{21}\})} < 1 \end{gathered} \tag{3.125}$$

where the $\kappa_{m,n}$ are elements of a Y, Z, H, or G matrix.

More information regarding external stability can be obtained by viewing the mappings between $\Gamma_{L,s}$ and $\Gamma_{in,\,out}$. We begin by recognizing that the input reflection coefficient of a terminated two-port, (3.107), can be written

$$\Gamma_{in} = \frac{\Delta\Gamma_L - S_{11}}{S_{22}\Gamma_L - 1} \tag{3.126}$$

and a similar expression can be written for Γ_{out},

$$\Gamma_{out} = \frac{\Delta\Gamma_s - S_{22}}{S_{11}\Gamma_s - 1} \tag{3.127}$$

These have the general form

$$f(s) = \frac{as + b}{cs + d} = \frac{1}{c}\left(a + \frac{bc - ad}{cs + d}\right) \tag{3.128}$$

where s is a complex variable, in this case corresponding to Γ_s, and a, b, c, d are constants, which can be equated to terms in (3.126) and (3.127) by inspection. It is well known from the theory of complex variables and conformal transformations that terms of the form $1/(cs + d)$, on the right of (3.128), map into a circle in the $f(s)$ plane as the phase of s is varied from 0 to 2π [3.3]. The remaining constants simply modify the diameter and center of that circle.

Knowing this, we can create the locus of Γ_L where $|\Gamma_{in}| = 1$. Plotted on the Γ_L plane, that locus is also a circle, and it represents the boundary between values of Γ_L that might cause instability (the *unstable region*) and those that cannot. The expression for that circle is found from yet another tedious but straightforward derivation; the result is

$$\Gamma_L = \frac{(S_{22} - \Delta S_{11}^*)^*}{|S_{22}|^2 - |\Delta|^2} + \frac{|S_{21}S_{12}|\exp(j\theta)}{|S_{22}|^2 - |\Delta|^2} \tag{3.129}$$

where θ varies from 0 to 2π. From symmetry, we find a similar expression for the locus of Γ_s where $|\Gamma_{out}| = 1$:

$$\Gamma_s = \frac{(S_{11} - \Delta S_{22}^*)^*}{|S_{11}|^2 - |\Delta|^2} + \frac{|S_{21}S_{12}|\exp(j\theta)}{|S_{11}|^2 - |\Delta|^2} \tag{3.130}$$

The first term in (3.129) or (3.130) is the center of the circle and the coefficient of the $\exp(j\theta)$ term is the radius. These are sometimes called *load* and *source stability circles*, respectively.

These expressions describe boundaries between stable and unstable regions in the Γ_L or Γ_s planes. If at least part of an unstable region falls within either the $\Gamma_s < 1$ or $\Gamma_L < 1$ plane, it is possible to find a pair of passive terminating impedances, at the two ports, that could cause oscillation. Of course, in that situation, oscillation can still be prevented by the selection of appropriate source and load impedances, so a circuit that fails the test in (3.123) and (3.124), or has a source or load in an unstable region defined by (3.129) or (3.130), will not necessarily oscillate. Such a circuit is called *conditionally stable*. If it passes the tests in (3.123) and (3.124), the unstable region is entirely outside the Γ_s and Γ_L planes; then it is *unconditionally stable*.

Figure 3.8 shows an example of a set of source and load stability circles for a high-frequency field-effect transistor (FET) from 2 to 10 GHz in 2-GHz steps. The plot indicates that an inductive source having a large reflection coefficient could cause instability; similarly, an inductive load could cause instability, especially at low frequencies. The plots show immediately that this device, like virtually all modern microwave FETs, is conditionally stable.

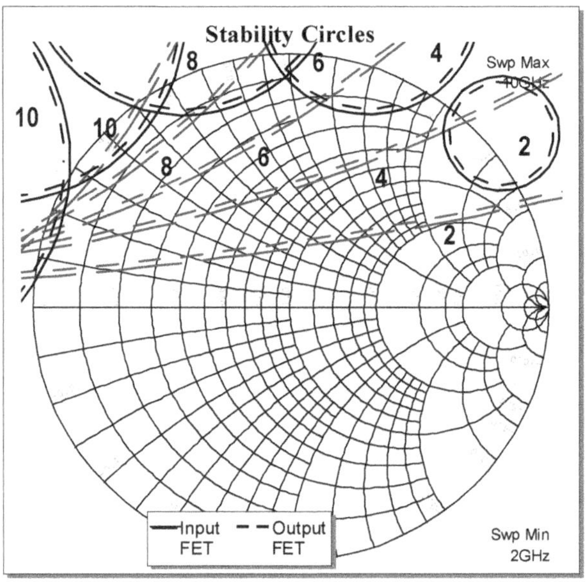

Figure 3.8 Example of stability circles for a high-frequency field-effect transistor from 2 to 10 GHz in 2-GHz steps. The short dashed lines identify the unstable region in each circle. The load circles (representing Γ_L) are dashed, and each circle is labeled with its frequency.

It is important to recognize that these stability tests are valid only at the frequency for which the S parameters are defined. To guarantee that a termination does not cause a situation that could lead to oscillation, it is necessary to evaluate the stability in terms of both Γ_s and Γ_L over the entire frequency range for which the circuit is conditionally stable. The source and load circles of (3.129) and (3.130) give a clear and immediate indication of how close to disaster a circuit is placed by some particular load. They can also be helpful in avoiding cases where Γ_{in} or Γ_{out} are close to unity, which results in difficult matching and high sensitivity to variations in the source or load impedances.

In a multistage amplifier, this method must be applied individually to every active stage in the cascade. The source and load impedances seen by some stage embedded in a cascade depend on the other stages in the cascade and only indirectly on the cascade's source and load. Thus, the cascade's source and load impedances must be transformed through all intervening stages to determine the source and load impedances seen by any one stage. This is a laborious process.

A final consideration is that the location of the unstable region defined by (3.129) or (3.130) is not always obvious; it could be either inside or outside of the stability circle. The simplest way to determine which region is the stable one is to select some convenient value of Γ_s or Γ_L and to evaluate the resulting Γ_{in} or Γ_{out}. For example, $\Gamma_L = 0$ gives $\Gamma_{\text{in}} = S_{11}$, so a quick look at the S parameters should identify the unstable region immediately; that is, if $|S_{11}| < 1$, $\Gamma_L = 0$ is in the stable region.

3.4.3 General Linear Circuit Stability

While the view of stability in Section 3.4.2 has practical value, it also has obvious limitations. The first, already stated, is that it isn't really a generally valid stability analysis, as it does not indicate the presence of right-plane poles. A second problem is its applicability only to two-ports. In this section we dispense with some of those limitations.

3.4.3.1 A More General View of External Stability

Suppose we have a multiport circuit whose ports are terminated by a set of impedances, Z_n, where n indicates the port number. The vector of voltages across the terminations, $\mathbf{V}$, is

$$\mathbf{V} = -\mathbf{Z}_L \mathbf{I} \tag{3.131}$$

where $\mathbf{Z}_L$ is a diagonal matrix whose elements are the Z_n and $\mathbf{I}$ is the vector of currents in the network ports (which is the negative of the currents in the loads; thus the negative sign). If we describe the network by a Y matrix,

$$\mathbf{I} = \mathbf{Y}\mathbf{V} \tag{3.132}$$

For oscillation to occur, port voltages must exist in the absence of excitation, so we do not include sources. Substituting (3.132) into (3.131) gives

$$\mathbf{V} = -\mathbf{Z}_L\mathbf{Y}\mathbf{V} \tag{3.133}$$

If we have a solution with nonzero $\mathbf{V}$, the circuit oscillates.

This is a classical eigenvalue problem. It indicates that oscillation will occur when the matrix $-\mathbf{Z}_L\mathbf{Y}$ has a real eigenvalue equal to one, indicating the existence of a pole on the $j\omega$ axis. A real eigenvalue greater than one indicates a right-plane pole. In either case, the associated eigenvector is the oscillatory mode of the circuit.

Unlike the stability criteria in Section 3.4.2, this criterion doesn't simply show that oscillation is possible; when satisfied, it indicates that oscillation *will* occur. Thus, it is useful in the design of oscillators as well as in the stability assessment of amplifiers; indeed, the oscillatory mode, which is provided by this method, is a good starting point for the nonlinear analysis of an oscillator. It does have the disadvantage, however, of requiring a search over a broad space of frequency and load impedance. Much of this can be simplified by the use of numerical optimization techniques, which home in on the values of frequency and load impedance that cause oscillation.

A related approach is to include the loads in the network, creating an augmented Y matrix, $\mathbf{Y}_a$. Then

$$\mathbf{Y}_a = \mathbf{Y} + \mathbf{Y}_L \tag{3.134}$$

where $\mathbf{Y}_L = \mathbf{Z}_L^{-1}$ is the matrix of load admittances, again a diagonal. For oscillation to occur, we must have

$$\mathbf{Y}_a\mathbf{V} = 0 \tag{3.135}$$

with some nonzero $\mathbf{V}$. This can happen only when the determinant of $\mathbf{Y}_a$ is zero; that is, $\mathbf{Y}_a$ is singular. This idea is the basis of many approaches to stability analysis, most notably the normalized determinant function, which we describe below. Like the eigenvalue method, however, it requires a broad search over frequency and load-admittance values.

3.4.3.2 Internal Stability

We now turn to internal stability, the stability of the circuit as a whole, not simply that of a reduced N-port in which we have access only to external ports. In this case we need to find the poles of a network function and ascertain whether any are in the right half of the s plane.

Methods for assessing the stability of active circuits have existed for many decades. The problem is not to find one, but to find ones that are appropriate for microwave circuits. Many are based on the idea of examining the behavior of a network function over values in the s plane. In microwave technology, however, we rarely have that information; usually we have information about its behavior on the $j\omega$ axis of the s plane alone. Fortunately, in most cases it is possible to extract the needed data from evaluations along the $j\omega$ axis only, but some easily met conditions apply. Some come from the form of the equations, while others come from the inherent characteristics of high-frequency circuits.

Consider again a multinode circuit described by a Y matrix. One can determine a network function as

$$N(s) = \frac{V_j}{I_k} = \frac{(-1)^{j+k}\Delta_{jk}}{\Delta} \tag{3.136}$$

where Δ is the Y matrix determinant and Δ_{jk} is the determinant of a minor of Y, the Y matrix with row j and column k removed. The signed minor is called a *cofactor*. It is clear from (3.136) that the zeros of the determinant are the poles of the function. Thus, our stability criterion is that no zeros of Δ can exist in the right half of the s plane.

A time-honored way to determine the presence of singularities in the right half plane is to use Nyquist's criterion. It is based on Cauchy's principle of the argument from complex variable theory [3.3]. It involves the evaluation of a function $F(s)$ along a closed path in the s plane. $F(s)$ must be *analytic*; that is, infinitely differentiable everywhere within the region of interest, a requirement met by virtually all practical linear circuits. Then, if we traverse a closed path in the s plane, the path of the function $F(s)$ encircles the origin N times in the same direction, where

$$N = Z - P \tag{3.137}$$

and Z and P are the number of zeros and poles, respectively, within the s-plane path. The situation is illustrated in Figure 3.9. In this case $F(s) = \Delta(s)$, so we are looking for right-half-plane zeros of $F(s)$.

Now, imagine that we traverse a path in the s plane that includes the entire right half of the plane. We begin at $s = 0$, then continue northward along the $j\omega$ axis toward $\omega \to \infty$. We then must convert to a polar representation, $s = |s|\exp(j\phi)$, where $|s| \to \infty$, and let ϕ vary from $\pi / 2$ to $-\pi / 2$. That brings us to the point (if we may call it that) on the $j\omega$ axis where $\omega \to -\infty$. From there, we return to $s = 0$. After this effort, the number or encirclements of the origin of $F(s) = \Delta(s)$ would equal the number of zeros minus the number of poles. A good start, but not what we need.

We have three problems remaining: first, there may have been poles or zeros on the $j\omega$ axis with which we would have collided. We can imagine avoiding them by circling around them on an infinitesimally small path. As a practical matter, however, if we encountered any zeros on that axis, we would have found a troublesome zero and the game would be over. Secondly, we need to do something about any right-plane poles, either determine the number of them or come up with an expression that retains the zeros but has no right-plane poles. The third problem is the evaluation of the function along the part of the path other than the $j\omega$ axis.

The latter two problems can be solved in the following manner. We define another function, $\Delta_0(s)$, the determinant of the matrix with all controlled sources turned off. Such a determinant has the same degree as $\Delta(s)$ and the same poles, but, as it represents a passive circuit, can have no right-plane zeros. We then form

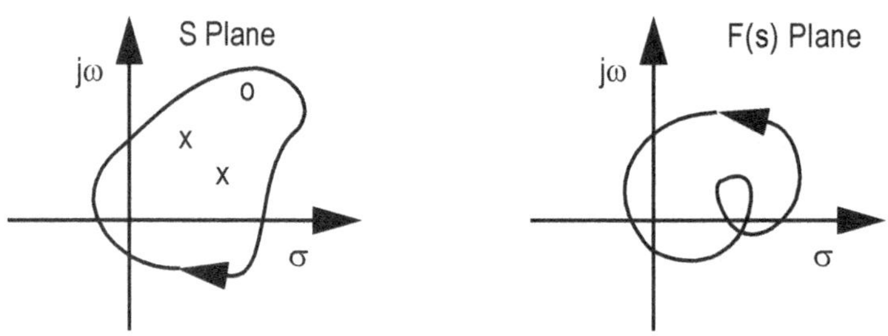

Figure 3.9 This figure illustrates a mapping of a closed path in the s plane to one in the $F(s)$ plane. The s-plane path encircles two poles and one zero, so from (3.137), $F(s)$ must circle the origin −1 times, or once in a direction opposite to the s-plane path.

$$F_n(s) = \frac{\Delta(s)}{\Delta_0(s)} \tag{3.138}$$

Dividing these two quantities cancels the poles of Δ and Δ_0. Although we have introduced new poles, the zeros of Δ_0, those are of no consequence, as they are guaranteed to be in the left half plane. Finally, since both Δ and Δ_0 are of the same degree, $F_n(s)$ reduces to a constant over the part of the s plane path where $|s| \to \infty$. Being a constant, that part of the path cannot cause any encirclement of the origin, so we can ignore it. Thus, we need only evaluate $F_n(s)$ along the $j\omega$ axis, in practice, from some frequency well below any that might cause oscillation to a high frequency that is well above the point where the circuit might oscillate. The mapping along the negative frequency part of the $j\omega$ axis is simply the mirror of the positive-frequency part, so it need not be evaluated explicitly. Then, any encirclement of the origin by the $F_n(s)$ mapping indicates a right-plane pole.

Equation (3.138) has been called the *normalized determinant function*. The idea is most directly from [3.4], but it derives from early work by Bode [3.5]. Other methods for both internal and external stability are similarly related to the use of Nyquist's criterion, sometimes with determinants of network matrices. Often it is assumed that evaluation along the $j\omega$ axis is adequate to evaluate stability, but rarely is that assumption clearly justified. Users of any stability evaluation based on Nyquist should ascertain that the evaluation along the $j\omega$ axis alone is valid; the best way is to be sure that $|F(s)|$ varies as $1 / |s|$ or becomes a constant as $|s| \to \infty$. This condition is met by most electronic circuits [3.5].

A limitation of this method is the need to turn off controlled sources. If solid-state devices are represented by equivalent circuits, that is easy to do. In many amplifier designs, however, the devices are represented by a set of S parameters, so turning off the controlled sources isn't possible. In that case, other methods must be used.

3.4.3.3 Interface Stability

With an understanding of Nyquist's criterion, we can return to a problem left open at the end of Section 3.4.2: determining external stability for a cascade of stages. The approach described here is that of Jackson [3.6]. It has been implemented in a number of circuit simulators.

The transfer function for a general feedback system can be written

$$F(s) = \frac{G(s)}{1 + G(s)H(s)} \tag{3.139}$$

where $G(s)$ is the transfer function of the active stage and $H(s)$ is the feedback. The stability of such a system can be examined by Nyquist's criterion, as it is clear that the poles of the system occur when $1 + G(s)H(s) = 0$ or, equivalently, $G(s)H(s) = -1$. As noted in Section 3.4.3.2, a plot of the denominator function, as s circles the right half of the s plane, circles the origin $Z - P$ times. Equivalently, we can examine only $G(s)H(s)$ and note whether it circles the $\sigma = -1$ point. We need only develop an expression in this form and in terms of quantities we are likely to have available to us.

Suppose we inject a signal, b_i, into the interface between two cascaded stages as shown in Figure 3.10. We can determine the wave variables at the interface as follows:

$$a = \Gamma_1(s)(b_i + b) = \Gamma_1(s)b_i + \Gamma_1(s)\Gamma_2(s)a \tag{3.140}$$

or

$$F(s) = \frac{a}{b_i} = \frac{\Gamma_1(s)}{1 - \Gamma_1(s)\Gamma_2(s)} \tag{3.141}$$

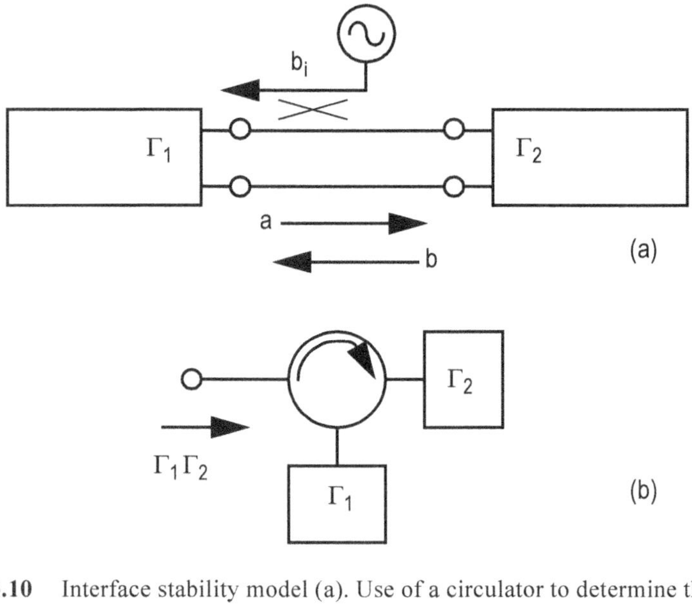

Figure 3.10 Interface stability model (a). Use of a circulator to determine the quantity $\Gamma_1\Gamma_2$ from a single measurement (b).

which has the same form as (3.139). Just as with the general feedback network, we apply Nyquist's criterion to the denominator of (3.141). It is simplest to map $\Gamma_1(s)\Gamma_2(s)$ along the path from $\omega = 0$ to $\omega \to \infty$ and to mirror that plot through the σ axis to obtain the negative-frequency part of the curve. Because of the negative sign in the denominator of (3.141), a path that circles the $\sigma = +1$ point indicates instability. We need not evaluate the path along $|s| \exp(j\phi)$, where $|s| \to \infty$ and ϕ varies from $\pi / 2$ to $-\pi / 2$, because the reflection coefficients reduce to a constant over that part of the s path. We show this in the following manner. The reflection coefficient is given by

$$\Gamma(s) = \frac{Z(s) - Z_0}{Z(s) + Z_0} \tag{3.142}$$

where

$$Z(s) = \frac{N(s)}{D(s)} \tag{3.143}$$

and $N(s)$ and $D(s)$ are polynomials in s. Substituting these into (3.142) gives

$$\Gamma(s) = \frac{N(s) - Z_0 D(s)}{N(s) + Z_0 D(s)} \tag{3.144}$$

The degree of the numerator and denominator are the same, so as $|s| \to \infty$, $\Gamma(s)$ approaches a constant, and there is no chance that $\Gamma_1(s)\Gamma_2(s)$ could circle the +1 point over this part of the path.

Another potential problem is the possibility of right-plane poles in $\Gamma(s)$. Such poles occur only when $Z(s) = -Z_0$, an unlikely condition, but one that can be checked easily.

This method still depends on the terminating impedances of the cascade, so it still must be applied to every interface and for all possible terminations. This is often less burdensome than it might appear, as deep in a cascade of stages, the sensitivity of Γ_1 and Γ_2 to source and load terminations is likely to be negligible.

3.5 Transfer Scattering Matrix

Just as it is possible to create a two-port matrix in terms of V, I that cascades (the ABCD matrix), it is possible to create a cascading matrix in wave quantities. That matrix is called a *transfer scattering matrix,* or *T matrix*. It has the form

$$\begin{bmatrix} b_1 \\ a_1 \end{bmatrix} = \begin{bmatrix} T_{11} & T_{12} \\ T_{21} & T_{22} \end{bmatrix} \begin{bmatrix} a_2 \\ b_2 \end{bmatrix} \tag{3.145}$$

There seems to be no standard form of the T matrix. It is sometimes written

$$\begin{bmatrix} a_1 \\ b_1 \end{bmatrix} = \begin{bmatrix} T_{11} & T_{12} \\ T_{21} & T_{22} \end{bmatrix} \begin{bmatrix} a_2 \\ b_2 \end{bmatrix} \tag{3.146}$$

or sometimes in inverse form, with (a_2, b_2) on the left. Throughout this book, we use the first version, (3.145), as it turns out to be more practical.

Suppose we have a pair of two-ports in cascade, described by T matrices $\mathbf{T}_m$ and $\mathbf{T}_n$, as shown in Figure 3.11. As long as the two matrices have the same normalizing impedances at their mating ports,

$$\begin{bmatrix} a_{2m} \\ b_{2m} \end{bmatrix} = \begin{bmatrix} b_{1n} \\ a_{1n} \end{bmatrix} = \mathbf{T}_n \begin{bmatrix} a_{2n} \\ b_{2n} \end{bmatrix} \tag{3.147}$$

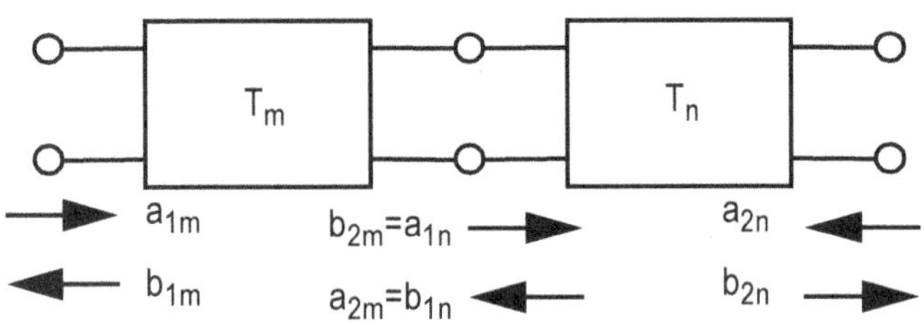

Figure 3.11 Two cascaded two-ports, illustrating the use of the transfer scattering matrix.

and

$$\begin{bmatrix} b_{1m} \\ a_{1m} \end{bmatrix} = \mathbf{T}_m \begin{bmatrix} a_{2m} \\ b_{2m} \end{bmatrix} = \mathbf{T}_m \mathbf{T}_n \begin{bmatrix} a_{2n} \\ b_{2n} \end{bmatrix} \tag{3.148}$$

so the T matrix of the cascade is simply the product of the individual stages' T matrices.

It is simple to convert the T matrix to an S matrix. As a general process, we begin by premultiplying the left side by **1**:

$$\begin{bmatrix} 1 & 0 \\ 0 & 1 \end{bmatrix} \begin{bmatrix} b_1 \\ a_1 \end{bmatrix} = \begin{bmatrix} T_{11} & T_{12} \\ T_{21} & T_{22} \end{bmatrix} \begin{bmatrix} a_2 \\ b_2 \end{bmatrix} \tag{3.149}$$

Recognizing that each column of a matrix multiplies its respective b or a term, we rearrange the (3.149) so that the b terms are on the left and a on the right:

$$\begin{bmatrix} 1 & -T_{12} \\ 0 & -T_{22} \end{bmatrix} \begin{bmatrix} b_1 \\ b_2 \end{bmatrix} = \begin{bmatrix} 0 & T_{11} \\ -1 & T_{21} \end{bmatrix} \begin{bmatrix} a_1 \\ a_2 \end{bmatrix} \tag{3.150}$$

Multiplying both sides by the inverse of the matrix on the left gives

$$\begin{bmatrix} b_1 \\ b_2 \end{bmatrix} = \frac{1}{T_{22}} \begin{bmatrix} T_{12} & T_{11}T_{22} - T_{12}T_{21} \\ 1 & -T_{21} \end{bmatrix} \begin{bmatrix} a_1 \\ a_2 \end{bmatrix} \tag{3.151}$$

so

$$\mathbf{S} = \frac{1}{T_{22}} \begin{bmatrix} T_{12} & T_{11}T_{22} - T_{12}T_{21} \\ 1 & -T_{21} \end{bmatrix} \tag{3.152}$$

A similar derivation gives the T matrix in terms of the S:

$$\mathbf{T} = \frac{1}{S_{21}}\begin{bmatrix} S_{11}S_{22} - S_{21}S_{12} & S_{11} \\ -S_{22} & 1 \end{bmatrix} \tag{3.153}$$

How do we cascade a pair of two-ports when the normalizing impedances at the interface are unequal? One option is to renormalize one of the two-ports as described in Section 3.1.3. A simpler option, however, is to insert a two-port consisting of a simple "through" connection between the two cascaded ports. The S parameters of the connection block are simply renormalized to those of the two cascaded two-ports. Specifically, the connection block has the S parameters,

$$\mathbf{S} = \begin{bmatrix} 0 & 1 \\ 1 & 0 \end{bmatrix} \tag{3.154}$$

To make things easy, we assume that these S parameters are normalized to Z_{01n}, the normalizing impedance of port 1 of the nth two-port. We then renormalize the S parameters to make the input normalizing impedance equal to Z_{02m}, the normalizing impedance of port 2 of the mth two-port. The S parameters of the resulting block are

$$\mathbf{S} = \frac{1}{Z_{01n} + Z_{02m}}\begin{bmatrix} Z_{01n} - Z_{02m} & 2\sqrt{Z_{01n}Z_{02m}} \\ 2\sqrt{Z_{01n}Z_{02m}} & Z_{02m} - Z_{01n} \end{bmatrix} \tag{3.155}$$

and its T matrix is

$$\mathbf{T} = \frac{1}{2\sqrt{Z_{01n}Z_{02m}}}\begin{bmatrix} -(Z_{01n} + Z_{02m}) & Z_{01n} - Z_{02m} \\ Z_{01n} - Z_{02m} & Z_{01n} + Z_{02m} \end{bmatrix} \tag{3.156}$$

The transfer scattering matrix is especially important in circuit analysis, as it allow simple *de-embedding*, or removal of the effects of unwanted elements. Suppose we have a component that can be described as a cascade of two-ports, and we have a characterization of the entire component but we want to know the S parameters of only one of the two-ports. This situation might occur, for example, if we have S-parameter measurements of a transistor chip that include bond wires, but we want the S parameters of the chip alone. If we can generate a characterization of the bond wires, perhaps through an electromagnetic simulation, we simply pre- and postmultiply the T matrix of the entire structure by the *inverses* of the wires' T matrices. This gives us the T matrix of the transistor alone, which can be converted easily to S parameters. See Section 6.1.4 for more information.

References

[3.1] Balabian, N., T. A. Bickart, and S. Seshu, *Electrical Network Theory*, New York: Wiley, 1969.

[3.2] Gonzalez, G., *Microwave Amplifiers: Analysis and Design*, Englewood Cliffs, NJ: Prentice-Hall, 1984.

[3.3] LePage, W. R., *Complex Variables and the LaPlace Transform for Engineers*, New York: Dover, 1961.

[3.4] Struble, W., and A. Platzker, "A Rigorous yet Simple Method for Determining Stability of Linear N-Port Networks," *IEEE GaAs IC Symposium Digest*, 1993, p. 251.

[3.5] Bode, H., *Network Analysis and Feedback Amplifier Design*, New York: Van Nostrand, 1945.

[3.6] Jackson, R. W., "Criteria for the Onset of Oscillation in Microwave Circuits," *IEEE Trans. Microwave Theory Tech.*, Vol. 40, 1992, p. 566.

Chapter 4

Matching Circuits

Much of the design of high-frequency components involves the use of matching circuits to interconnect stages, with a goal of maximizing power transfer. In this regard, RF and microwave design is fundamentally different from analog electronic design, which largely involves a focus on voltage (or occasionally current) gain, and devices are operated within their "midfrequency" range, where their parasitic capacitances are negligible. In microwave circuits, we rarely have that luxury; devices are operated near the upper limits of their useful frequency ranges, and power gain is the important criterion. We must achieve optimum performance under these difficult conditions, and that invariably requires impedance matching.

While narrowband matching is straightforward, broadband matching circuit design is neither straightforward nor particularly satisfying. There usually exists no straightforward, elegant technique for designing a broadband matching circuit, and it is often impossible to create an optimum design. Indeed, it is often impossible to know how close any design approaches some optimum, or even to have a clear idea of what that optimum is. In such a situation, it is not surprising that many different approaches to the impedance-matching problem have been developed. In the past, reactive synthesis dominated, but today, methods that are somewhat less elegant but more practical, involving computer optimization, are favored. In this chapter, we begin with the former and end with the latter.

4.1 Fundamentals

4.1.1 Power Transfer and Port Impedances

The problem we address is illustrated in Figure 4.1, which shows a source having an impedance Z_S and a load Z_L. The source and load may be real impedances, such as the ubiquitous 50Ω standard, or may be complex. The reactive part of the source or load could be a lumped inductor or capacitor, or it might be some more complex structure whose reactance does not vary in a simple way with frequency.

If we can find a lossless two-port network whose input impedance $Z_1 = Z_S^*$, when terminated by Z_L, the power transferred into Z_L is equal to the available power of the source. This is proven not only by (3.52) but by the simple fact that power transferred to the input of a lossless matching circuit cannot be dissipated anywhere except the real part of the terminating impedance. Furthermore, by the same reasoning, when the input port is conjugate matched, the output must be conjugate matched as well, so $Z_2 = Z_L^*$. The circuit must be equivalent to a source having an impedance Z_2 and the same available power as V_S and Z_S. Mathematically,

$$P_{av} = \frac{|V_S|^2}{8\mathrm{Re}\{Z_S\}} = \frac{|\hat{V}_S|^2}{8\mathrm{Re}\{Z_2\}} \tag{4.1}$$

where the circumflex indicates the value of the transformed voltage source.

The above argument is the formal justification for determining the gain of a two-port in terms of source and load impedances or reflection coeffi-

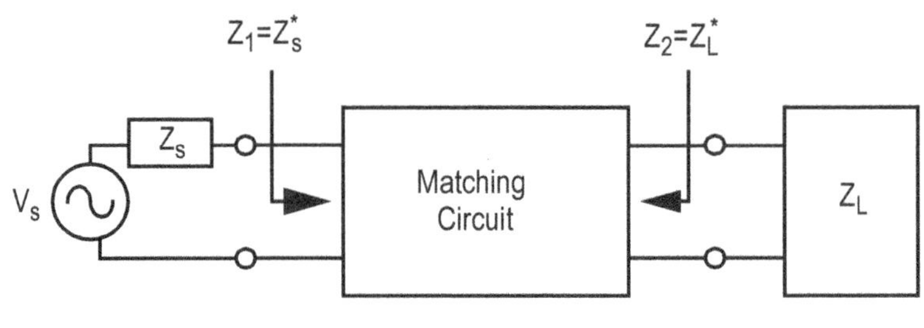

Figure 4.1 An illustration of the matching problem. The matching circuit is a lossles two-port having, ideally, a conjugate match at both ports.

cients, as we did in Section 3.3. When those impedances are realized by a lossless matching circuit, the available and delivered powers are the same, and so must be the gain.

The matching circuit can be realized by any kind of lossless elements that are practical, realizable, and provide the necessary impedance transformation. Depending on the frequency and the application, the circuit may consist of lumped capacitors, inductors, or transmission-line segments. The circuit also may be sufficiently narrowband that a design at center frequency is adequate, or the bandwidth may be great enough that a more complex design is necessary. In the latter case, we shall see that it is impossible to match a load perfectly over a finite bandwidth, and the greater the bandwidth, the worse the match. There are fundamental reasons for this.

4.1.2 Impedance Normalization

In much of this chapter, we deal with normalized impedance and admittance. To normalize impedances and admittances, we divide impedances by R_n and admittances by G_n, where $G_n = 1 / R_n$. R_n is invariably either the real part of the source impedance, $\mathrm{Re}\{Z_s\}$, or the real part of the load impedance, $\mathrm{Re}\{Z_L\}$. For LC elements,

$$\hat{B} = \frac{\omega C}{G_n} \quad \text{or} \quad \hat{B} = \frac{1}{G_n \omega L}$$
$$\hat{X} = \frac{\omega L}{R_n} \quad \text{or} \quad \hat{X} = \frac{1}{R_n \omega C} \tag{4.2}$$

We use the circumflex to indicate a normalized quantity. While normalization might at first seem like a complication, it allows us to use a Smith chart based on 1Ω for any matching problem, and, as we shall see, allows the use of prototype circuits based on $\mathrm{Re}\{Z_L\} = 1\Omega$ for creating matching circuits for any source and load impedance.

4.2 Narrowband Matching

The narrowband matching approaches we discuss in this section are inherently bandpass matching circuits. Here we assume that the bandwidth of interest is narrow enough that it need not be taken into consideration, and a simple circuit designed to operate at the bandcenter is adequate. This approach can be used when the Q of the load is low, relative to the bandwidth of interest. As the load Q increases, the bandwidth of the matched load may become impractically narrow; in that case, a broadband matching approach must be used. This is covered more completely in Sections 4.4 to 4.6.

4.2.1 L-Section Matching Circuits Using LC Elements or Stubs

In Chapter 3 we saw that achieving a certain value of gain required that we provide some particular source and load impedances to a two-port. We shall designate these Z_{s2} and Z_{L2}, respectively. These were not necessarily conjugate-match values. In the simplest case, the external source and load are the standard impedance, Z_0. With reference to Figure 4.1, our problem is to synthesize a lossless matching circuit at the input having $Z_s = Z_0$ and $Z_L = Z_{s2}^*$. This way, when the input matching circuit has been synthesized, $Z_2 = Z_{s2}$. A similar process is used for the output matching circuit.

A simple approach is to select a series reactive element X that causes the real part of the input admittance to be some $Y_0 + jB$, where $Y_0 = 1 / Z_0$ and B is some resulting susceptance. We then add a shunt element of susceptance $-B$, making the input admittance simply Y_0. The value of X can be found algebraically or by trial and error. A better way is to use a Smith chart, as illustrated in Figure 4.2, curve (a). In the chart, $\hat{Z}_L = Z_{s2}^* / Z_0 = 0.2 - j0.4$ is the normalized load impedance; we use $R_n = Z_0$ as the normalization quantity and the circumflex indicates a normalized value. Adding a series reactance (in this case, an inductance) moves the impedance along a constant-resistance line to the point where it intersects the $\hat{Y}_1 = 1$ line, where $\hat{Y}_1 = 1 / \hat{Z}_1$ is the normalized input admittance per Figure 4.1. That point is $\hat{Y}_1 = 1.0 - j2.0$ or equivalently $\hat{Z}_1 = 0.2 + j0.4$. The value of the normalized series reactance is the difference between the reactances at the starting and ending points of that path, in this case +0.8, or an actual reactance of $+0.8\ Z_0$. The normalized shunt susceptance necessary to cancel that of $\hat{Y}_1$ is simply 2.0, or an actual susceptance of $2.0\ Y_0$.

The matching elements can be realized in any manner that is appropriate to the type of circuit and its frequency. In many cases a lumped series inductor and shunt capacitor can be used. At higher frequencies, it may be necessary to use a high-impedance transmission-line segment to approximate the series inductor and a low-impedance shunt stub to realize the capacitance. Many options are possible, and selecting the correct one is an exercise in engineering judgment.

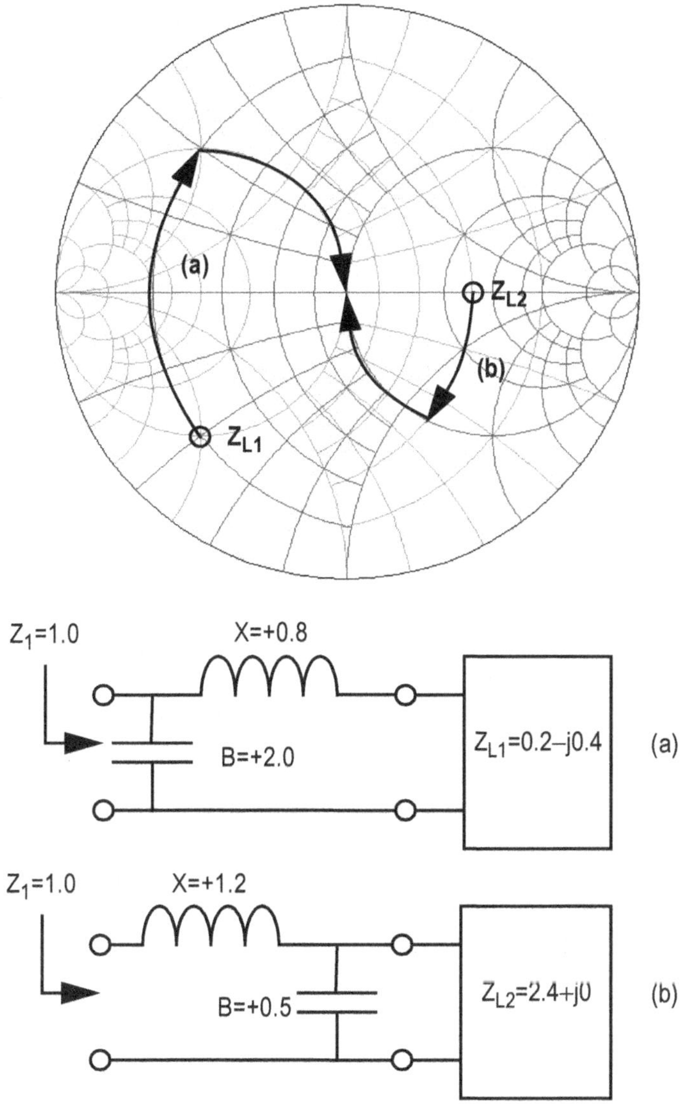

Figure 4.2 Two impedances, Z_{L1} and Z_{L2}, and their respective matching circuits, (a) and (b). The immittances shown in the circuits are normalized. It should be clear from the Smith chart that the series-L, shunt-C network, as used in (a), cannot be used for the impedance in (b). The Smith chart is useful not only for realizing the matching circuit, but for determining quickly what kind of matching network can be used for a particular load.

A series-L, shunt-C matching circuit cannot match any possible load. An example is shown in Figure 4.2, curve (b). In this case it should be clear that no series reactance exists that results in $\hat{Y}_1 = 1 + j\hat{B}$. It is possible, however, to use a shunt susceptance and series reactance to match the load; a shunt-C, series-L circuit is shown in the figure, but one could also use a shunt-L, series-C circuit. The choice depends largely on practical considerations such as circuit layout. For example, a shunt inductance could serve as a convenient bias path as well as a matching element.

Instead of a series-L element in Figure 4.2, curve (a), we could use a series transmission line of characteristic impedance Z_0. The load's reflection coefficient of $\hat{Z}_{L1}$ is −136 degrees; this would have to be shifted to +136 degrees at $\hat{Y}_1 = 1.0 - j2.0$, or a change of 88 degrees. This corresponds to a transmission-line length of 44 degrees. One could also use a series transmission line and shunt C to match Z_2, but the line would have to be more than one-quarter wavelength long. A transmission line with a shunt inductance would also work and would result in a smaller circuit.

In simple matching circuits, matching a highly reactive load may be difficult. As the Q of the load increases, the sensitivity of the input impedance to the element values increases, so the tuning of the circuit may be "touchy." Furthermore, losses in the circuit elements may make the matching circuit loss uncomfortably high. One way around this dilemma is to use more than two elements in the circuit, so a complete match is not achieved in a single LC section. This also provides greater bandwidth in matching high-Q loads.

4.2.2 Realization of L and C Elements with Transmission Lines

It is well known that a shorted or open-circuit stub can realize an inductance or capacitance. From (1.52), with $Z_L = 0$, we find the input reactance of a lossless short-circuit stub to be

$$X = Z_0 \tan(\beta l) \tag{4.3}$$

where l is the line length and β is the phase-shift constant in rad/m. As long as $\beta l < \pi/2$, the stub appears inductive. For an open-circuit stub, we obtain

$$B = Y_0 \tan(\beta l) \tag{4.4}$$

and for $\beta l < \pi/2$ the stub is capacitive.

It is important to note that the behavior of the reactance or susceptance of a stub as a function of frequency is different from that of a lumped inductor or capacitor. The approximation is best when the stub is short, Z_0 is high for the inductive stub, and Y_0 is high for the capacitive.

These stubs work best in microstrip as shunt elements, where the stub consists of a single strip conductor over a ground plane. Realization as series elements would require two parallel conductors, which would in fact realize coupled lines over a ground plane. This is not the same as a simple stub.

A series inductor is often approximated by a series, high-impedance transmission-line section. The reactance is approximately (and coincidentally) given by (4.3). In this case it is essential that Z_0 be as high as practically possible and $\beta l \ll \pi / 2$; then $X \approx Z_0 \beta l$. This is not a very accurate approximation.

A series capacitor is sometimes realized as an "interdigital" structure, a short multistrip set of coupled lines. These are adequate only for very small values of capacitance, and they are difficult to characterize. They also have substantial fringing capacitance, which creates a shunt capacitance component. Such capacitors are best evaluated with an electromagnetic field simulator, a computationally expensive process. For these reasons, they are not used commonly in microstrip circuits.

Because of the ease of realization, series transmission lines, inductive, high-impedance series lines, and shunt capacitive stubs are the most commonly used matching-circuit structures in microwave circuits. For RF circuits, LC matching is often used; small, low-parasitic chip inductors and capacitors are readily available for such circuits.

4.2.3 Series-Line Matching

Series-line matching is illustrated in Figure 4.3. In this case, we use a single series line of characteristic impedance Z_n to perform the matching. We normalize the source and load impedances to that impedance; that is, Z_0 / Z_n and Z_L / Z_n. Z_n is selected so that a constant-radius curve on the Smith chart intersects the real axis at Z_0 / Z_n.

The selection of Z_n is the tricky part of the process. It requires a certain amount of trial and error, but once found, creates a matching circuit that is simple and usually is easily realizable.

4.2.4 Quarter-Wave Transformer Matching

We saw in Section 1.4 that a multisection quarter-wave transformer could provide a remarkably broadband match to a resistive load. It is a simple matter to add a shunt or series reactive element to cancel the load's reactance or susceptance, making the load resistive, and to use a transformer to match that resistive load to the source. This can result in a large matching circuit, but one that is fairly broadband. Transformer matching often results in a circuit that is larger, however, than those realized in the more sophisticated ways that we describe later in this chapter.

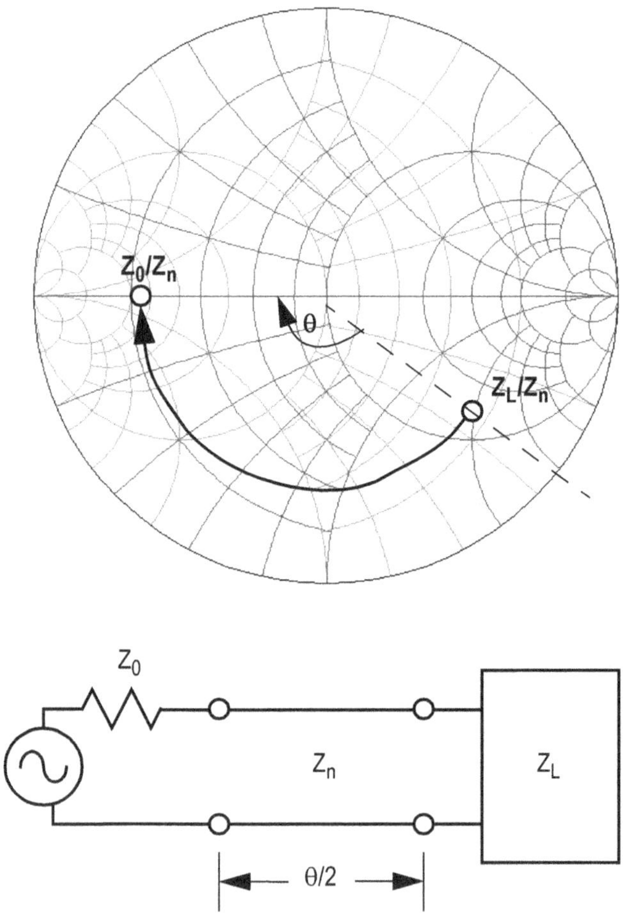

Figure 4.3 An example of series-line matching. The source and load are normalized to the characteristic impedance of the series line, Z_n. Z_n is selected so that the constant-radius curve intersects the real axis at Z_0 / Z_n.

4.2.5 Simple Broadbanding Technique

A simple technique can be used in combination with single-frequency matching to produce a good match over a moderate bandwidth. Figure 4.4 illustrates the technique. It is most useful when the impedance locus as a function of frequency, $Z_L(\omega)$, is oriented so that the high-frequency end of the locus is counterclockwise from the low-frequency end.

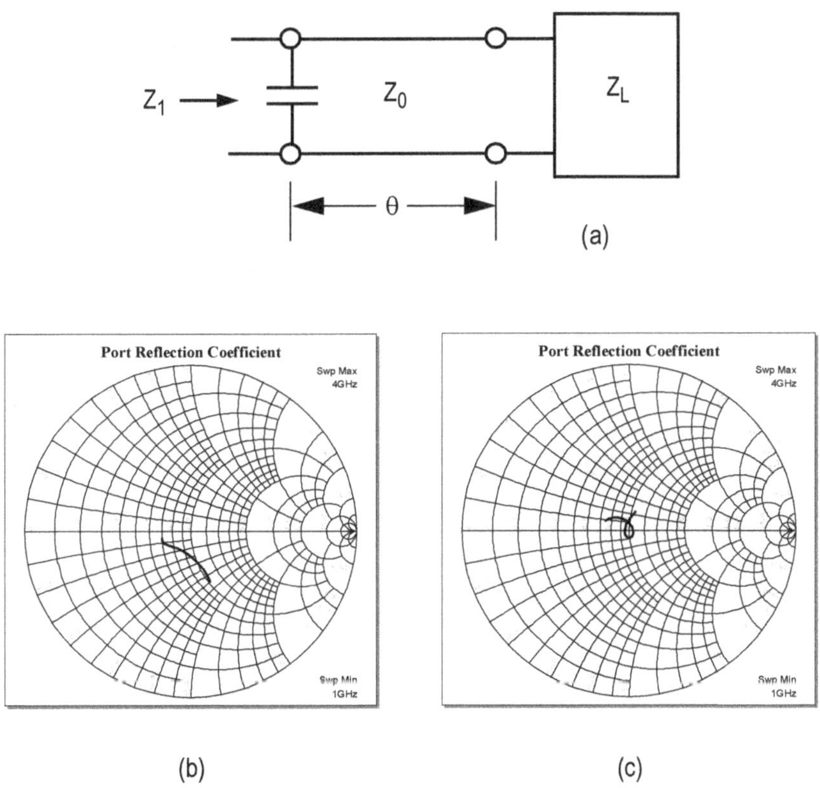

Figure 4.4 Illustration of a simple broadbanding method (a). The impedance $Z_L(\omega)$ from 1.0 to 4.0 GHz is shown in (b); when a transmission line is added, the high-frequency end of the impedance locus undergoes more phase shift than the low end, so the locus eventually wraps up on itself. A shunt capacitance then moves the locus to the center of the reflection-coefficient plane (c).

If we connect a transmission line to the load, the input-impedance locus wraps up on itself as the line is lengthened. This occurs because the phase change along the line is greater at higher frequencies. The line length is selected so that the wrapped locus is centered on the $\hat{Y}_1 = 1 + jB$ curve of the chart, and then a shunt susceptance of $-B$ is used to center the locus on the $\hat{Z}_1 = 1$ point.

4.3 Transmission-Line Transformers

Transmission-line transformers are used extensively in RF circuits, especially power amplifiers. They can be realized in a number of ways. An obvious one is to wind a conventional transformer on a toroidal core, but this approach usually does not provide good performance, as it is difficult to achieve high coupling between the coils, and interwinding capacitance restricts the high-frequency performance. A better approach is to wind a twisted-pair transmission line on a toroidal ferrite form. We can view this as a pair of coupled lines, where the toroidal core has a minimal effect on the odd mode but increases the even-mode inductance substantially. This causes the even-mode characteristic impedance to be high, so a kind of parallel-wire balun (Section 2.3.2) can be realized successfully. Interconnections of such baluns can then provide impedance transformation. With such transformers, multioctave bandwidths can be achieved.

Another type of transmission-line transformer is a rather odd type of twisted-pair structure, analogous to an autotransformer. This turns out to be very useful and to have some interesting properties.

4.3.1 Wirewound Impedance Transformer

A more or less conventional transformer can be realized at RF frequencies. Figure 4.5 shows such a transformer; it consists of a trifilar winding on a toroidal ferrite core. The design of such transformers largely follows that of low-frequency transformers; the magnetic coupling between the coils must be maximized (thus the trifilar winding), and the inductive reactance of the windings must be large relative to the source and load impedances. Because all windings have the same number of turns, the selection of impedance transformations is limited. Attempting to achieve other transformations by using windings having different numbers of turns is usually impractical, as it increases leakage inductance beyond a tolerable level. The greatest problem with such transformers, however, is bandwidth limitation caused by magnetic field leakage and capacitance between the coils' windings.

Sometimes no attempt is made to maximize the coupling in a transformer, and instead the leakage inductance of both coils is resonated by capacitors. This realizes a narrowband transformer, which could be viewed as a pair of coupled, parallel-LC resonators. Such "tuned interstage" transformers were once used commonly between amplifier stages of a radio receiver to provide filtering as well as coupling. They are largely obsolete today, as they are invariably too large for modern radios, and better interstage coupling and filtering techniques have been devised.

Conventional transformers are used occasionally in RF applications. The availability of a center tap and the easy accommodation of an unbalanced input are useful in mixers, for example. The designs we discuss later, however, are generally superior and are much more widely employed in RF circuitry.

4.3.2 Toroidal Balun

In Section 2.3.2 we examined a simple, parallel-strip balun and concluded that it doesn't work very well unless the even-mode characteristic impedance is made quite high. The even-mode impedance is given by

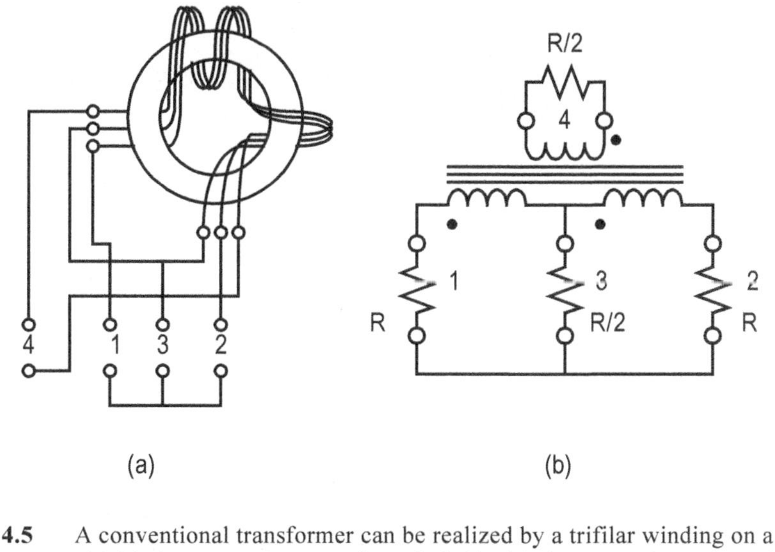

Figure 4.5 A conventional transformer can be realized by a trifilar winding on a toroidal inductor (a). It can realize a hybrid with the unequal port impedances, as shown in (b).

$$Z_{0e} = \sqrt{\frac{L_e}{C_e}} \tag{4.5}$$

where L_e is the even-mode inductance and C_e is the even-mode capacitance. It is usually impossible to achieve an adequate even-mode impedance simply by minimizing the even-mode capacitance, say, by elevating the strips above a ground plane, but substantially increasing the even-mode inductance is often possible.

Figure 4.6 shows a twisted pair of wires wound on a toroidal ferrite core. The small-diameter wires are usually insulated by varnish or by a thin plastic layer. Viewed as a pair of coupled lines, the lines' odd-mode fields are concentrated between the wires and are minimally affected by the core, but the even-mode magnetic field couples to the core as strongly as would the field from a single wire. This increases its even-mode inductance substantially over its free-space value.

The characteristic impedance of the twisted pair, viewed as a simple balanced transmission line, is given by

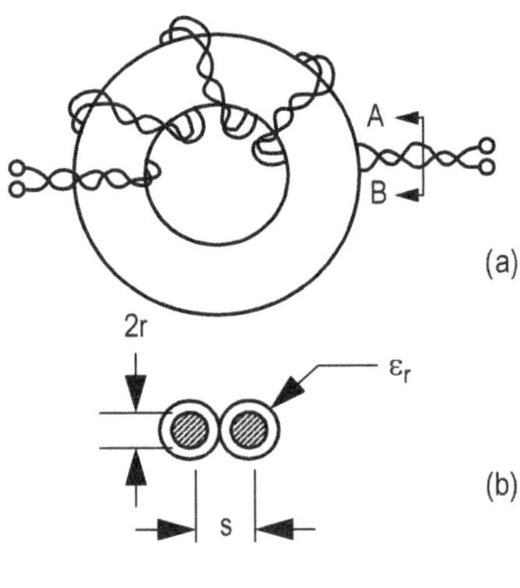

Figure 4.6 Toroidal core wrapped with a twisted pair of wires (a). The wires can be treated as a transmission line; cross-section AB, shown in (b), can be used to determine the characteristic impedance. The wire covering is usually a thin layer of varnish, which affects the characteristic impedance minimally.

$$Z_0 = 2Z_{0o} \approx \frac{120}{\sqrt{\varepsilon_{eff}}} \cosh^{-1}\left(\frac{s}{2r}\right) \tag{4.6}$$

where s is the wire spacing, r is its radius, and ε_{eff} is an effective dielectric constant (Section 1.2.2.4). Unfortunately as $s \to 2r$, the value of Z_0 given by (4.6) varies rapidly, and a value of ε_{eff} is difficult to estimate accurately. Furthermore, the type and thickness of the wires' insulation, the coupling of fringing fields to the ferrite core, and the tightness of the twist also affect the impedance. These characteristics are difficult to quantify, so determining the characteristic impedance accurately is difficult. However, a pair of varnish-insulated wires of a size appropriate for a small, low-power transformer, twisted moderately tightly, has a characteristic impedance in the range of 50Ω to 70Ω. For more precise estimation of the impedance, see [4.1].

The even-mode impedance is difficult to estimate accurately. The even-mode inductance per length is not difficult to determine, but the even-mode capacitance is problematical, as the transmission line, wrapped on a toroidal form, is not uniformly spaced above a ground plane. Usually it is adequate to make sure that the total inductance of the twisted pair, viewed as a single wire, is high enough so that the reactance is much larger than the load or source impedance at the lowest frequency of interest. It is usually easy to achieve adequate inductance. The inductance of a toroidal inductor is a textbook electromagnetics problem; it is given by

$$L = \frac{\mu N^2 d}{2\pi} \ln\left(\frac{b}{a}\right) \tag{4.7}$$

where b and a are, respectively, the outer and inner diameters of the toroid, d is the cross-sectional diameter, N is the number of turns, and μ is the permeability.

Other requirements are as described in Section 2.3.2. The lower end of the balun's passband is established by the even-mode inductance, and the high end by the match between the load and the parallel-wire transmission line's characteristic impedance. At frequencies above that lower limit, the balun can be treated as a simple, floating, balanced transmission line.

A pair of toroidal baluns can realize a 4:1 impedance transformer. This is accomplished by connecting the ports in parallel at one end and in series at the other, as shown in Figure 4.7. This halves the impedance at the parallel-connected port and doubles it at the series-connected port; additionally, the resulting structure is balanced.

At high power levels, thin, varnish-insulated wire may not be adequate to carry the required current and a small core may saturate magnetically. One solution is to use larger-diameter, plastic- or varnish-insulated wire with larger toroidal cores. Another is to use a section of small-diameter coaxial line either wrapped around a toroid or loaded with a cylinder of ferrite or a series of large ferrite beads. In the latter case, the coaxial line can be wound into a loop or kept straight, depending on the frequency and the necessary line length.

4.3.3 Transmission Line "Autotransformer"

This type of transformer is analogous to a lumped-element autotransformer and is compared to one in Figure 4.8. At low frequencies, it is, in fact, equivalent to an autotransformer, but at high frequencies, the transmission-line characteristics dominate in establishing its behavior. This way, the capacitance of the windings, which is a deleterious parasitic in a conventional transformer, is part of the transmission line and thus an integral part of the transformer.

By itself, the structure realizes a 4:1 impedance transformation, but two or more such structures can be interconnected to form other transformation ratios. As with other transmission-line transformers (and autotransformers, for that matter), its operation at lower frequencies depends on its conductors' even-mode inductances, which are equivalent to the self inductances of the coils in the autotransformer.

At low frequencies, the electrical length of the transmission line is small, so it is effectively a simple autotransformer. The low-frequency cutoff then is determined, as in any conventional transformer, by the self inductances of the windings. The structure behaves like a transformer as long as those reactances are large relative to the source and load resistances.

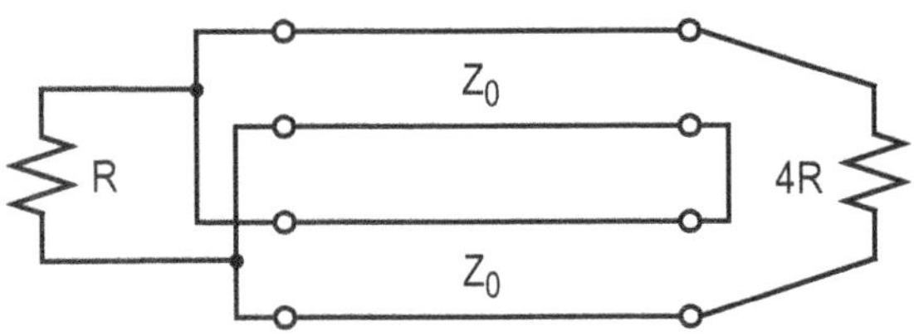

Figure 4.7 A 4:1 balanced transformer can be realized by two parallel-strip baluns. The transmission lines can be realized by two twisted pairs wound on a single toroidal core.

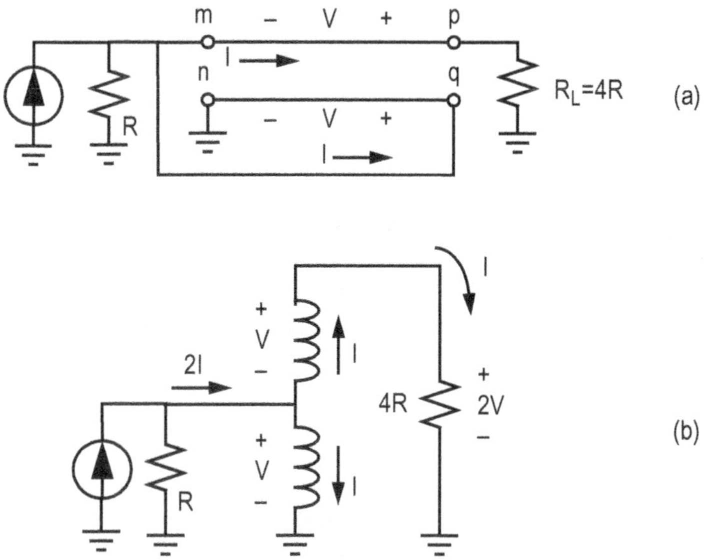

Figure 4.8 Transmission-line transformer (a) and an autotransformer (b). The (*m*, *n*, *p*, *q*) node numbering is used in the derivation of its operation. The transmission line transformer in (a) operates in a manner identical to that of the autotransformer, as illustrated by the voltages and currents in (b).

This is, in turn, a function of the length of the wires and the toroidal core's dimensions, as in (4.7).

At high frequencies, the analysis of the transformer is straightforward but algebraically messy. We begin with the assumption that the transmission line is lossless; then its Y matrix, from (1.51), is

$$Y = \frac{j}{Z_0 \sin(\theta)} \begin{bmatrix} -\cos(\theta) & 1 \\ 1 & -\cos(\theta) \end{bmatrix} \tag{4.8}$$

where θ is the electrical length of the line. A two-port matrix can be converted into a nodal matrix via the following pattern:

$$\begin{bmatrix} I_m \\ I_n \\ I_p \\ I_q \end{bmatrix} = \begin{bmatrix} Y_{11} & -Y_{11} & Y_{12} & -Y_{12} \\ -Y_{11} & Y_{11} & -Y_{12} & Y_{12} \\ Y_{21} & -Y_{21} & Y_{22} & -Y_{22} \\ -Y_{21} & Y_{21} & -Y_{22} & Y_{22} \end{bmatrix} \begin{bmatrix} V_m \\ V_n \\ V_p \\ V_q \end{bmatrix} \tag{4.9}$$

where the Y_{ij} are the two-port Y parameters. This is formally an *indefinite admittance matrix*, which we shall study in greater detail in Chapter 5. Its columns and rows are designated, in order, *m*, *n*, *p*, and *q*, consistent with Figure 4.8. V_m is the voltage between node *m* and ground, I_n is the current at node *n*, and so on.

In the transformer, node *n* is grounded. Grounding node *n* involves merely deleting column *n*, as it multiplies V_n, which is zero. Row *n* can also be eliminated, as it describes I_n, which is of no interest. Connecting nodes *m* and *q* is a bit more complicated, but by expanding the linear equations with $V_m = V_q$, noting that $Y_{11} = Y_{22}$ and $Y_{21} = Y_{12}$, and finally including R_L, we obtain the nodal Y matrix of the entire circuit, Y_n:

$$Y_n = \begin{bmatrix} 2(Y_{11} - Y_{12}) & Y_{12} - Y_{11} \\ Y_{12} - Y_{11} & -Y_{11} + \dfrac{1}{R_L} \end{bmatrix} \tag{4.10}$$

Substituting the values of Y_{ij} from (4.8) gives

$$Y_n = \frac{j}{Z_0 \sin(\theta)} \begin{bmatrix} -2(\cos(\theta) + 1) & \cos(\theta) + 1 \\ \cos(\theta) + 1 & \dfrac{Z_0 \sin(\theta)}{jR_L} - \cos(\theta) \end{bmatrix} \tag{4.11}$$

To find the input impedance, we must invert the Y matrix in (4.11) to obtain a Z matrix; Z_{11} then becomes the input impedance. This is a messy job, best accomplished with symbolic manipulation software. We show the results in Figure 4.9. One surprising result is that the performance is optimal as $\theta \to 0$. As we noted in Chapter 2, a pair of coupled lines devolves into a simple, balanced transmission line when the even-mode characteristic impedance is infinite. By treating the wires as a simple transmission line, we have implicitly assumed that the line's even-mode impedance is infinite, and, therefore, that its even-mode inductance is infinite. This means that the lines have an infinite even-mode input impedance, even as frequency approaches zero.

Figure 4.9 shows that the operation of the structure, viewed as a transmission line, converges to that of the autotransformer at low frequencies. At high frequencies, the core has little effect on the operation of the transformer. Therefore, such a transformer should have smooth response from its low-frequency cutoff to a high frequency where $\theta \sim 1$ rad.

The widest bandwidth is achieved with $Z_0 \approx R_L/2$. At lower values of Z_0, the real part of the input impedance rolls off early; at higher values, the imaginary part increases. At the optimum value, the input return loss is be-

$$Z := 50$$

$$R := 100$$

$$Z11(x) := -Z \cdot j \cdot \sin(x) \cdot \left(\frac{R^2 \cdot \cos(x) - R^2 \cdot \cos(x)^2 - 2 \cdot Z^2 \cdot \sin(x)^2 + R \cdot Z \cdot \sin(x) \cdot i + R \cdot Z \cdot \cos(x) \cdot \sin(x) \cdot i}{R^2 \cdot \cos(x)^3 - R^2 \cdot \cos(x)^2 - R^2 \cdot \cos(x) + R^2 + 4 \cdot Z^2 \cdot \cos(x) \cdot \sin(x)^2 + 4 \cdot Z^2 \cdot \sin(x)^2} \right)$$

$$x := 0.01, 0.02 .. 1.57$$

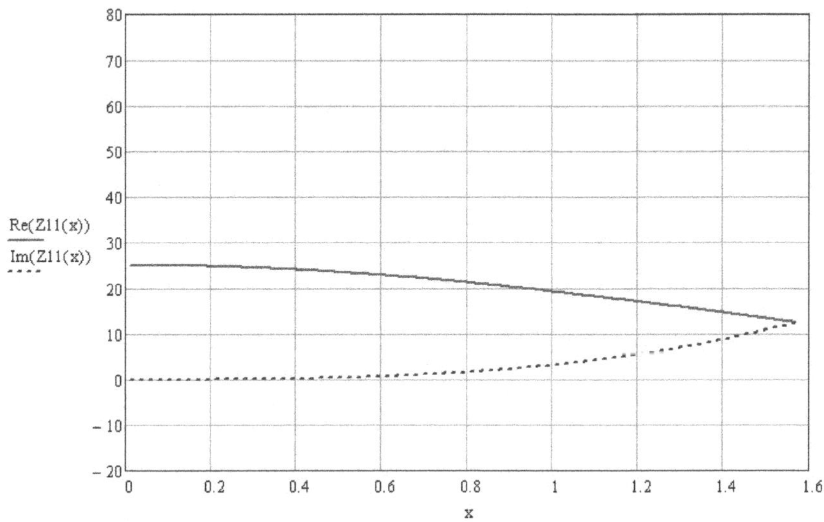

Figure 4.9 The real and imaginary parts of the input impedance of the transmission-line transformer in Figure 4.8(a). The variables have been abbreviated in the calculation; *Z* is the characteristic impedance, *R* is R_L, *x* is the electrical length, θ.

low 15 dB for $\theta \leq 1.0$ rad. It is likely that the bandwidth could be extended somewhat with capacitive tuning.

A number of similar transformers can be realized. A few of these are shown in Figure 4.10. See [4.2] for more such circuits and an analysis of them. The low-frequency analysis of such structures can often be performed by treating the coupled lines as transformers. At high frequencies, the analysis is more complex. They can be analyzed in a manner similar to that of the transformer in Figure 4.8(a) or simply by means of conventional circuit-analysis software.

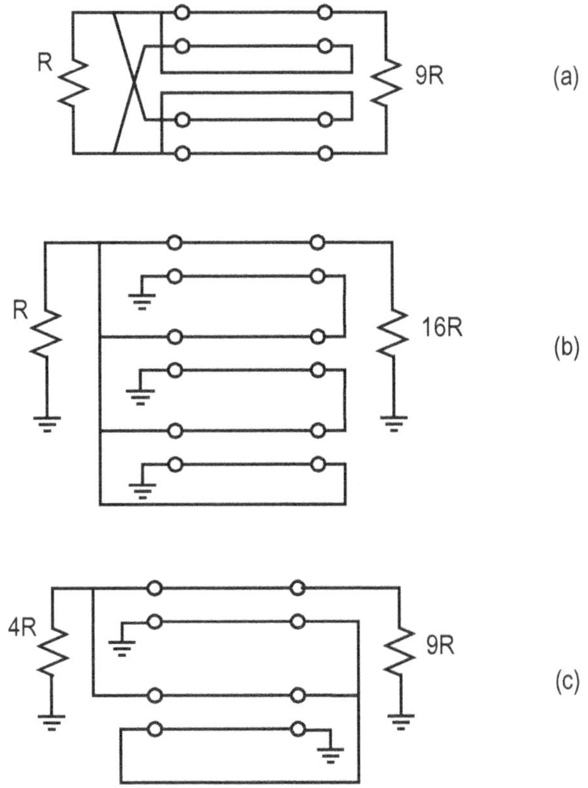

Figure 4.10 Three transmission-line transformer configurations: (a) 9:1; (b) 16:1; and (c) 9:4.

4.4 Classical Synthesis

Over the years, a number of matching-circuit synthesis methods have been developed. These are based on the idea of creating a normalized low-pass prototype network and scaling it to the operating frequency and impedance level. The matching circuit that results can retain the low-pass structure or be transformed into a bandpass structure. A high-pass structure is also possible, but that is generally less practical and therefore rarely used.

These method have two disadvantages. First, the source resistance is real and is treated as a free variable in the synthesis, so the process rarely results in a termination having the desired Z_s. It is sometimes possible, however, to shift that resistance to the desired value. The second problem is that the resulting element values are sometimes not realizable. These difficulties are consequences of the methods' main advantage, which is the realization of an optimum network. Even with the disadvantages, however, the process can be used to create an initial matching design, which can be optimized on the computer to obtain the desired Z_s and practical element values, although often at some cost to the network's performance.

4.4.1 Matching Limitations

Before discussing specific techniques, we must recognize that an ideal match over a finite bandwidth cannot be achieved. For a given load, there is always a trade-off between the quality of the match, as indicated by the reflection coefficient at the input of the matching circuit, and the bandwidth. This limitation was examined by Bode [4.3] and later by Fano [4.4], and is often called *Fano's limit*. Fano found that the best reflection coefficient achievable by a lossless matching circuit of any type is given as follows:

$$\int_0^\infty \ln\left(\frac{1}{|\Gamma(\omega)|}\right) d\omega = \frac{\pi}{RC} \tag{4.12}$$

This includes an implicit assumption that the load consists of a parallel resistor and capacitor, but one can view this more generally by substituting Q/ω_0 for RC, where ω_0 is the frequency at which Q is defined. To make this a bit more concrete, suppose we could create a matching circuit with $|\Gamma| = 1$ out-of-band and a constant $|\Gamma|$ in-band, from some ω_1 to ω_2. Then (4.12) becomes

$$\int_{\omega_1}^{\omega_2} \ln\left(\frac{1}{|\Gamma|}\right) d\omega = \frac{\pi}{RC} \tag{4.13}$$

This can be solved easily, with $Q = \omega_0 RC$, to obtain

$$|\Gamma| = \exp\left(\frac{-\pi\omega_0}{(\omega_2 - \omega_1)Q}\right) \tag{4.14}$$

While this result was derived for a specific case, its implications are generally valid. It shows that the best input reflection coefficient achievable is limited by the bandwidth and the Q of the load; the greater the bandwidth or the higher the Q, the worse the reflection coefficient must be. In real circuits, of course, $|\Gamma|$ is not constant, but (4.12) still tells us that a good match in some part of the band can be achieved only at the expense of a bad one elsewhere, so ideally a matching circuit must roll off quickly out of band, placing the required bad match outside the band of interest. From this argument, it is clear that an efficient matching circuit must be some type of filter. In fact, as we shall see in the next few sections, an effective matching circuit is simply a well-chosen filter circuit.

4.4.2 Prototype Networks

A matching circuit is built around a low-pass prototype network. This network usually has a Chebyshev characteristic, as it provides an input reflection coefficient closest to the limit defined by (4.12). The use of a Butterworth characteristic, also called a *maximally flat* characteristic, is also possible and may be preferred when a lack of ripple or a low group-delay variation near cutoff is needed, and the poorer in-band match is acceptable.

The prototype network is synthesized for a 1Ω load impedance and radian frequency at cutoff of 1 rad/sec. In a Chebyshev matching circuit, that cutoff frequency is usually the highest-frequency point at which the loss equals the maximum ripple. For a Butterworth circuit, it is usually the 3-dB point.

4.4.2.1 Series RL or Shunt RC

Figure 4.11 illustrates the matching problem that is addressed here. The load is reactive, modeled by either a series resistor and inductor or a shunt resistor and capacitor, the first elements of a low-pass structure. Since they are normalized to 1 rad/sec, the series reactance or shunt susceptance are equal to the inductance or capacitance, as appropriate to the filter structure. The remaining elements are the inductors and capacitors of the matching circuit and the source resistance, whose value is determined by the design process; it cannot be specified independently. When the design is complete, some type of modification or a transformer can be used to achieve the desired source resistance.

The g_n are called *immittance parameters*, as they can represent either impedances or admittances according to their use in the circuit. The g_n values for shunt elements are susceptances or conductances and for series elements are reactances or resistances. The treatment of the resistors at the ends of the network depends on the adjacent element; if it is a shunt element, the resistor is treated as a series element, and its value is a resistance. If the adjacent element is a series element, the resistor is treated as a shunt element, and its value is a conductance. The networks in Figure 4.11 are duals; they have identical responses and the g_n values for both are the same.

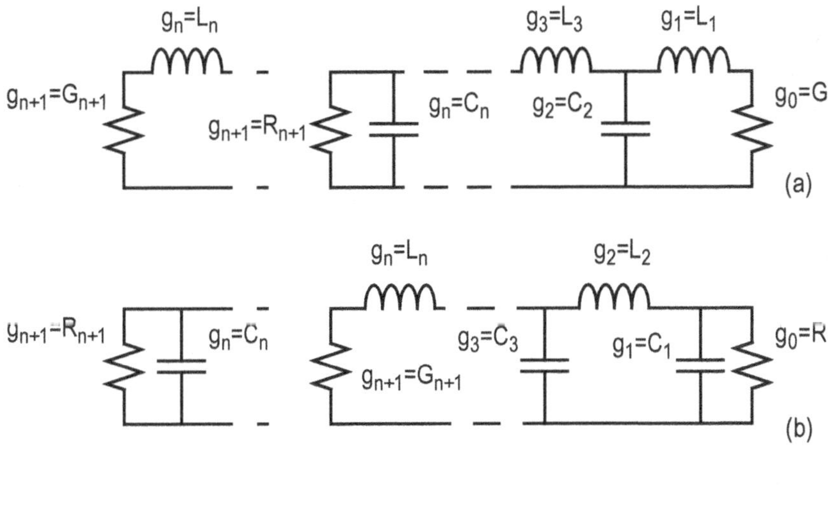

Figure 4.11 Low-pass prototype networks for matching-circuit design, (a) for a load that is inductive and (b) capacitive. The g_n are immittance parameters, which are numerically equal to the normalized R, L, C, and G values. L_1 and C_1 are part of the load. The alternative end sections at the left are used when n is odd.

4.4.2.2 Shunt RL or Series RC Loads

It is possible that the load may best approximate a shunt RL or series RC combination. In that case, the prototype network has a high-pass structure. The immittance parameters are identical to those of the low-pass case, but their interpretation is as series capacitances or shunt inductances. Since this case is the dual of the low-pass case, it is a straightforward extension of the following theory, so we shall not examine it specifically.

4.4.3 Normalization and Frequency Scaling

As with narrowband matching, we normalize impedances as in (4.2). In all cases, $R_n = R$ and $G_n = G$, where R or G is the real part of the load impedance or admittance for which we wish to create a matching circuit. We assume that R, G are constant with frequency. More modern methods that we discuss later in this chapter do not have that requirement.

Frequency can be scaled in a couple of different ways. For low-pass structures, the scaling is very simple: we use $\Omega = \omega / \omega_1$, where Ω is the normalized frequency for the prototype network, ω is the actual, unscaled frequency, and ω_1 is the edge of the passband. For a bandpass structure, the scaling is a bit more complicated, as we must map the passband of a low-pass filter into that of a bandpass. This allows us to begin with a low-pass design and create a bandpass matching circuit from it. The normalized low-pass design is called a *prototype* network. The mapping is

$$\Omega = \frac{1}{B}\left(\frac{\omega}{\omega_0} - \frac{\omega_0}{\omega}\right) \tag{4.15}$$

where ω_0 is the geometrical center of the band and B is the fractional bandwidth. Specifically,

$$\omega_0 = \sqrt{\omega_l \omega_h} \tag{4.16}$$

$$B = \frac{\omega_h - \omega_l}{\omega_0} \tag{4.17}$$

where ω_l, ω_h are the low and high band edges, respectively, of the bandpass circuit. For small fractional bandwidths, we can approximate

$$\omega_0 \approx \frac{\omega_l + \omega_h}{2} \tag{4.18}$$

This maps $\Omega \to \infty$ of the low-pass prototype circuit into $\omega = 0$ and $\omega \to \infty$ of the real circuit. It similarly maps $\Omega = 0$ of the prototype into ω_0 and $\Omega = 1$, the prototype's cutoff frequency, into ω_l and ω_h.

4.4.4 Load Scaling and the Decrement

For low-pass matching networks, the load must be scaled in frequency and impedance so that it can be part of a prototype network with $R = 1$ and $\Omega = \omega / \omega_1$. To do this, we first multiply the load capacitances or inductances by ω_1 to obtain the frequency-normalized value:

$$\begin{aligned} X &= L\omega = (L\omega_1)\frac{\omega}{\omega_1} = (L\omega_1)\Omega \\ B &= C\omega = (C\omega_1)\frac{\omega}{\omega_1} = (C\omega_1)\Omega \end{aligned} \tag{4.19}$$

The load must also be scaled according to the resistance. Invariably the real part of the load resistance is the appropriate scaling factor, as tables of prototype element values are derived for $\hat{R} = 1$, where the circumflex indicates a normalized quantity. Then

$$\begin{aligned} \hat{X} &= \frac{X}{R} = \left(\frac{L\omega_1}{R}\right)\Omega \\ \hat{B} &= \frac{B}{G} = BR = (RC\omega_1)\Omega \end{aligned} \tag{4.20}$$

where $G = 1 / R$. Then $g_1 = RC\omega_1$ for a shunt RC load or $g_1 = L\omega_1 / R$ for a series RL load. The normalized resistive part of the load is, of course, 1Ω.

This normalization applies also to other elements in the prototype matching network, so it can be inverted to determine the values of the real network's elements. Thus, when the prototype network has been designed, we find the real element values to be

$$L_n = \frac{g_n R}{\omega_1}$$
$$C_m = \frac{g_m}{R\omega_1} \quad (4.21)$$

The load's *decrement*, δ, is defined as follows:

$$\delta = \frac{1}{R\omega_1 C} = \frac{1}{g_1} \quad \text{or} \quad \delta = \frac{R}{\omega_1 L} = \frac{1}{g_1} \quad (4.22)$$

when the load is normalized as we have discussed. The decrement also can be viewed as the inverse of the load's Q at the band edge, so clearly it affects how well the load can be matched over a bandwidth. The maximum in-band loss or, alternatively, the maximum reflection coefficient, can be determined as a function of δ alone.

Figure 4.12 shows the maximum loss of a Chebyshev matching circuit as a function of δ, when it is designed to minimize the maximum loss. In this case, the ripple is not specified; the resulting ripple is given by Figure 4.13. Any attempt to reduce the ripple inevitably increases the maximum loss; this is guaranteed by (4.12). It is clear that more complex matching circuits reduce the maximum loss, but, as n increases, one quickly reaches a point of diminishing returns. There is little point in using $n > 3$, and in most cases $n = 2$ reaches a practical optimum; because of element parasitics or the inaccuracies of realizing lumped elements by distributed structures, the small benefits of greater n may not be realized in practice. This is especially the case when δ is large, on the order of 1.0 or more.

Other criteria, such as design for a specified maximum ripple, are possible and give different results. When the ripple, ε, is specified, the maximum transducer gain G_{tm} of a matching circuit having an RC or RL load is given by [4.5]

$$(1 - G_{tm})^{1/2} > \varepsilon \sinh\left\{ n \operatorname{asinh}\left[\sinh\left(\frac{1}{n} \operatorname{asin}\left(\frac{1}{\varepsilon}\right)\right) - 2\delta \sin\left(\frac{\pi}{2n}\right)\right]\right\} \quad (4.23)$$

For bandpass networks, the frequency normalization is a little more complicated. As before, the fundamental principle is that the branch reactances or susceptances of the prototype and scaled networks are the same, for corresponding frequencies. For the series inductive branch we need

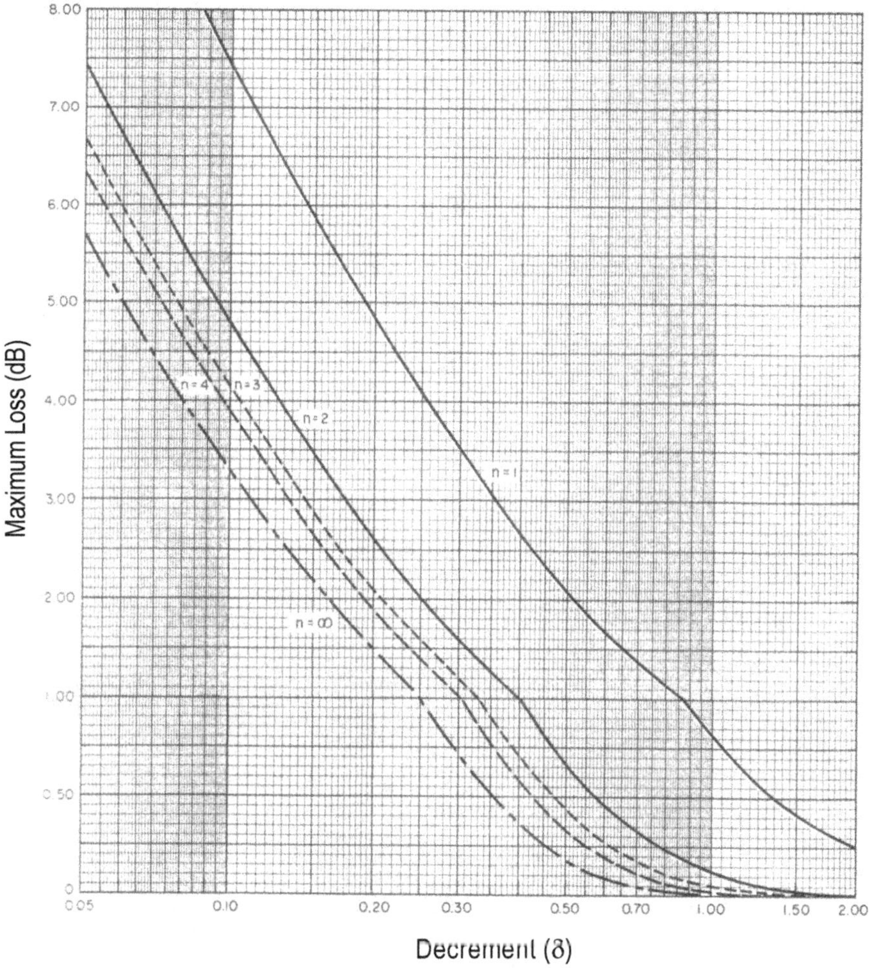

Figure 4.12 Maximum insertion loss of a Chebyshev matching network as a function of the number of elements n and the decrement δ. Reproduced by permission from [1.9] © 1980 by Artech House, Inc.

$$\begin{aligned} \hat{L}\Omega &= \frac{L}{B}\left(\frac{\omega}{\omega_0} - \frac{\omega_0}{\omega}\right) \\ &= \frac{L}{\omega_h - \omega_l}\left(\omega - \frac{\omega_0^2}{\omega}\right) \\ &= \frac{\omega L}{\omega_h - \omega_l} - \frac{1}{\omega C(\omega_h - \omega_l)} \end{aligned} \tag{4.24}$$

This implies that the inductance is scaled in frequency as $\omega_h - \omega_l$ and is resonated at ω_0 with a series capacitor. The decrement becomes

$$\delta = \frac{R}{(\omega_h - \omega_l)L} \tag{4.25}$$

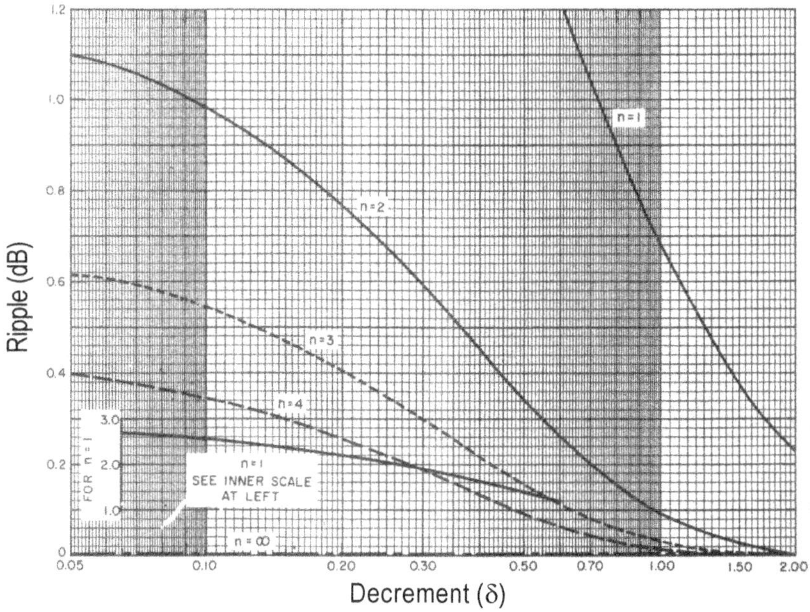

Figure 4.13 Maximum ripple in a Chebyshev matching circuit as a function of the decrement, when the circuit is designed to minimize the maximum loss. Reproduced by permission from [1.9] © 1980 by Artech House, Inc.

which is invariably larger than the low-pass case (4.22). This is not unexpected, as the narrower bandwidth should allow a better match.

The scaling of shunt capacitances is analogous to the inductances. Impedance normalization for the bandpass network is the same as for the low-pass: multiply all inductances by R_n and susceptances by G_n. Specifically,

$$L_n = \frac{g_n}{(\omega_h - \omega_l)} R_n$$
$$C_m = \frac{g_m}{(\omega_h - \omega_l)} G_n \tag{4.26}$$

4.4.5 Examples

We wish to design a matching circuit for a load consisting of a 70Ω real part in parallel with a 7.5-pF capacitor. The band of interest is 400 to 600 MHz. We shall realize it by both low-pass and bandpass matching circuits.

4.4.5.1 Low-Pass Matching Circuit

In this case the cutoff frequency is 600 MHz and $G = 1 / 70$. From (4.22) we find the decrement to be

$$\delta = \frac{G}{\omega_1 C} = \frac{1}{g_1} = 0.51 \tag{4.27}$$

From Figure 4.12, we find the maximum insertion loss to be 0.65 dB for $n = 2$, 0.45 dB for $n = 3$, and 0.35 dB for $n = 4$. We select $n = 3$.

The values of the g_n can be obtained from graphs of low-pass prototype element values. In Figures 4.14 and 4.15 we reproduce graphs from [1.9] for $n = 2$ and $n = 3$. (The case of $n = 1$ is not included, as it is, in effect, no matching circuit at all. Similarly, $n = 4$ has not been reproduced, as it is rarely used.) From the graphs, we obtain

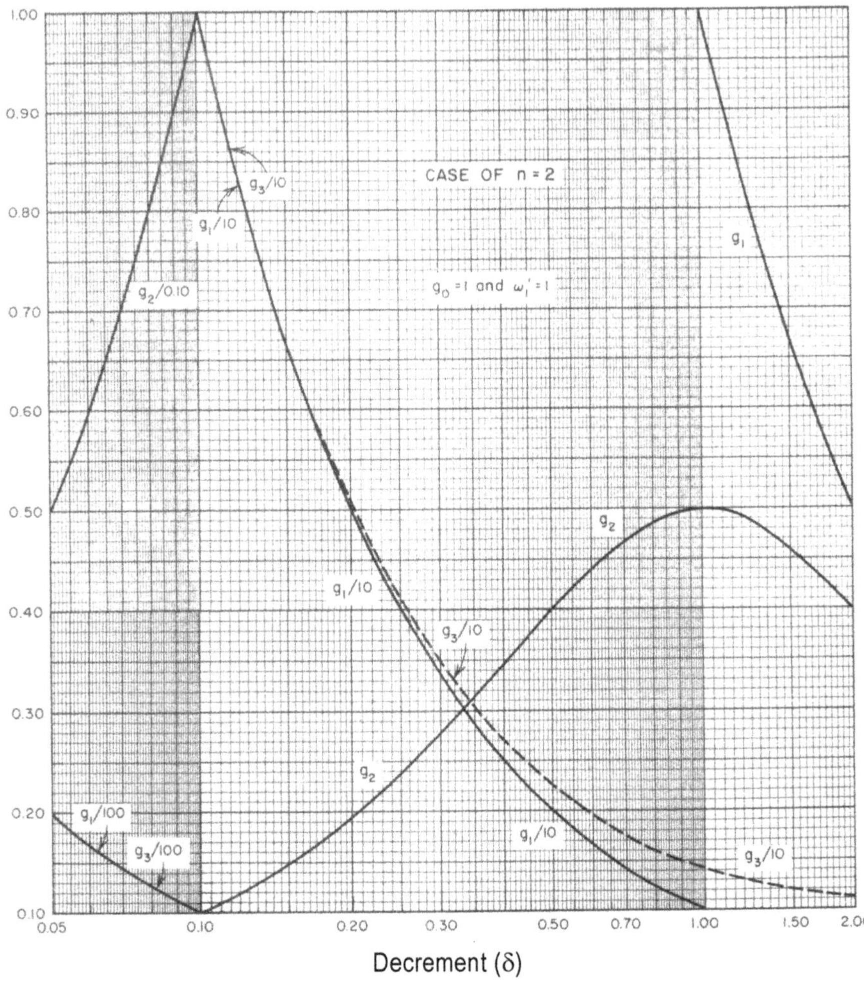

Figure 4.14 Values of g_n for an $n = 2$ matching network. Reproduced by permission from [1.9] © 1980 by Artech House, Inc.

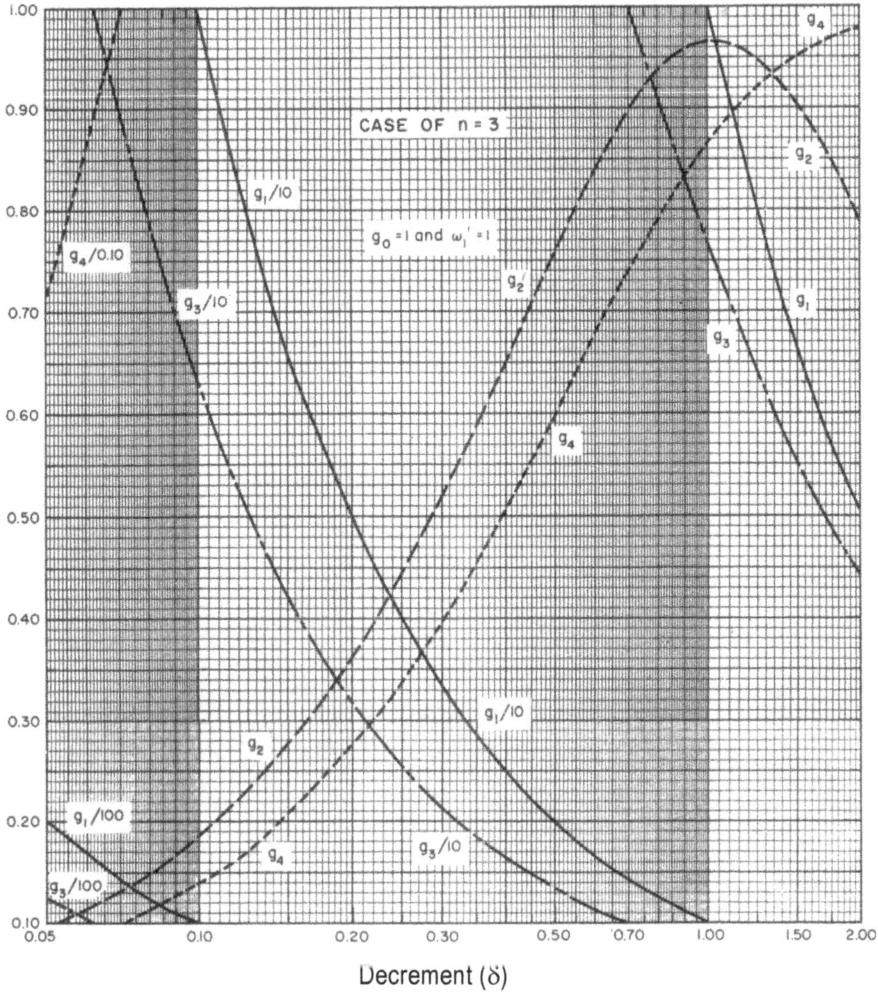

Figure 4.15 Values of g_n for an $n = 3$ matching network. Reproduced by permission from [1.9] © 1980 by Artech House, Inc.

$$g_2 = 0.76$$
$$g_3 = 1.35 \quad (4.28)$$
$$g_4 = 0.60$$

and, of course, $g_1 = 1.98$ from (4.27). This value is not needed in the design. From (4.21) we obtain the filter's element values,

$$L_2 = 14.1\,\text{nH}$$
$$C_3 = 5.15\,\text{pF} \quad (4.29)$$
$$R_4 = 42\Omega$$

Note that the prototype element g_4 is adjacent to a shunt element, so it represents a terminating resistance. Figure 4.16 shows the matching circuit and its response.

4.4.5.2 Bandpass Matching Circuit

From (4.25) the decrement is

$$\delta = \frac{R}{(\omega_h - \omega_l)L} = \frac{1}{g_1} = 1.52 \quad (4.30)$$

From Figure 4.12, the maximum insertion loss for an $n = 2$ filter is only 0.05 dB, which usually is less than the loss from circuit-element parasitics and uncertainties in their values. There is no point in using a more complex structure.

From Figure 4.14 we obtain

$$g_2 = 0.45$$
$$g_3 = 1.2 \quad (4.31)$$

In this case, g_3 represents a conductance, as it is adjacent to a series element. From (4.26),

$$
\begin{aligned}
L_2 &= 25.1 \text{ nH} \\
G_3 &= 0.0172
\end{aligned}
\tag{4.32}
$$

The terminating resistance is $1 / G_3$, which is 58.3Ω. We now add elements to resonate the shunt capacitance of the load and the series inductance. To do this, we must first determine the center frequency, which is the geometric mean of the high- and low-frequency limits:

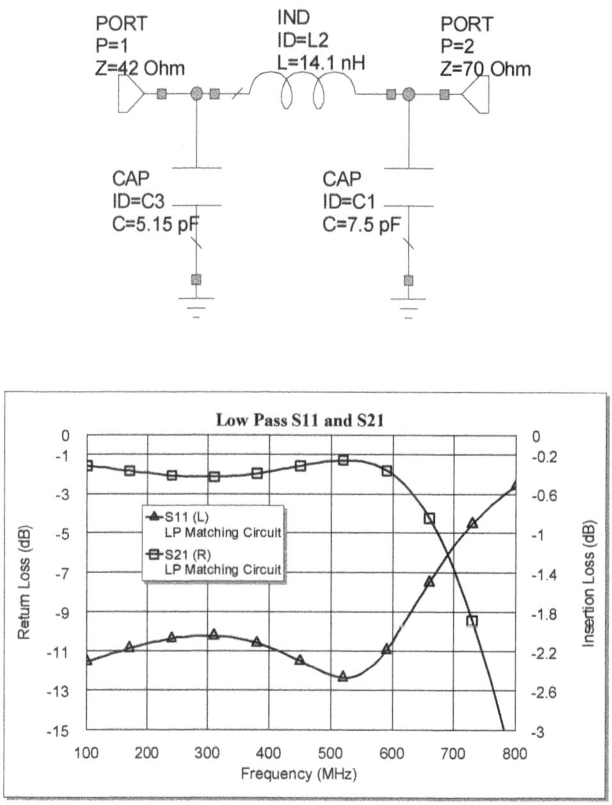

Figure 4.16 Low-pass matching circuit of the example. Because the matching circuit is designed for an unnecessarily broad range of frequencies, the match over the entire band is mediocre.

$$\omega_0 = \sqrt{\omega_l \omega_h} \tag{4.33}$$

which, in temporal frequency, is 490 MHz. The resonating elements are

$$\begin{aligned} L_1 &= 14.1 \text{ nH} \\ C_2 &= 4.21 \text{ pF} \end{aligned} \tag{4.34}$$

Figure 4.17 shows the resulting circuit and its response.

In both matching circuits, the source resistance is unlikely to be what the designer desires, as it is determined in the design process and cannot be independently specified. There are a few ways to accommodate this. One is simply to adjust the bandwidth and center frequency to give the desired $g_{n+1} = 1$. This process is necessarily iterative and not terribly elegant. If the resistance is close to the desired value, however, a little numerical optimization on the computer is usually enough to provide an acceptable result. If the desired resistance is very different from that determined in the design, it may be necessary to use some kind of transformer. A final shunt LC circuit in a bandpass design, for example, could have a tapped shunt inductor. The impedance transformations described in Section 4.4.6 may also be useful.

Another problem in this approach is that nonrealizable element values sometimes result. This often happens when the initial lumped-element circuit is converted to a distributed form. An obvious difficulty is the series LC resonator, which cannot be realized accurately by a microstrip resonator. We address some of these problems later in this chapter.

4.4.6 Impedance Transformations

We noted earlier that a problem in the synthesis approach is the inability to specify the source resistance independently. That problem can sometimes be solved through the use of impedance transformations.

Impedance transformations involve changing an inductive or capacitive L section at the source end of the circuit into a tee or pi section. In the process, the terminating resistance can be shifted. Figure 4.18 shows inductive and capacitive transformations that perform this function. The values of the inductances and capacitances of the transformed networks are as follows:

$$
\begin{aligned}
L_{t1} &= n^2(L_p + L_s) - nL_p \\
L_{t2} &= (1 - n)L_p \\
L_{t3} &= nL_p
\end{aligned}
\tag{4.35}
$$

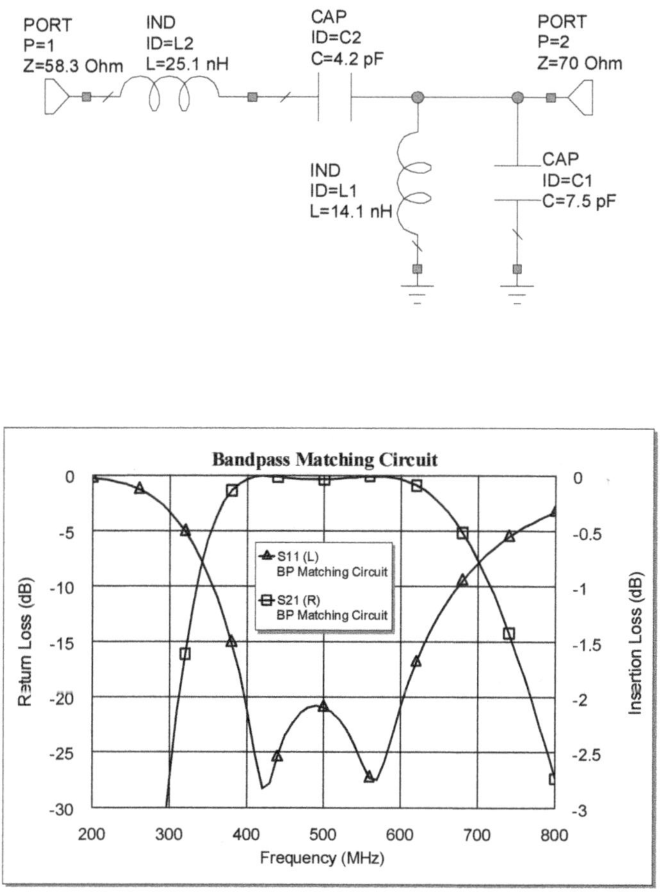

Figure 4.17 Bandpass matching circuit of the example.

$$
\begin{aligned}
L_{p1} &= \frac{n^2 L_s}{1-n} \\
L_{p2} &= nL_s \\
L_{p3} &= \frac{nL_s L_p}{n(L_s + L_p) - L_p}
\end{aligned}
\tag{4.36}
$$

$$
\begin{aligned}
C_{t1} &= \frac{C_s C_p}{n^2(C_p + C_s) - nC_s} \\
C_{t2} &= \frac{C_p}{1-n} \\
C_{t3} &= \frac{C_p}{n}
\end{aligned}
\tag{4.37}
$$

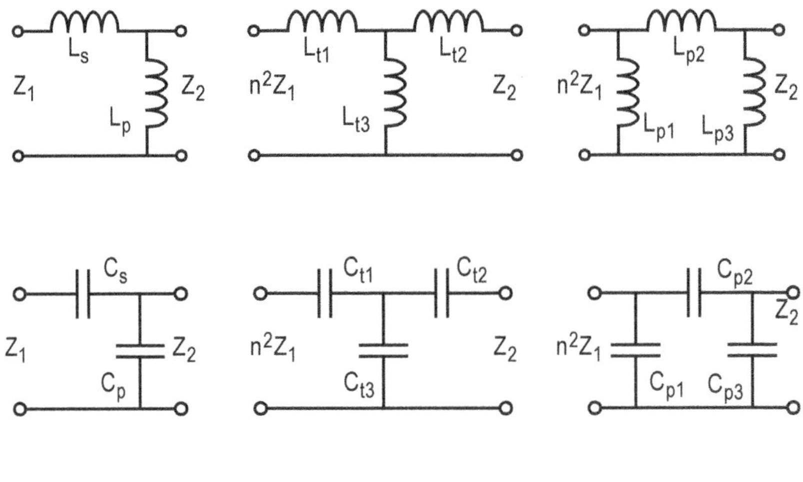

Figure 4.18 These networks can be used to convert the impedance at the source end of a matching network when the synthesis procedure does not allow it to be specified, and the end of the matching circuit has a capacitive or inductive L section. See (4.35) through (4.39) for the element values.

$$
\begin{aligned}
C_{p1} &= \frac{(1-n)C_s}{n^2} \\
C_{p2} &= \frac{C_s}{n} \\
C_{p3} &= C_s + C_p - \frac{C_s}{n}
\end{aligned}
\tag{4.38}
$$

with the constraints

$$
\begin{aligned}
&\frac{1}{1 + L_s/L_p} < n < 1 \\
&\frac{1}{1 + C_p/C_s} < n < 1
\end{aligned}
\tag{4.39}
$$

Since $n < 1$, this technique appears to provide only a downward shift in impedance level. The circuits, however, can be reversed to provide an upward shift, as long as the original L section is also reversed.

While these are not applicable to the low-pass LC ladder circuits we have been considering, they are applicable to LC bandpass matching circuits and often to other kinds of matching circuits, which may be generated by circuit-synthesis software.

4.5 Distributed Networks

The 600-MHz matching circuits of the example in Section 4.4.5 probably could not be realized precisely as designed. In practice, chip inductors and capacitors would be used, and those have parasitics that the designer would have to account for. Chip capacitors have series inductance, for example, which might necessitate modification of the resonating inductances. The matching circuits usually would be treated as initial designs, subject to optimization on the computer with models of real elements.

Inevitably, however, we need to create a matching circuit at frequencies where such elements simply cannot be used, as the values are not realizable and the parasitics are too great. In that case, a distributed circuit can be employed, but we need some way to convert the lumped-element design to a distributed one. This process is necessarily approximate, as there are no exact equivalents between distributed and lumped circuit elements. Nevertheless, within some limits on bandwidth, a distributed equivalent usually can be devised. For bandwidths beyond those limits, the distributed realization can be treated as an initial design subject to computer optimization.

4.5.1 Simple Resonator Equivalents Based on Slope Parameters

From the discussions and examples in Section 4.4, it should be clear that bandpass matching is inherently superior to low-pass matching. A lumped-element design from the network synthesis approach consists of a number of resonators, and we would like to convert these to the closest possible distributed equivalents.

It is well known that a short-circuit, quarter-wavelength stub qualitatively approximates a parallel LC resonator and an open quarter-wavelength stub approximates a series LC resonator. We can select the characteristic impedances of such stubs to realize a close equivalence of the two resonators. It seems logical to select the stub so the variation of its input impedance, at its resonant frequency, matches that of the LC resonator. For a shorted stub, we want $dB/d\omega$ of both resonators to be the same at ω_0. For the lossless shorted stub,

$$B = Y_0 \tan(\beta l) \tag{4.40}$$

where β is the phase shift per meter and l is the length. From Chapter 1, $\beta = \omega / v_p$ and the stub is resonant when $l = \lambda / 4$. Differentiating gives

$$\left.\frac{dB}{d\omega}\right|_{\omega = \omega_0} = Y_0 \frac{\pi}{2\omega_0} \tag{4.41}$$

For reasons that will be clear momentarily, we normalize this to $\omega_0 / 2$ and call it the *susceptance slope parameter*, b:

$$b = \left.\frac{\omega_0}{2}\frac{dB}{d\omega}\right|_{\omega = \omega_0} = Y_0 \frac{\pi}{4} \tag{4.42}$$

Applying the same process to a parallel LC resonator gives

$$b = \omega_0 C \tag{4.43}$$

and equating these results gives an equivalence between the stub and LC resonators. Their reactance slopes, at resonance, are identical when

$$Y_0 = \frac{4}{\pi}\omega_0 C \tag{4.44}$$

Following the same process for the open stub and series LC resonator gives the reactance slope parameter, x:

$$x = \frac{\omega_0}{2}\frac{dX}{d\omega}\bigg|_{\omega = \omega_0} = Z_0\frac{\pi}{4} \tag{4.45}$$

and the equivalence,

$$Z_0 = \frac{4}{\pi}\omega_0 L \tag{4.46}$$

To show how well this works, we return to the bandpass example in Section 4.4.5. As a practical matter, only the series resonator can be replaced by a resonant stub because the capacitance of the parallel resonator is part of the load. However, to illustrate the equivalence, we can ignore this little detail and replace the shunt resonator by a stub as well. The circuit and performance are shown in Figure 4.19. It is clear that, over this 40% bandwidth, the equivalence is quite good. Over a very broad bandwidth, however, it might not be adequate, and for bandwidths approaching an octave it is unlikely to be workable.

4.5.2 Converting Series Elements to Shunt

A problem with the circuit in Figure 4.19 is the open-circuit, series stub that replaces the series resonator. Such a stub is not realizable in microstrip or other strip transmission media, as it would in fact be a pair of coupled lines having both odd- and even-mode impedances. Precise realization would require an infinite even-mode impedance, and that is clearly impossible.

A better option is to convert the series section into a parallel one. This can be accomplished through impedance inversion. Suppose we replace the series stub in Figure 4.19 by a short-circuit, shunt stub followed by a quarter-wavelength transformer, as shown in Figure 4.20. We can show that this is equivalent to the open-circuit, series stub as follows. The admittance looking into the right end of the transformer is

$$Y_1 = \frac{Y_0^2}{G + jY_{0p}\tan(\theta)} \tag{4.47}$$

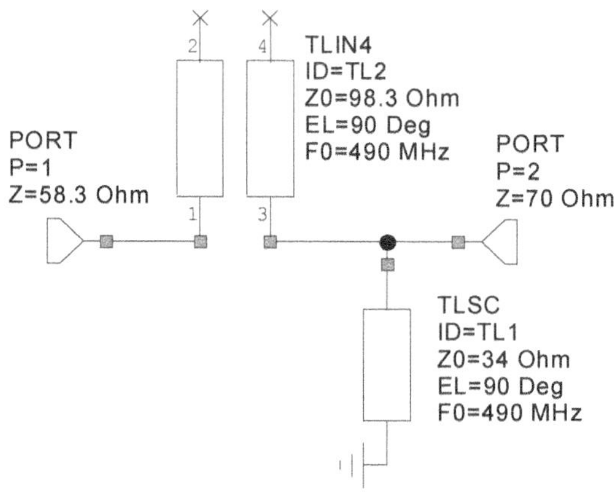

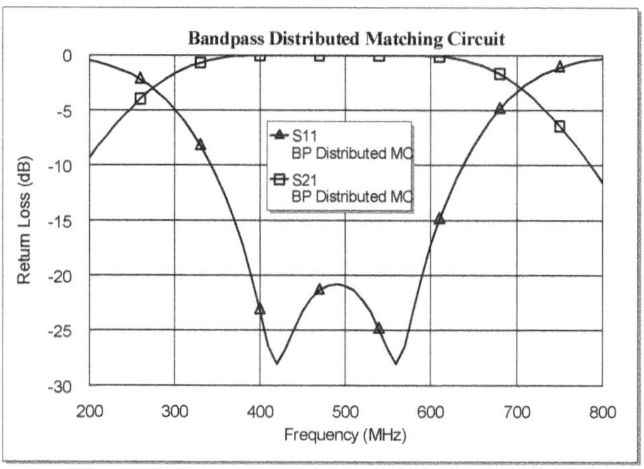

Figure 4.19 Bandpass matching circuit of Figure 4.17 realized with stubs instead of lumped-element resonators. The return loss and maximum in-band insertion loss are similar, but the bandwidth is somewhat greater.

where Y_0 is the characteristic admittance of the quarter-wave transformer, G is the source admittance (which need not equal the inverse of the source resistance in Figure 4.19), Y_{0p} is the characteristic admittance of the shunt stub, and θ is the electrical length of the stub. We assume that losses are negligible. This gives

$$Z_1 = \frac{1}{Y_1} = \frac{G}{Y_0^2} + \frac{jY_{0p}\tan(\theta)}{Y_0^2} \tag{4.48}$$

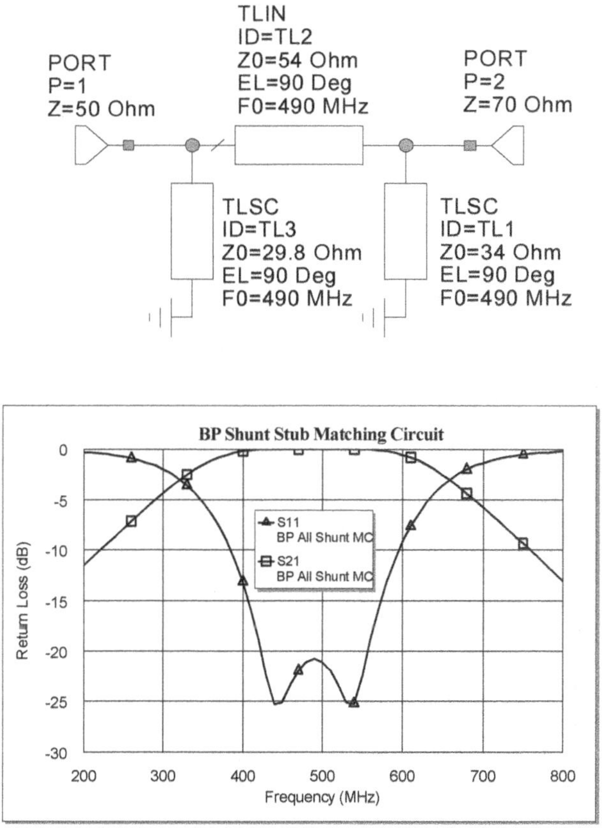

Figure 4.20 Matching circuit consisting entirely of shunt stubs. This is more easily and accurately realizable in strip transmission media than the circuit in Figure 4.19.

For the series stub in Figure 4.19,

$$Z_1 = R + jZ_{0s}\tan(\theta) \tag{4.49}$$

where Z_{0s} is the characteristic impedance of the original, series stub. These are equivalent when

$$\begin{aligned} R &= \frac{G}{Y_0^2} \\ Z_{0s} &= \frac{Y_{0p}}{Y_0^2} \end{aligned} \tag{4.50}$$

So, finally,

$$\begin{aligned} G &= RY_0^2 \\ Y_{0p} &= Z_{0s}Y_0^2 \end{aligned} \tag{4.51}$$

This has the additional advantage of an additional degree of freedom, Y_0, which can be selected to provide the desired value of G.

4.5.2.1 Example: Conversion of a Series Resonator to Shunt

To continue our example, we convert the 98Ω stub in Figure 4.19 to a shunt representation. To obtain a 50Ω source instead of the 58Ω one, the first equation of (4.50) gives $Z_0 = 1 / Y_0 = 54\Omega$. Using that in the second equation, with $Z_{0s} = 98\Omega$, we obtain $Y_{0p} = 1 / 29.8\Omega$. Figure 4.20 shows the circuit and its response. It is interesting that the inversion worked well over the entire band, although the transformer is an ideal inverter only at its center frequency.

It is important to note that we needed only one transformer because the series stub was at the input end of the matching structure. If it were in the center, we would need transformers on both sides. The determination of their values follows the same approach as for a single one.

4.5.2.2 Impedance and Admittance Inverters

Other types of inverters can be realized, and, in fact, many are superior to a quarter-wave transformer. Figure 4.21 shows four such immittance inverters; those of Figure 4.21(a) and (b) can be used to convert series resonators to shunt; (c) and (d) to convert shunt to series, should that be necessary. A straightforward analysis shows that, when terminated by Y_L, the input admittance, Y_{in}, of (a) and (b) is

$$Y_{in} = \frac{B^2}{Y_L} \tag{4.52}$$

where B is the branch susceptance; that is, $C\omega$ for (a) and $1 / L\omega$ for (b). Similarly, for circuits (c) and (d),

$$Z_{in} = \frac{X^2}{Z_L} \tag{4.53}$$

where Z_{in} is the input impedance, X is the branch reactance, and Z_L is the load impedance. In filter parlance, (a) and (b) are called *J inverters* and (c) and (d) *K inverters*. *J* inverters are admittance inverters and *K* inverters are impedance inverters. They are defined as

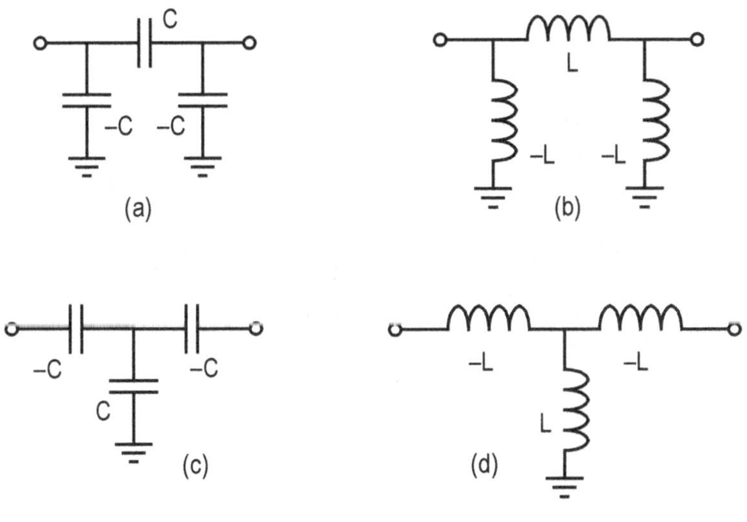

Figure 4.21 Examples of immittance inverters. Circuits (a) and (b) are useful for converting series resonators to shunt; (c) and (d) for converting shunt elements to series. The negative capacitances and inductances are realized by reducing the values of inductance or capacitance in adjacent resonators.

$$Y_{in} = \frac{J^2}{Y_L} \tag{4.54}$$

and

$$Z_{in} = \frac{K^2}{Z_L} \tag{4.55}$$

So, for transmission-line inverters, $J = Y_0$ and $K = Z_0$. For the inverters of Figure 4.21, $J = B$ and $K = X$.

The negative capacitances and inductances in Figure 4.21 are not as much of a difficulty as they might seem, as they can be absorbed into the adjacent LC resonators. Then, only a coupling capacitor or inductor remains, which can be realized easily and has the advantage of being much smaller than the quarter-wavelength line in Figure 4.20. These structures are best used to convert series LC resonators to shunt in the lumped-element version of the circuit; the resulting shunt resonators can then be converted to distributed ones.

A greater problem is the frequency sensitivity of the inversion, which, for the inverters in Figure 4.21, limits accurate design to a bandwidth of approximately 20%. The bandwidths of circuits using transmission-line inverters are considerably less, however, so the immittance inverters of Figure 4.21 are usually preferred. Of course, for wider bandwidths, the approach can be viewed as an initial estimate that must be completed by numerical optimization.

Figure 4.22 illustrates the conversion of an internal series resonator to shunt. As we noted earlier, this requires a pair of admittance inverters, one on either side of the shunt resonator. We do this, as before, by equating the input impedances of the two structures. In the figure, Y_L is the admittance of the right half of the matching circuit. This admittance, transferred through J_2, is Y_{L2}:

$$Y_{L2} = \frac{J_2^2}{Y_L} \tag{4.56}$$

Adding the shunt resonator's admittance and transforming it through J_1 gives

$$Z_i = \frac{j\omega C + \dfrac{1}{j\omega L} + Y_{L2}}{J_1^2} \tag{4.57}$$

The corresponding impedance of the original network, $\hat{Z}_i$, is

$$\hat{Z}_i = j\omega\hat{L} + \frac{1}{j\omega\hat{C}} + \frac{1}{Y_L} \tag{4.58}$$

where the circumflex indicates a quantity from the original network. Equating these gives

$$C = \hat{L}J_1 \tag{4.59}$$

$$L = \frac{\hat{C}}{J_1} \tag{4.60}$$

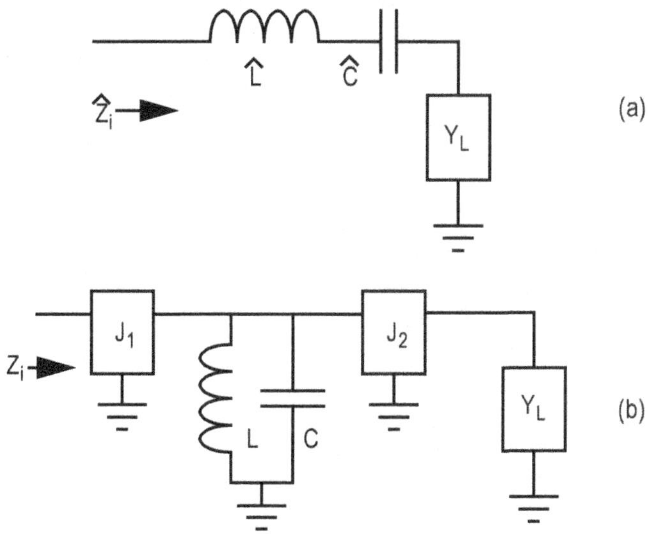

Figure 4.22 An internal series resonator (a) and its conversion to a parallel resonator plus admittance inverters (b).

and, with the substitution of (4.56),

$$\frac{1}{Y_L} = \frac{J_2^2}{Y_L J_1^2} \tag{4.61}$$

so

$$J_1 = J_2 \tag{4.62}$$

In this case, we have used the inverters to transform the central resonator to shunt with no change in impedance level. Allowing such changes provides and extra degree of freedom in generating a circuit having easily realizable values. It also allows the realization of a bandpass matching circuit directly from the immittance parameters and to use resonators having convenient (even identical) slope parameters. While this property is most valuable for filter design, where the designer may be forced to use a single type of cavity resonator for all the resonances, it may still be useful on occasion in matching circuits. This more general process can be found in [1.9].

4.5.2.3 Example: Use of Lumped-Element Inverters

We convert the central resonator in the matching circuit of Figure 4.23(a) to a shunt one plus the two lumped-element J inverters of Figure 4.21(a). Because we must absorb the negative capacitances into the resonators adjacent to the inverters, we perform the conversion on the lumped-element, bandpass matching circuit. If a distributed circuit is desired, the resulting lumped-element circuit consisting of only shunt resonators can be converted into a distributed one by the methods we have already discussed.

The load consists of a 60Ω resistance in parallel with a 14.7-pF shunt capacitance. We wish to match it over 150 MHz at a center frequency of 1.0 GHz. This combination has a decrement of $\delta = 1.2$; from Figure 4.15 we obtain the immittance parameters and synthesize the matching network. The result is shown in Figure 4.23(a). The element values for the shunt resonator are given by (4.59) and (4.60); after some experimentation, we find that $J = 1 / 70$ gives useful values for the inductance and capacitance. From the comments following (4.52), we use $C\omega = J = 1 / 70$ or $C = 2.27$ pF. To account for the negative capacitances in the J inverter, we must subtract that capacitance from the original parallel resonators. The new, central resonator has two such capacitances in parallel with it, however, one from each inverter, so we must subtract twice the capacitance from that resonator. The result is shown in Figure 4.23(b). Figure 4.23(c) compares the input reflection coefficients of the two circuits.

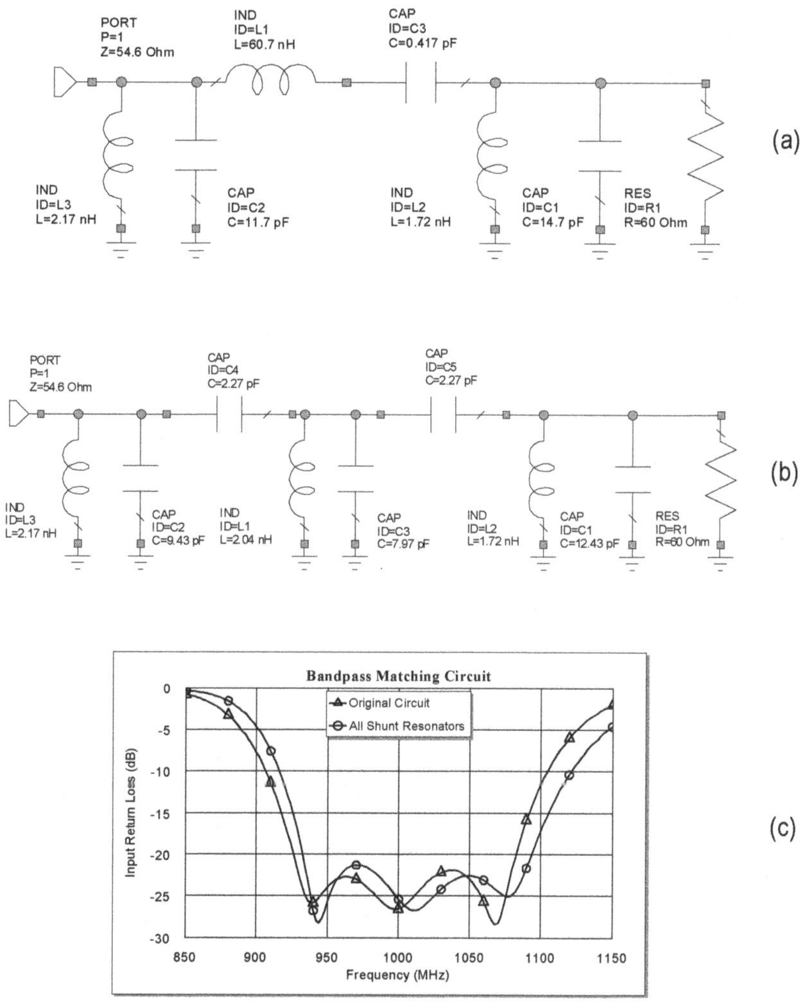

Figure 4.23 Bandpass matching circuit (a) and its shunt-resonator realization (b). The input reflection coefficients of the two circuits are compared in (c).

4.5.3 Richards' Transformation

Richards' transformation is a mapping between the LaPlace transform variable, s, to a second variable, which we denote S, that is applicable to distributed circuits. As such, it allows direct transformation between a lumped-element circuit and a distributed one. The response of the resulting distributed circuit is not identical to that of the lumped circuit, but the transformation does preserve bandwidth and, within limits, provides a good approximation to the lumped-element response.

The transformation is

$$S = \tanh(Ts) \tag{4.63}$$

where S is sometimes called the *Richards' variable*; $S = \Sigma + j\Omega$, analogous to $s = \sigma + j\omega$; and T is a constant having units of time, which we shall discuss shortly. When $s = j\omega$,

$$S = \tanh(Tj\omega) = j\tan(\omega T) = j\tan(\theta) \tag{4.64}$$

where θ is the electrical length of the line. The input admittance, Y_{in}, of an open-circuited line is

$$Y_{in} = jY_0\tan(\theta) = jY_0S \tag{4.65}$$

and the input impedance, Z_{in}, of a short-circuited line is

$$Z_{in} = jZ_0\tan(\theta) = jZ_0S \tag{4.66}$$

so an open-circuited line behaves like a capacitor of capacitance Y_0 and a shorted line like an inductor of inductance Z_0. We can replace capacitors in a lumped-element matching circuit with open-circuited lines having $Y_0 = C$ and inductors with short-circuited lines having $Z_0 = L$, and the response in S is the same as that in s.

The quantity T is a free variable, which can be chosen to be anything convenient, but certain values are likely to be more convenient than others. When $\theta = \pi / 4$, $\Omega = \omega$, so the responses of the distributed and lumped-element circuits are equivalent at that point. The most useful value of T is therefore

$$T = \frac{\theta}{\omega} = \frac{\pi}{4\omega} \tag{4.67}$$

The use of Richards' transformation is best illustrated by an example.

4.5.3.1 Example: Low-Pass Matching Circuit

We wish to generate a matching circuit for a load consisting of a 25Ω resistor in shunt with a 2.5-pF capacitor. We shall create a low-pass matching circuit with a cutoff of 2.0 GHz.

From (4.22) we find $\delta = 0.6$ and from Figure 4.15 we obtain $g_1 = 1.65$, $g_2 = 0.84$, $g_3 = 1.13$, and $g_4 = 0.675$. Scaling these to 25Ω and 2 GHz, we obtain the complete matching circuit of Figure 4.24(a). From (4.67), $T = 7.956 \cdot 10^{-11}$, so the lines are 45 deg long at 2 GHz. The characteristic impedance of the series stub is just the series inductor's reactance, $g_2 \cdot 25\Omega$, or 21Ω. The characteristic admittance of the shunt stub is similarly the capacitor's admittance, which is 1 / 22.1Ω. The resulting distributed matching circuit is shown in Figure 4.24(b) and the response of both circuits in (c). Converting the series stub to a shunt stub can be accomplished by the methods described in previous sections.

4.6 Modern Methods

The direct synthesis methods of the previous sections have a number of disadvantages. We have noted a couple of them: the inability to specify the source impedance independently and, frequently, unrealizability. We can add a few more. First, we often need a gain slope in the matching circuit; in an amplifier; for example, a slope is needed to compensate for the transistor's natural gain roll-off. Second, more often than not, the load impedance or admittance is provided as a table of values that cannot be approximated adequately by a simple RL or RC circuit. Finally, in the case where both terminations are reactive, there is no known method for synthesizing an optimum network. In practice, numerical optimization usually provides a better circuit than one based on necessarily nonoptimum analytic solutions.

More modern methods can accommodate these cases, but they often are not so straightforward, depending on numerical optimization or on the designer's judgment. Usually, they must be implemented in some kind of software, as they are too complex to use otherwise. Still, because of their greater versatility, they are the dominant methods for microwave and RF circuit design.

4.6.1 Direct Optimization

Especially for simple tasks, one should not underestimate the capabilities of modern numerical optimizers. Figure 4.25 show the results of a matching circuit designed by direct optimization, using the pointer-robust optimizer in Microwave Office [2.13]. The passband was specified by

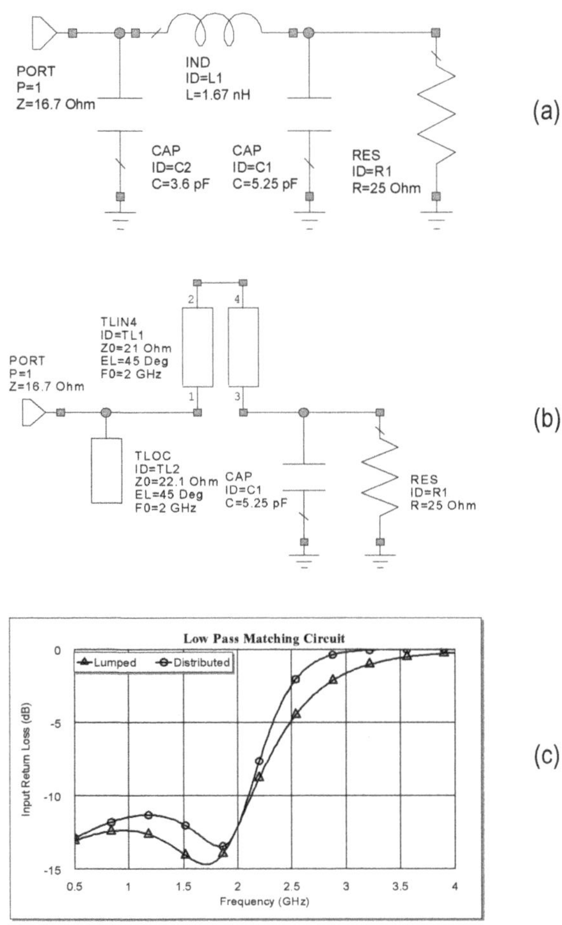

Figure 4.24 Matching circuit (a) realized by distributed elements (b). The plots of their input reflection coefficients (c) show that they are not identical, but are close enough for most practical purposes.

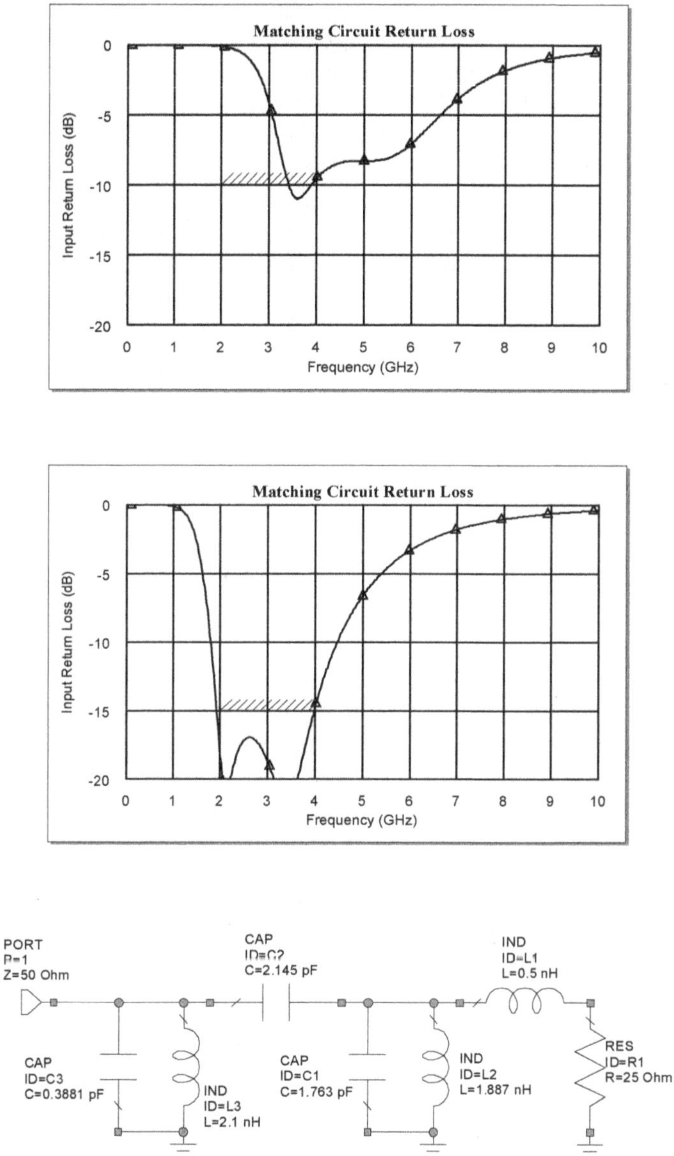

(a)

(b)

(c)

Figure 4.25 A matching circuit designed by direct optimization. Initial response (a), final response (b), and the final matching circuit (c).

optimizing the input return loss to a value less than –15 dBm, and a minimax error function was used, in which the maximum deviation from a goal is minimized. This tends to create a Chebyshev response. The initial element values were the circuit simulator's default values, 1 nH and 1 pF. No constraints were applied to the variables' values.

This example shows that modern optimizers, in contrast to the conventional wisdom, are often capable of designing matching circuits without a good initial design. Even so, one should be careful not to bring unreasonable expectations to this process. While direct optimization often works well for single matching circuits, designing an entire circuit, consisting of several matched stages, might not be successful. Worse, it may not provide the kinds of impedances desired (for example, for stability) at each interface. We examine these requirements more completely in Chapter 7.

4.6.2 Real Frequency Method

The real frequency method, originally proposed by Carlin [4.6], has become one of the dominant methods for matching-circuit design, and, since other methods use many of the same principles, it is worthwhile to examine it closely. Describing it in great detail, however, requires a sophisticated knowledge of general circuit theory, which is beyond the scope of this book, so we discuss it here in fairly general terms. Most users of this technique will employ it in the form of commercial software. For more specific information, it is best to consult [4.6–4.8].

The real frequency method is useful for a variety of matching situations. It can accommodate a complex source and load and an arbitrary $Z_L(\omega)$ function, although it is still subject to the broadband matching limitations described in Section 4.4.1. The source or load impedance is described by a table of values as a function of frequency and need not be modeled as a resistor and lumped-element inductor/capacitor, or for that matter, by any kind of circuit model. The real part of the load impedance or admittance need not be constant, and in fact, generally is not. While it is not simply an optimization approach, numerical optimization is an essential part of the method.

The method is based on the fact that a form of the Hilbert transform can be used to determine the reactive part of a lossless ladder network if the resistive part is known. The port reactance, $X_2(\omega)$, is

$$X_2(\omega) = \frac{1}{\pi}\int_{-\infty}^{\infty}\frac{R_2(\xi)}{\xi - \omega}d\xi \tag{4.68}$$

where ξ is a variable of integration, representing radian frequency, and $Z_2(\omega) = R_2(\omega) + jX_2(\omega)$ as in Figure 4.1. $Z_2(\omega)$ may be a conjugate match to $Z_L(\omega)$ or it may be chosen according to other criteria, such as achieving a specific value of gain or gain-slope compensation.

We begin by assuming that $R_2(\omega) = R_L(\omega)$, where $R_L(\omega)$ is the resistive part of the load impedance (Figure 4.1). Having a table of $R_2(\omega)$, we first approximate $R_2(\omega)$ by a piecewise linear approximation and $X_2(\omega)$ is found from (4.68), where the linearized $R_2(\omega)$ is used in the expression. The resistance and reactance functions can then be expressed as a linear combination of basis functions:

$$\begin{aligned} \bar{R}_2(\omega) &= \sum_{n=0}^{N} r_n \alpha_n(\omega) \\ \bar{X}_2(\omega) &= \sum_{n=0}^{N} r_n \beta_n(\omega) \end{aligned} \tag{4.69}$$

where the overbar indicates the approximate quantity. The basis functions, $\alpha_n(\omega)$ and $\beta_n(\omega)$, are linear ramps, where $\alpha_n(\omega)$ are even functions of ω and $\beta_n(\omega)$ are odd. The coefficients of the two expressions, r_n, are the same. $\bar{X}_2(\omega)$, as found from (4.68), may not be the desired value, so both $R_2(\omega)$ in the system (4.68) and (4.69) must be adjusted to achieve the desired reflection coefficient. Both optimization and least-squares methods have been used for this step. The criterion for the reflection coefficient may be a gain slope, to compensate for natural gain roll-off in the device, or simply minimum reflection coefficient over the band of interest. This method allows any such matching criterion.

At this stage, we have a set of values for $R_2(\omega)$ and $X_2(\omega)$ that satisfy (4.68) and thus can be realized by a minimum-phase network, particularly a ladder network. The task now is to synthesize a two-port ladder network having resistance and reactance functions matching (4.69). The resistance function must have the form

$$R_2(\omega) = \frac{A\omega^{2m}}{B_0 + B_1\omega^2 + \ldots + \omega^{2n}} \tag{4.70}$$

where $m \le n$. One can easily write a set of linear equations which, when solved, guarantees that (4.70) agrees with the $R_2(\omega)$ and $X_2(\omega)$ data at the frequency increments. However, this may result in a network of higher order than desired. Furthermore, even with a high order, the result may still have large errors in the regions between those points, and usually it is better to tolerate some error at the increments in exchange for reduced error overall. Thus, an optimization may be necessary to determine the final values of the A and B_n coefficients.

The final step is to synthesize a representation of $Z_2(s)$ from (4.70) in the form

$$Z_2(s) = \frac{N(s)}{D(s)} \tag{4.71}$$

where $N(s)$ and $D(s)$ are polynomials in the LaPlace transform variable, s. A process for doing this is called the *Gewertz procedure*, which we shall not describe here. In short, one must substitute $-s^2$ for ω^2 in (4.70) to obtain the denominator polynomial, and the numerator polynomial is then found in a fairly complex manner. The ladder network is then synthesized by a conventional technique, such as partial-fraction expansion.

In the years since its inception, this method has been modified somewhat. Its fundamental concepts have also been used to develop other modern methods. Readers interested in more detail can find a straightforward method for implementing this type of matching process in [4.8].

4.6.3 Synthesis and Parasitic Absorption

A simple yet effective synthesis method involves the design of an LC matching circuit in which the load reactance can be "absorbed" into the network. For example, if the load is a shunt RC, and the matching-circuit element adjacent to the load is a capacitor whose capacitance is greater than that of the load, the load capacitance can be made part of that network capacitance. To do this, we must design a matching circuit between the real source and load resistances in which (1) the gain and ripple satisfy (4.23), and (2) the load-end element (L or C) is appropriate to absorb the load reactance.

A problem in the use of (4.23) is that it does not provide the minimum value of the matching circuit's optimized gain; it provides only a gain that corresponds to a certain value of ripple, which in general is not optimum. It is conceivably possible, however, to find the optimum gain iteratively, by adjusting the ripple, ε, or simply by optimizing (4.23) over ε. Because there is only one variable, given a value of decrement δ, this is a simple, if inelegant, process.

A well-established procedure for synthesizing such a network is called *Darlington synthesis*. It is based on the synthesis of the impedance $Z_2(\omega)$ in Figure 4.1. The procedure is as follows:

1. Develop an expression for the network's transducer gain, $G_t(\omega)$, including only even powers of ω.
2. Replace all ω^2 terms by $-s^2$.
3. Determine the product $\Gamma(s)\Gamma(-s)$ from

$$\Gamma(s)\Gamma(-s) = 1 - G_t \tag{4.72}$$

   where $\Gamma(s)$ is the source reflection coefficient.

4. To find $\Gamma(s)$, first assign all the left-half-plane (LHP) poles of $\Gamma(s)\Gamma(-s)$ to $\Gamma(s)$. Assigning zeros is a bit more problematical, as there are benefits to assigning right-half-plane (RHP) or LHP zeros to $\Gamma(s)$. In parasitic absorption, it is best to assign the RHP zeros to $\Gamma(s)$.
5. $\Gamma(s)$ can have a leading (+) or (–) sign, and the selection is usually important. Since

$$\Gamma(s) = \frac{Z_2(s) - R_L}{Z_2(s) + R_L} \tag{4.73}$$

   the sign must be consistent with the network at $s = 0$. In a low-pass matching circuit, for example, $Z_2(0) = R_s$, so the sign must be negative if $R_s < R_L$. Similarly, at $s \to \infty$, $\Gamma(s)$ will be +1 or –1; the former implies a series inductor as the element adjacent to the load, and the latter a shunt capacitor. The sign must therefore be consistent with the type of element that is to be absorbed.

6. Find $Z_2(s)$ as

$$Z_2(s) = R_L \frac{1 + \Gamma(s)}{1 - \Gamma(s)} \tag{4.74}$$

7. Use continued fractions to synthesize the network. This process is described in any ordinary circuit-theory text.

This is a simple method, one that is easy to understand and to implement in software. It has the advantage of producing a network that includes the correct source resistance. It also can result in a bandpass network consisting of elements other than series or shunt LC resonators, so it is often simpler and more practical than the methods in Sections 4.4 and 4.5. The disadvantage is its requirement of an RL or RC load model, although sometimes more complex RLC loads can be accommodated. A disadvantage, as with other synthesis methods, is its tendency occasionally to produce unrealizable networks, either in LC form or when converted to a distributed realization.

References

[4.1] Lefferson, P., "Twisted Magnet Wire Transmission Line," *IEEE Trans. Parts, Hybrids, and Packaging*, Vol. PHP-7, 1971, p. 148.

[4.2] Abrie, P. L. D., *The Design of Impedance-Matching Networks for Radio-Frequency and Microwave Amplifiers*, Dedham, MA: Artech House, 1985.

[4.3] Bode, H. W., *Network Analysis and Feedback Amplifier Design*, New York: Van Nostrand, 1945.

[4.4] Fano, R. M., "Theoretical Limitations on the Broadband Matching of Arbitrary Impedances," *Jour. Franklin Inst.*, Vol. 249, 1950, p. 57.

[4.5] Youla, D. C., "A New Theory of Broadband Matching," *IEEE Trans. Circuits Syst.*, Vol. CT-11, 1964, p. 30.

[4.6] Carlin, H. J., "A New Method of Broad-Band Equalization Applied to Microwave Amplifiers," *IEEE Trans. Microwave Theory Tech.*, Vol. MTT-27, 1979, p. 93.

[4.7] Cuthbert, T., *Broadband Direct-Coupled and Matching RF Networks*, Greenwood, AR: TRCPEP, 1999.

[4.8] Abrie, P. L. D., *Design of RF and Microwave Amplifiers and Oscillators*, Dedham, MA: Artech House, 1985.

Chapter 5

Circuit Analysis

We can discuss matters like the properties of scattering matrices until the reader's eyes glaze over, but inevitably we must create a circuit and analyze it. That is, after all, the point of what we have been discussing. We have already described one method, the direct manipulation of S matrices, in Chapter 3. Two more methods, which we examine in this chapter, are more systematic and useful for analyzing large networks. These are network graphs and the indefinite admittance matrix. Network graphs are a useful way to generate a network function analytically. The indefinite admittance matrix, or nodal matrix, is a method that is best suited to circuit analysis by computers, although it can be used analytically as well. Understanding these methods is essential for the analysis of high-frequency circuits.

5.1 Network Graph Analysis

Network graph analysis is a relatively simple technique that allows even a fairly complex network function to be generated in a systematic manner. Generating network functions in this way is usually simpler than direct manipulation of S or Y matrices and provides an intuitive sense of the circuit, which is missing from algebraic formulations. The process involves first setting up a network graph, then reducing it in a systematic manner to generate the network function. Much of this reduction can be performed graphically. Another technique, used traditionally with S-parameter formulations, involves Mason's rule, sometimes called the *nontouching loop rule*. It provides a general technique for finding a network function, and it does not require graphical reduction. It can be used with all kinds of network graphs and is especially useful when graph reduction is not successful.

5.1.1 General Network Graphs

A network graph is a symbolic representation of a set of network equations, consisting of a number of nodes and directed branches. Each node is a network variable, and each branch represents the relationship between those variables. While most graph analysis of microwave circuits is applied to S matrices, it can be applied to any set of linear equations.

For example, the equation

$$I_1 = Y_{11}V_1 + Y_{12}V_2 \tag{5.1}$$

can be expressed by the network graph shown in Figure 5.1. It has three nodes, I_1, V_1, and V_2, and two branches, Y_{11} and Y_{12}, which are directed from V_1 and V_2, respectively, toward I_1. The branch direction shows the dependency of the variables; as (5.1) indicates that I_1 is directly dependent on V_1 and V_2, the branch directions are both from the voltage nodes to I_1. The values of the branches are called *transmittances*; Y_{11} and Y_{12} are the transmittances of their respective branches. Some of the nodes can be called *sources* and others *sinks*. If all branches connected to a node are directed away from it, that node is a source; if all connected branches are directed toward the node, it is a sink. We sometimes say that a branch directed toward a node is *incident* on it; if it is directed away from a node, it *emerges* from it. In the above example, there are two sources, V_1 and V_2, and one sink, I_1. The branches emerge from V_1 and V_2 and are both incident on I_1.

More complex sets of equations can be described by network graphs. Suppose we have a set of equations of the form

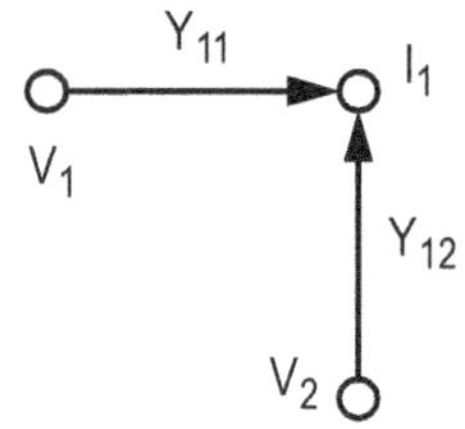

Figure 5.1 This graph represents the expressions in (5.1). The nodes of the graph are the variables and the branches are the coefficients. The branch direction is from the independent to the dependent variable.

$$\begin{aligned} x_2 &= a_{21}x_1 + a_{23}x_2 \\ x_3 &= a_{31}x_1 + a_{32}x_2 \\ x_4 &= a_{43}x_3 + a_{42}x_4 \end{aligned} \tag{5.2}$$

where the x_n can represent any type of quantity. In this case, we have one source, x_1, which can be viewed as the independent variable in the system of equations, while the remaining three variables, x_2, x_3, and x_4, are dependent, directly or indirectly, on node x_1. Node x_4 is a sink. The resulting graph is shown in Figure 5.2; it should be straightforward to relate the nodes and branches to the terms in (5.2).

An important advantage of graph analysis lies in the fact that graphs often can be reduced in a straightforward manner, leading to a transfer function between a pair of quantities represented by the nodes. Some possible reductions are summarized in Figure 5.3. The simplest is probably two parallel branches, Figure 5.3(a); in that case,

$$x_2 = a_{21}x_1 + b_{21}x_1 = (a_{21} + b_{21})x_1 \tag{5.3}$$

so the two are reduced to a single branch whose transmittance is the sum of the transmittances of the original branches. In Figure 5.3(b), two cascaded branches are shown, and it should not require any great effort to prove that they are equivalent to a single branch whose transmittance is the product of the branch transmittances. Figure 5.3(c) shows a number of branches joining to form a third. The equations it represents are

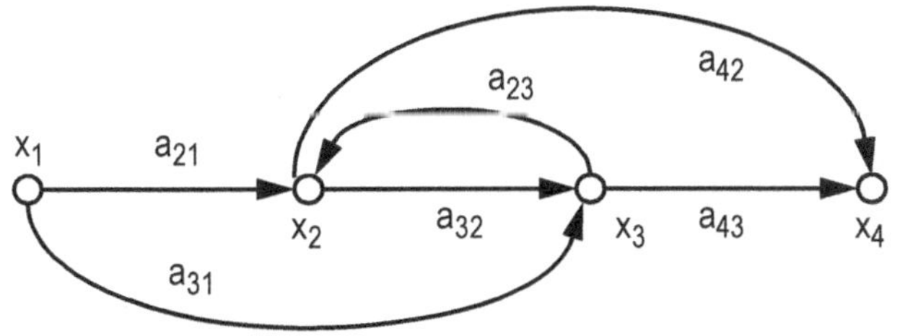

Figure 5.2 This network graph represents the set of equations in (5.2). The branches' directions are from the right-side variables in (5.2) to the left-side one, and the coefficients are the branch transmittances.

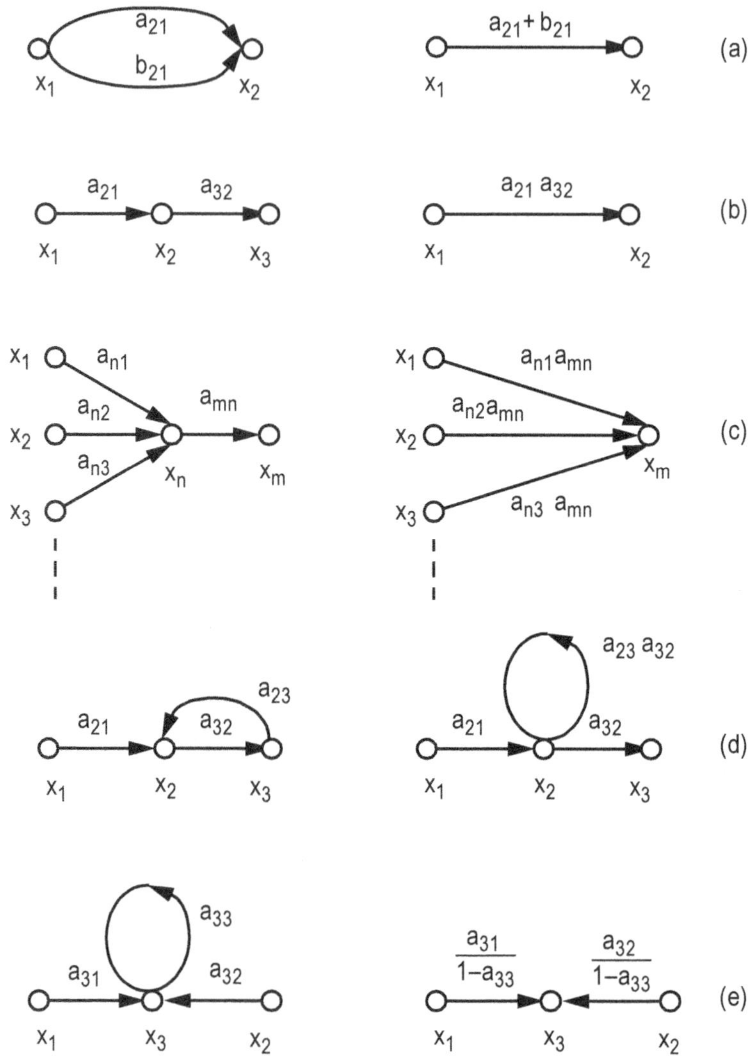

Figure 5.3 Equivalent circuits of certain graph structures can be used to reduce the complexity of a circuit graph. The derivations of these equivalents are given in the text. Shown are equivalents for parallel branches (a), series branches (b), summed branches (c), a feedback loop (d), and removal of a self loop (e). Other equivalents are possible, as well [5.1].

$$\begin{aligned} x_n &= a_{n1}x_1 + a_{n2}x_2 + \ldots \\ x_m &= a_{mn}x_n \end{aligned} \tag{5.4}$$

so

$$x_m = a_{mn}a_{n1}x_1 + a_{mn}a_{n2}x_2 + \ldots \tag{5.5}$$

and the equivalent is shown in the figure.

A more interesting case is shown in Figure 5.3(d), where the graph includes a feedback loop. The equations are

$$\begin{aligned} x_2 &= a_{21}x_1 + a_{23}x_3 \\ x_3 &= a_{32}x_2 \end{aligned} \tag{5.6}$$

By substitution,

$$x_2 = a_{21}x_1 + a_{32}a_{23}x_2 \tag{5.7}$$

The result is a *self loop* at x_2. Finally, Figure 5.3(e) shows how that self loop can be removed. We have

$$x_3 = a_{31}x_1 + a_{32}x_2 + a_{33}x_3 \tag{5.8}$$

Collecting x_3 terms on the left and dividing gives

$$x_3 = \frac{a_{31}}{1 - a_{33}}x_1 + \frac{a_{32}}{1 - a_{33}}x_2 \tag{5.9}$$

These five equivalents are not the only ones that could be derived, but they are probably the most useful. See [5.1] for other equivalents, or simply derive them yourself!

A few notes on network graphs are in order. First, the equivalents in Figure 5.3 cannot be applied mindlessly. Consider, for example, the case illustrated in Figure 5.4(a). This is essentially that of Figure 5.3(d), but with an extra node, x_4, and a branch incident on x_3. If one were mindlessly to convert the feedback loop as in Figure 5.3(d), the path from node x_4 to node

x_2 would be lost, and the graph would no longer represent the circuit. We can find the correct equivalent circuit in the same manner as finding the original one. The equations for the unreduced graph in Figure 5.4 are

$$\begin{aligned} x_2 &= a_{21}x_1 + a_{23}x_3 \\ x_3 &= a_{32}x_2 + a_{34}x_4 \end{aligned} \tag{5.10}$$

Substituting x_3 into the first of the two equations gives

$$x_2 = a_{21}x_1 + a_{32}a_{23}x_2 + a_{34}a_{23}x_4 \tag{5.11}$$

The resulting graph is shown in Figure 5.4(b). In general, such problems can arise when a branch is incident on one of the nodes in the part of the circuit to be reduced; branches emerging from a node usually have no effect, but the transformation must, of course, preserve that node.

A second consideration is the uniqueness of graphs. Graphs of a particular set of equations are not unique, and the way that they are formulated can strongly affect the ease of the analysis. As we shall see in Section 5.1.2, it is possible to create a graph that simply cannot be reduced, while putting the equations in a different form can make a significant reduction

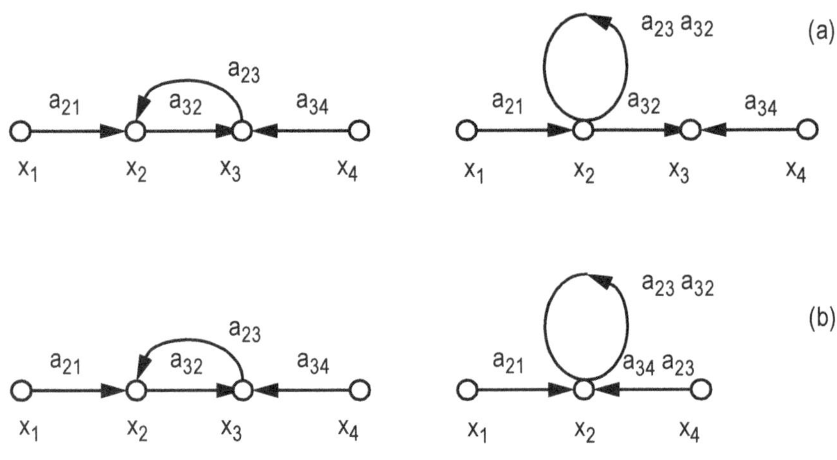

Figure 5.4 Incorrect graph (a) resulting from a naive use of the equivalence in Figure 5.3(d). Correct graph (b) according to (5.11).

possible. Finally, while graphs usually ease the analysis of most circuits, some are not easily amenable to graph analysis. In those, straightforward algebraic manipulation of the original circuit equations may be preferable.

5.1.2 Example: A Terminated Two-Port

The reduction of a graph to determine a network function is best described by an example. Figure 5.5 shows a terminated two-port. The two-port is described by its Y matrix, and it is excited by a current source I_S. We wish to find the quantity V_1 / I_S. Knowing that, we can find other important quantities, such as gain and input admittance. Others, such as reverse gain and output admittance, can be found by symmetry.

We begin with the circuit equations:

$$\begin{aligned} I_1 &= Y_{11}V_1 + Y_{12}V_2 \\ I_2 &= Y_{21}V_1 + Y_{22}V_2 \\ I_1 &= I_s - G_sV_1 \\ I_2 &= -G_LV_2 \end{aligned} \tag{5.12}$$

Figure 5.6(a) shows a graph of these equations. Right away, we encounter a problem: in the graph, V_1 and V_2 are sources, but should not be; they should be dependent variables, so there must be at least one branch incident on each of them. I_s should be the only source. Reformulating the last two equations of (5.12) solves the problem. We use

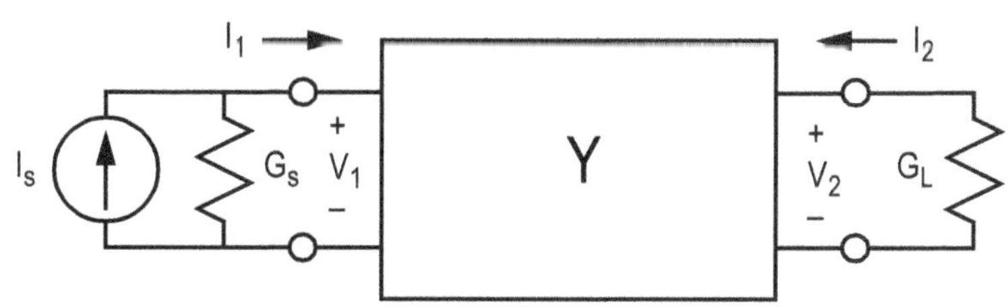

Figure 5.5 A terminated two-port. As an example of network graph analysis, we create a graph and compute V_1 / I_S.

$$
\begin{aligned}
I_1 &= Y_{11}V_1 + Y_{12}V_2 \\
I_2 &= Y_{21}V_1 + Y_{22}V_2 \\
V_1 &= R_s I_s - R_s I_1 \\
V_2 &= -R_L I_2
\end{aligned}
\tag{5.13}
$$

where $R_L = 1 / G_L$ and $R_s = 1 / G_s$. The resulting graph is shown in Figure 5.6(b). Clearly, graphs are not unique, and graph selection has a strong effect on the success of the analysis.

The reduction process is illustrated in Figure 5.7. We begin with the lower branches. We first eliminate the feedback loop between V_2 and I_2,

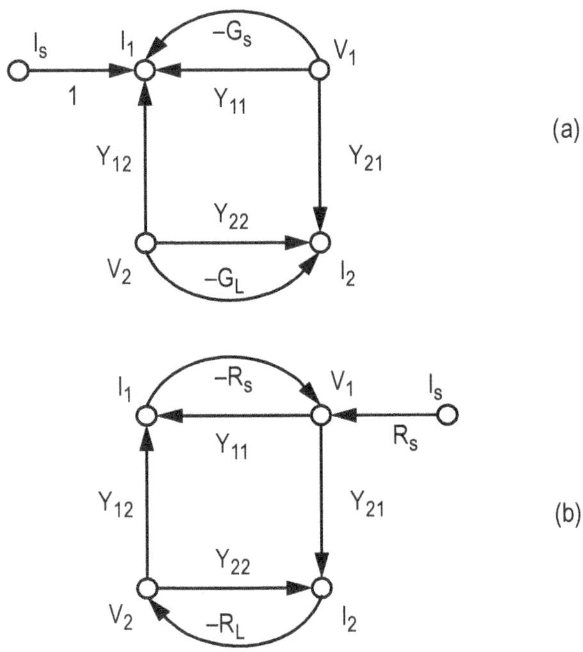

Figure 5.6 Two possible network graphs of the circuit of Figure 5.5. The representation in (a) is not useful, as it cannot be reduced to provide the desired network function. The graph in (b) is mathematically identical but much more amenable to analysis, as it can be reduced in a straightforward manner.

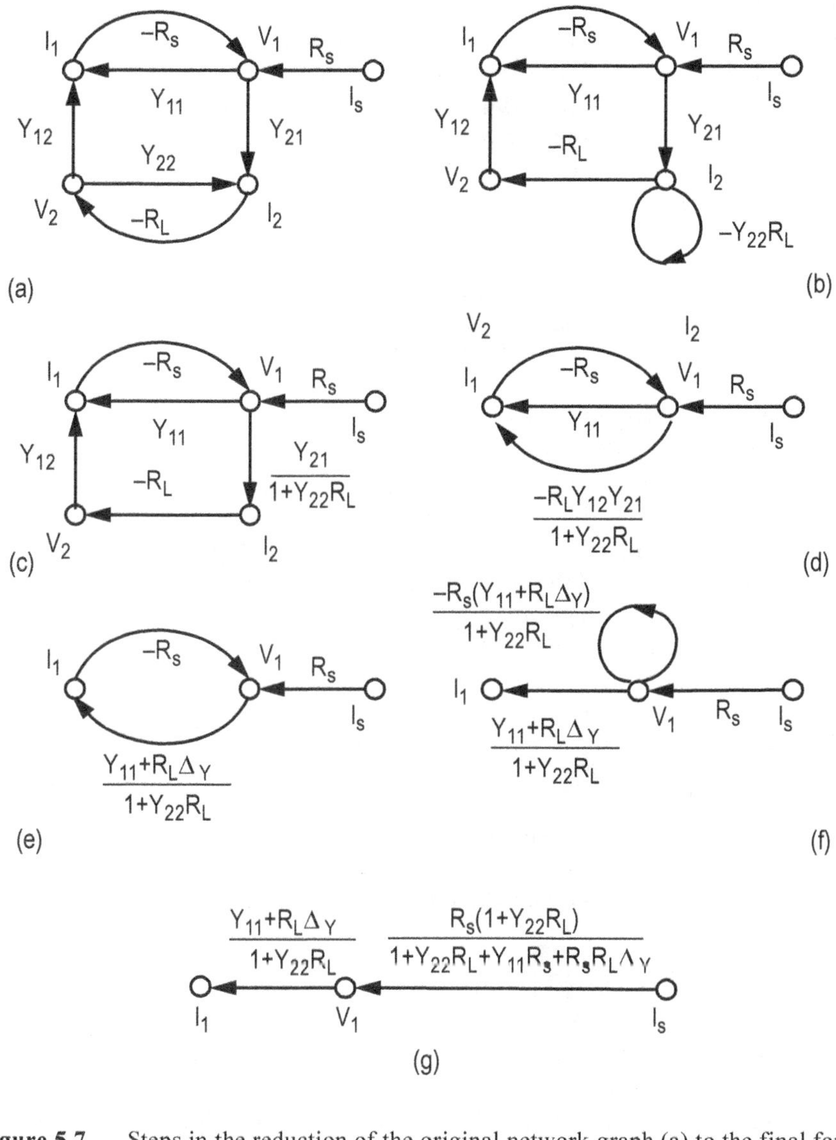

Figure 5.7 Steps in the reduction of the original network graph (a) to the final form, (g), where the V_1 / I_s network function is apparent. The steps are elimination of the lower feedback loop (b), elimination of the self loop (c), reduction of the lower branch (d), merging of the two lower parallel branches (e), conversion of the feedback loop to a self loop (f), and elimination of the self loop (g).

creating a self loop having transmittance $-Y_{22}R_L$. That self loop is then eliminated, resulting in a simple cascade of three branches, where the former Y_{21} branch is now $Y_{21}/(1 + Y_{22}R_L)$. The three-branch cascade is easily converted to a single branch and then added to the parallel Y_{11} branch. The resulting transmittance is $(Y_{11} + R_L\Delta_Y)/(1 + Y_{22}R_L)$, where Δ_Y is the determinant of the Y matrix. This leaves us with another feedback loop, which is converted into a self loop and eliminated in a manner identical to the one in the lower branches. The result for the branch between I_s and V_1 is

$$\frac{V_1}{I_s} = \frac{R_s(1 + Y_{22}R_L)}{1 + Y_{22}R_L + Y_{11}R_s + R_sR_L\Delta_Y} \tag{5.14}$$

This result was verified by direct algebraic manipulation of (5.13).

Note that the branch from V_1 to I_1 still exists, and we need not do anything with it. As long as we have a single path from the independent variable to the dependent variable of interest, nothing more needs to be done. We could still use this path, however, to find I_1 / I_s very easily, and that result might be useful for determining other circuit quantities.

From this result, other quantities can be found easily. For example, $I_2 = Y_{21}V_1$, so the current gain is

$$\frac{I_2}{I_s} = Y_{21}\frac{V_1}{I_s} \tag{5.15}$$

Other gains and port admittances are found similarly. To find output admittance and reverse gain, the forward results can be used with the input and output terms reversed; that is, exchange R_s and R_L, Y_{11} and Y_{22}, and Y_{21} and Y_{12}.

5.1.3 S Parameters and Mason's Rule

Direct reduction, as described in the previous section, can be used in S-parameter analysis. A second method, however, which may be more convenient is the application of Mason's rule. We illustrate its use by applying it to a two-port.

For a terminated two-port, we have

$$
\begin{aligned}
b_1 &= S_{11}a_1 + S_{12}a_2 \\
b_2 &= S_{21}a_1 + S_{22}a_2 \\
a_2 &= \Gamma_L b_2 \\
a_1 &= b_s + \Gamma_s b_1
\end{aligned}
\tag{5.16}
$$

The variables are as used in Chapter 3; in particular, Section 3.3.1 and Figure 3.6. Z_{01} is the normalizing impedance for port 1 and Z_{02} for port 2. The variable b_s requires some extra explanation. It represents the wave launched from a voltage source, V_s, with source impedance Z_s, onto a transmission line having characteristic impedance Z_{01}; from (1.41), with conversion of the voltage wave to a wave variable,

$$
b_s = \frac{V_s\sqrt{Z_{01}}}{Z_s + Z_{01}} \tag{5.17}
$$

The variables for the graph's nodes are now the a_n and b_n quantities, and the branch transmittances are the S_{ij} and $\Gamma_{L,s}$ values. The network graph is shown in Figure 5.8.

This graph can be reduced by application of the methods of Section 5.1.1, and the process is similar to that outlined in Figure 5.7. An alternative is to use Mason's rule [5.2]. The rule is given by the following rather complicated expression:

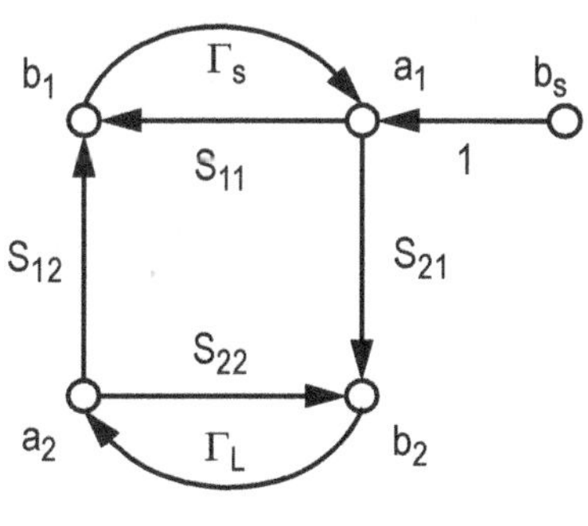

Figure 5.8 Network graph for a terminated two-port, in terms of S parameters, reflection coefficients, and wave variables.

$$T = \frac{P_1(1 - \sum L_{1,1} + \sum L_{2,1} - \dots) + P_2(1 - \sum L_{1,2} + \sum L_{2,2} - \dots) + P_3(1 - \sum L_{1,3} + \sum L_{2,3} - \dots) + \dots}{1 - \sum L_1 + \sum L_2 - \dots} \tag{5.18}$$

The equation is interpreted as follows:

1. T is a transfer function between any two node variables in the graph.
2. P_n is the nth path from the independent to the dependent variable. A path is the product of the transmittances along any unique set of branches between the node variables, in which no node is encountered more than once. The branches must be traversed in their defined directions.
3. L_n is an nth-order loop. A loop is a closed path from any node back to itself. As such, it includes what we have been calling feedback loops, and self loops but may also include other circular paths through the graph. A first-order loop is the product of the transmittances around such a loop. An nth-order loop is the product of any n *nontouching* first-order loops; that is, loops that do not have a common node.
4. $L_{n,m}$ is any nth-order loop that does *not* touch path m.

The summation signs in (5.18) indicate the summation of all such loops in the graph.

5.1.4 S-Parameter Examples

5.1.4.1 Input Reflection Coefficient

The use of (5.18) is best illustrated by a few examples. We derive the two-port gain and return loss, as we found by direct algebraic manipulation of (5.16) in Section 3.3.1. We begin with the input reflection coefficient. For this problem, we need to find $\Gamma_{in} = b_1 / a_1$, which is unaffected by Γ_s, so its branch need not be included; then $a_1 = b_s$, so b_s is also unnecessary. The resulting graph is shown in Figure 5.9.

We see that we have the following paths between b_1 and a_1:

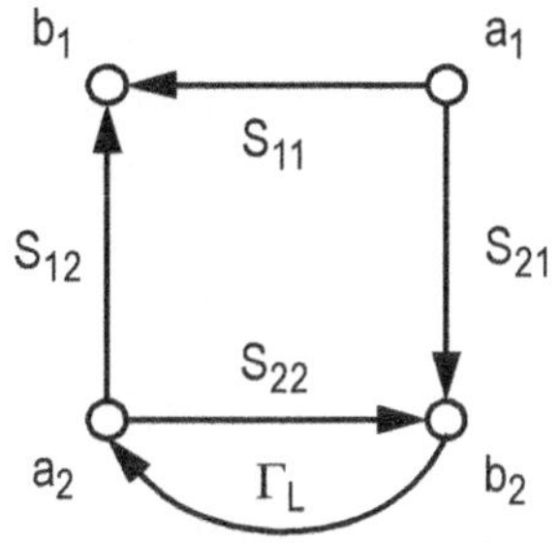

Figure 5.9 Network graph for the calculation of Γ_{in}. In this case, our desired quantity is not dependent on Γ_s, so we can eliminate it; then $a_1 = b_s$ so b_s is also unneeded.

$$\begin{aligned} P_1 &= S_{11} \\ P_2 &= S_{21}\Gamma_L S_{12} \end{aligned} \tag{5.19}$$

Only one loop exists, so no loops of higher order are possible. The loop is

$$\sum L_1 = \Gamma_L S_{22} \tag{5.20}$$

This loop does not touch P_1, but it does touch P_2. Thus,

$$\sum L_{1,1} = \Gamma_L S_{22} \tag{5.21}$$

Substituting these terms into (5.18) gives

$$\Gamma_{in} = \frac{b_1}{a_1} = \frac{S_{11}(1 - \Gamma_L S_{22}) + S_{21}\Gamma_L S_{12}}{1 - S_{22}\Gamma_L} \tag{5.22}$$

which is easily reduced to the more familiar form,

$$\Gamma_{in} = S_{11} + \frac{S_{21}S_{12}\Gamma_L}{1 - S_{22}\Gamma_L} \tag{5.23}$$

The output reflection coefficient, Γ_{out}, can be found by a similar graph. However, determining Γ_{out} is much simpler if we recognize that it is essentially the same problem, but reversed. Therefore, we exchange S_{11}, S_{22} and S_{12}, S_{21} and replace Γ_L with Γ_s in (5.23). Then

$$\Gamma_{out} = S_{22} + \frac{S_{21}S_{12}\Gamma_s}{1 - S_{11}\Gamma_s} \tag{5.24}$$

Γ_{in} could also be found from conventional graph reduction. Figure 5.10 shows the process, which is similar to that described in Figure 5.7. The lower loop is first converted to a self loop, the self loop is eliminated, and finally the resulting lower branch is added to the S_{11} branch. The result is the same as (5.23).

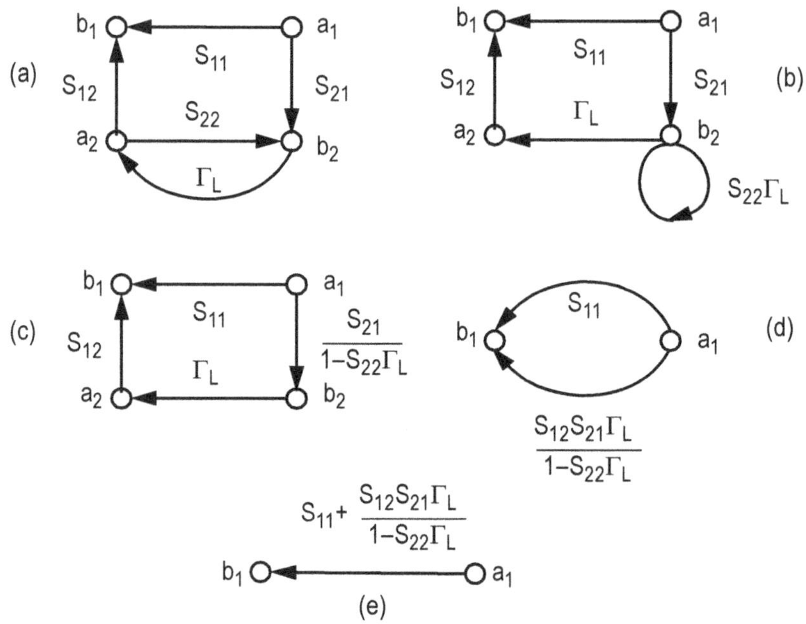

Figure 5.10 Determination of S_{11} from conventional graph reduction. The original graph is shown in (a); the lower loop is reduced to a self loop (b); the self loop is eliminated (c); the lower path is converted to a single path (d); and the two paths are added (e). The transmittance of the final, single path is S_{11}.

5.1.4.2 Transducer Gain

Next we find the transducer gain of the terminated two-port. This requires the full graph of Figure 5.8. We begin by recognizing that the gain cannot be expressed as a simple ratio of the graph's nodal variables, but some such ratio must be derived as part of the solution.

The transducer gain, G_t, is defined as

$$G_t = \frac{P_d}{P_{av}} \tag{5.25}$$

where P_{av} is the power available from the source and P_d is the power delivered to the load. From (3.92), (3.93), and (3.99),

$$\begin{aligned} P_{av} &= \frac{1}{2}\frac{|b_s|^2}{1-|\Gamma_s|^2} \\ P_d &= \frac{1}{2}|b_2|^2(1-|\Gamma_L|^2) \end{aligned} \tag{5.26}$$

So

$$G_t = \left|\frac{b_2}{b_s}\right|^2(1-|\Gamma_s|^2)(1-|\Gamma_L|^2) \tag{5.27}$$

Thus, our task is to determine b_2 / b_s from the graph and then to find G_t from (5.27). We first find the paths. There is only one path from b_s to b_2,

$$P_1 = S_{21} \tag{5.28}$$

Next we find the loops. The first-order loops are

$$\begin{aligned} L_1^a &= \Gamma_s S_{11} \\ L_1^b &= \Gamma_L S_{22} \\ L_1^c &= S_{21}\Gamma_L S_{12}\Gamma_s \end{aligned} \tag{5.29}$$

where the letter superscripts simply distinguish between the three loops. The only second-order loop is

$$L_2 = \Gamma_s S_{11} \Gamma_L S_{22} \tag{5.30}$$

There are no other second-order loops or third-order loops, as the third loop in (5.29) touches the other two. All of the loops touch path P_1. Substituting into (5.18) gives

$$\frac{b_2}{b_s} = \frac{S_{21}}{1 - (\Gamma_s S_{11} + \Gamma_L S_{22} + S_{21}\Gamma_L S_{12}\Gamma_s) + \Gamma_s S_{11}\Gamma_L S_{22}} \tag{5.31}$$

Substituting this into (5.27) and simplifying the expression gives the same result as in (3.103), which was derived algebraically:

$$G_t = \frac{|S_{21}|^2(1 - |\Gamma_s|^2)(1 - |\Gamma_L|^2)}{|(1 - S_{11}\Gamma_s)(1 - S_{22}\Gamma_L) - S_{12}S_{21}\Gamma_s\Gamma_L|^2} \tag{5.32}$$

5.1.4.3 Interface Mismatch in Cascaded Two-Ports

Figure 5.11(a) shows a cascade of two unilateral (i.e., $S_{12} = 0$) two-ports with a mismatched interface. We assume that the ports' normalizing impedances at the output of S_a and input of S_b are the same. Usually, we expect that cascading a number of two-ports, such as amplifiers, results in a gain that is the sum of the gains, in decibels, of the individual components. If the interface is mismatched, however, that is no longer the case.

The network graph is shown in Figure 5.11(b). While we could use Mason's rule to analyze this circuit, graph reduction can be accomplished almost by inspection. If it were analyzed by Mason's rule, we would have to deal with the fact that three first-order loops exist, three second-order, and one third-order. This would result in a messy denominator, which could be factored to obtain a simpler form, although with difficulty. The reduced graph is shown in Figure 5.11(c), and immediately we find the b_2 / b_s ratio to be

$$\frac{b_2}{b_s} = \frac{S_{21a}S_{21b}}{(1 - S_{11a}\Gamma_s)(1 - S_{22a}S_{11b})(1 - S_{22b}\Gamma_L)} \tag{5.33}$$

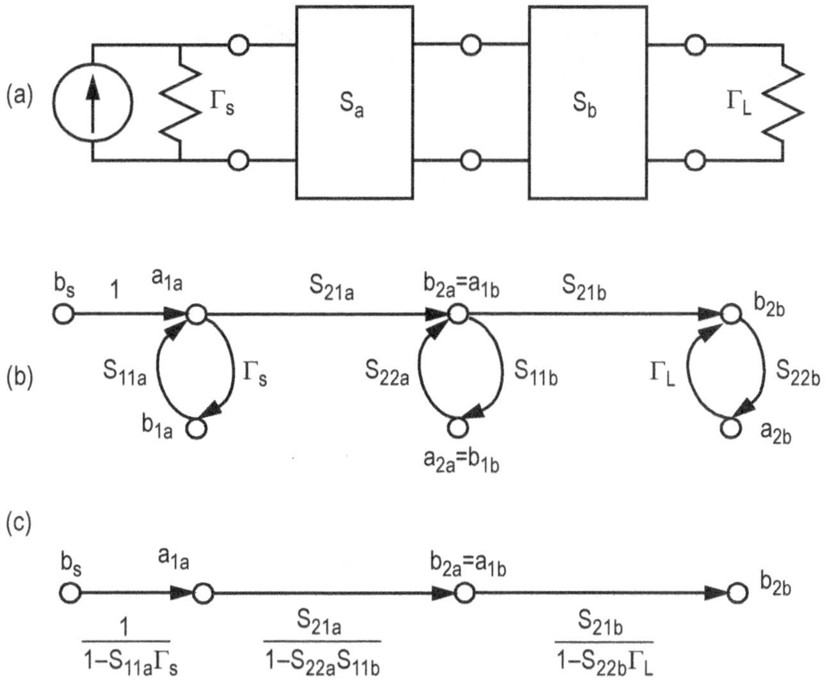

Figure 5.11 Cascaded two-ports with a mismatched interface (a), network graph (b), the reduced graph (c). The additional *a* or *b* in the subscripts refers to the first or second of the cascaded two-ports, respectively.

and from (5.27),

$$G_t = \frac{|S_{21a}S_{21b}|^2(1-|\Gamma_s|^2)(1-|\Gamma_L|^2)}{|(1-S_{11a}\Gamma_s)(1-S_{22a}S_{11b})(1-S_{22b}\Gamma_L)|^2} \qquad (5.34)$$

If the interface were matched, $S_{22a} = S_{11b} = 0$. The gain, G_{t0}, would be

$$G_{t0} = \frac{|S_{21a}S_{21b}|^2(1-|\Gamma_s|^2)(1-|\Gamma_L|^2)}{|(1-S_{11a}\Gamma_s)(1-S_{22b}\Gamma_L)|^2} \qquad (5.35)$$

so the interface mismatch adds a factor of $1/|1 - S_{22a}S_{11b}|^2$ to the gain expression. In most cases, we do not know the phases of S_{22a} or S_{11b}; we know only that the mismatch introduces an uncertainty in the cascade gain. The maximum value of the gain, G_{max}, is

$$G_{max} = G_{t0}/(1 - |S_{22a}S_{11b}|)^2 \tag{5.36}$$

and the minimum, G_{min}, is

$$G_{min} = G_{t0}/(1 + |S_{22a}S_{11b}|)^2 \tag{5.37}$$

It may be a little disconcerting to note that $G_{max} > G_{t0}$, so the gain can actually be greater than in the matched case. This happens because the "matched" case is one in which S_{22a} and S_{11b} are zero, not a conjugate match. The gain increases in (5.36) because the phases of S_{22a} and S_{11b} are conjugate, so there is actually some approximate interface matching taking place.

This result can be used to show how gain ripple, due to mismatched interfaces, can occur. Suppose we had a piece of ideal transmission line, whose characteristic impedance was equal to the port-normalizing impedance, between the components. The line would introduce forward and reverse paths of transmittance $\exp(-j\beta l)$ between the two-ports, where β is the phase-shift constant and l is the line length. This multiplies S_{11b} by $\exp(-j2\beta l)$, so the central self loop in Figure 5.11(b) would be

$$L = \exp(-j2\beta l)S_{22a}S_{11b} \tag{5.38}$$

and the central gain term in Figure 5.11(c) would become

$$\frac{a_{1b}}{a_{1a}} = \frac{S_{21a}}{1 - \exp\left(-j\frac{2\omega l}{v_p}\right)S_{22a}S_{11b}} \tag{5.39}$$

since $\beta = \omega/v_p$, where v_p is the phase velocity. The term in (5.39) causes the gain to vary cyclically with frequency between the values given by (5.36) and (5.37). The severity of that variation depends on the product $|S_{22a}S_{11b}|$.

It is also worthwhile to note that if *either* $S_{22a} = 0$ or $S_{11b} = 0$, the cascade gain of (5.35) is achieved. It is not necessary to match both ports;

only one matched port is needed. When one of those terms is set to zero, the middle self loop disappears, so it has no effect on the transfer function. This point is obvious from even the most casual inspection of the network graph, but it is not immediately evident from an algebraic treatment.

Much of the "conventional wisdom" on high-frequency circuit analysis states that graphs and Mason's rule are the only way to analyze microwave circuits described by S parameters. We have seen, however, that direct algebraic manipulation of the S matrices and wave variables is often a practical way to deal with such circuits, and even when graphs are used, Mason's rule isn't the only game in town. We shall see in the next section that nodal analysis, based on Y matrices, is yet another powerful method and is well suited to computer implementation.

5.2 Nodal Analysis

Nodal analysis, based on the concept of an indefinite admittance matrix, is a powerful technique for setting up circuit equations for solution either analytically or numerically. It is the preferred method, today, for computer analysis of all kinds of electronic circuits: digital, analog, RF, and microwave. We have already seen one example, in Section 4.3.3, that uses an indefinite admittance (or *nodal*) matrix. Here we discuss nodal analysis in greater detail.

5.2.1 Indefinite Admittance Matrix

Imagine that we have a circuit consisting of an arbitrary set of nodes and branches, in which all node voltages are referenced to a common ground node. The situation is illustrated in Figure 5.12; the terminals represent the nodes of the circuit. We can write a Y matrix describing this circuit:

$$\begin{bmatrix} I_1 \\ I_2 \\ \cdots \\ I_N \end{bmatrix} = \begin{bmatrix} Y_{1,1} & Y_{1,2} & \cdots & Y_{1,N} \\ Y_{2,1} & Y_{2,2} & \cdots & Y_{2,N} \\ \cdots & \cdots & \cdots & \cdots \\ Y_{N,1} & Y_{N,2} & \cdots & Y_{N,N} \end{bmatrix} \begin{bmatrix} V_1 \\ V_2 \\ \cdots \\ V_N \end{bmatrix} \tag{5.40}$$

This Y matrix has a number of interesting properties. First, since the network is floating, Kirchoff's current law requires that the sum of all the nodal currents be zero for all values of V_n. This means, in turn, that the

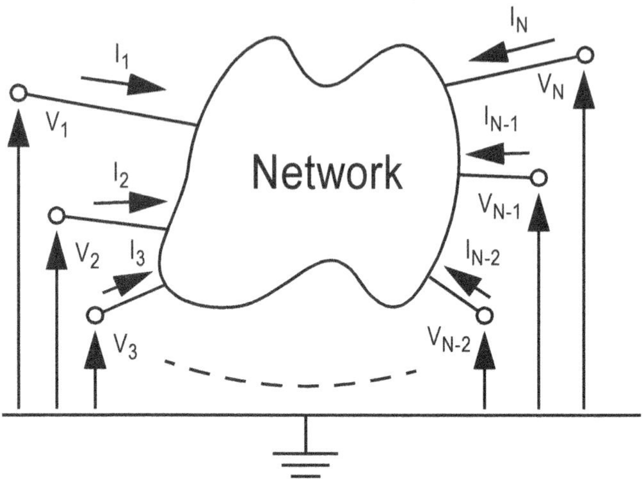

Figure 5.12 An N-node circuit represented by an indefinite admittance matrix. All node voltages are referenced to an arbitrary ground potential, and no terminals of the network are connected to ground.

columns of the matrix must sum to zero. Although it is less obvious intuitively, the rows must sum to zero as well. Second, grounding a node forces its voltage to zero, so the respective column of the matrix can be eliminated. Since the current in that node is usually not of interest, the respective row can be eliminated as well, preserving the convenient square form of the matrix. Eliminating a node by open-circuiting it is more complex, however. The matrix must be inverted, forming a Z matrix, then the corresponding current is set to zero. This eliminates a row and column of the inverted matrix, which can then be inverted again to return it to Y-matrix form.

5.2.1.1 Matrix Stamps

The importance of the indefinite admittance matrix is in the ease with which elements can be added to it. Invariably the process requires no user judgment, so it is well suited to computer implementation.

Suppose we connect a two-terminal element of admittance Y between nodes m and n, as shown in Figure 5.13. This changes the node currents as follows:

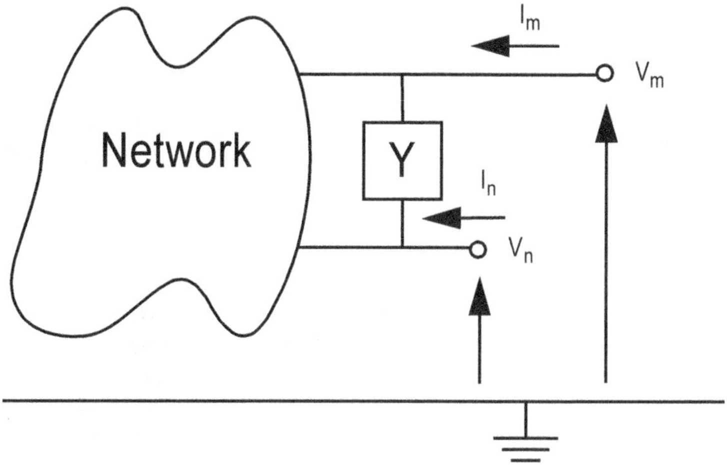

Figure 5.13 Adding an admittance to the network modifies the external node currents as in (5.41).

$$\begin{aligned} \Delta I_m &= Y(V_m - V_n) \\ \Delta I_n &= Y(V_n - V_m) \end{aligned} \tag{5.41}$$

so we add Y to Y_{mm} and Y_{nn} and subtract it from Y_{mn} and Y_{nm}. If the mth or nth row and column do not exist, we simply create them.

This pattern of adding the necessary values to the matrix is sometimes called a *stamp*, and adding the element values to the matrix is called *stamping the matrix*. The terminology reflects that fact that adding element values to circuit equations is quite automatic, requiring no thought or judgment. It involves merely stamping the matrix with the appropriate stamp (i.e., adding values to the appropriate matrix locations), and adding new row-column pairs as needed.

5.2.1.2 Voltage-Controlled Current Source

Similar stamps can be derived for other types of elements. For example, the voltage-controlled current source in Figure 5.14 has

$$\Delta I_p = G(V_m - V_n)$$
$$\Delta I_q = -G(V_m - V_n) \tag{5.42}$$

where (p, q) are the current-source nodes and (m, n) are the control-voltage nodes. G, the transconductance, is added to the (p, m) and (q, n) positions, while $-G$ is added to (p, n) and (q, m).

5.2.1.3 Grounded Elements

As another example, a grounded element, such as a two-port, can be added to the matrix as follows. We assume that port 1 is connected to node m and port 2 is connected to node n. The negative terminals of both ports are grounded. The two-port then changes the Y matrix as follows:

$$\Delta I_m = Y_{11} V_m + Y_{12} V_n$$
$$\Delta I_n = Y_{21} V_m + Y_{22} V_n \tag{5.43}$$

This shows that Y_{11} is added to the (m, m) position, Y_{12} to the (m, n) position, Y_{21} to the (n, m) position, and Y_{22} to the (n, n) position.

Conversely, the properties of the indefinite admittance matrix show how a grounded element can be "floated"; that is, the ground can be converted to an ordinary terminal. Suppose, for example, we have a bipolar transistor that is measured in a grounded-emitter test circuit with the base as port 1 and collector as port 2, resulting in two-port Y parameters. When the emitter is converted to a separate node, the transistor's Y matrix becomes an indefinite admittance matrix. This requires the addition of a third row and column, and we simply fill the new positions with values that cause all rows and columns to sum to zero. Thus,

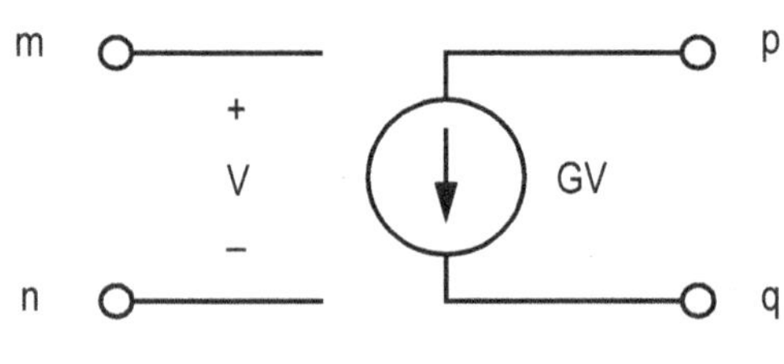

Figure 5.14 A voltage-controlled current source; its stamp is given by (5.42).

$$\begin{bmatrix} I_b \\ I_c \\ I_e \end{bmatrix} = \begin{bmatrix} Y_{11} & Y_{12} & -(Y_{11}+Y_{12}) \\ Y_{21} & Y_{22} & -(Y_{21}+Y_{22}) \\ -(Y_{11}+Y_{21}) & -(Y_{12}+Y_{22}) & Y_{11}+Y_{12}+Y_{21}+Y_{22} \end{bmatrix} \begin{bmatrix} V_b \\ V_c \\ V_e \end{bmatrix} \tag{5.44}$$

where the b, c, and e subscripts refer to the base, collector, and emitter respectively. To add this to the network's matrix, the terms in (5.44) are simply added to the respective node positions of the network matrix.

In general, any admittance element can be added to the matrix by an application of the above procedure. Suppose we are given the conventional, port-format admittance matrix of a multiport element and must add it to the indefinite admittance matrix. Figure 5.15 illustrates the case. The element's port no. 1 is connected to the network's node pairs (m, n), port 2 to (p, q), port 3 to (r, s), and so on. Thus, we have

$$\begin{aligned} I_m &= Y_{11}(V_m - V_n) + Y_{12}(V_p - V_q) + Y_{13}(V_r - V_s) + \dots \\ I_n &= -Y_{11}(V_m - V_n) - Y_{12}(V_p - V_q) - Y_{13}(V_r - V_s) - \dots \end{aligned} \tag{5.45}$$

and so on for the other port currents. This can be expressed in the form

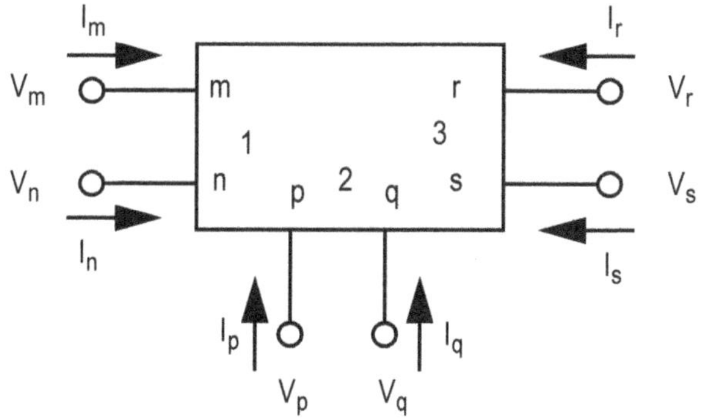

Figure 5.15 A multiport circuit that is added to the nodal matrix. We have assumed that it contains no grounded ports.

$$\begin{bmatrix} I_m \\ I_n \\ I_p \\ I_q \\ \cdots \end{bmatrix} = \begin{bmatrix} Y_{11} & -Y_{11} & Y_{12} & -Y_{12} & \cdots \\ -Y_{11} & Y_{11} & -Y_{12} & Y_{12} & \cdots \\ Y_{21} & -Y_{21} & Y_{22} & -Y_{22} & \cdots \\ -Y_{21} & Y_{21} & -Y_{22} & Y_{22} & \cdots \\ \cdots & \cdots & \cdots & \cdots & \cdots \end{bmatrix} \begin{bmatrix} V_m \\ V_n \\ V_p \\ V_q \\ \cdots \end{bmatrix} \tag{5.46}$$

which now has the structure of an indefinite admittance matrix. To add an element's matrix, in indefinite admittance matrix form, to the circuit matrix, we simply add it to the corresponding positions in the circuit's matrix.

For further examples of matrix stamps, see [5.3].

5.2.2 Matrix Reduction

Inevitably, we want to convert the indefinite admittance matrix to a port-format one, perhaps a two-port representing an amplifier or other component. To do so, we must begin by inverting the Y matrix to form a Z matrix. We assume that at least one node in the original circuit has been connected to ground and the corresponding row and column eliminated. If this is not done, the matrix is singular and cannot be manipulated. Attempting to analyze a floating circuit, with a singular admittance matrix, is a common error made by naive users of circuit-simulation software.

We first invert the matrix to obtain an impedance matrix:

$$\begin{bmatrix} V_1 \\ V_2 \\ \cdots \\ V_N \end{bmatrix} = \begin{bmatrix} Z_{11} & Z_{12} & \cdots & Z_{1N} \\ Z_{21} & Z_{22} & \cdots & Z_{2N} \\ \cdots & \cdots & \cdots & \cdots \\ Z_{N1} & Z_{N2} & \cdots & Z_{NN} \end{bmatrix} \begin{bmatrix} I_1 \\ I_2 \\ \cdots \\ I_N \end{bmatrix} \tag{5.47}$$

so we can excite the nodes with current sources and obtain the voltages of all the nodes that become ports in the reduced matrix. As illustrated in Figure 5.16, we first excite the node pair that will become port 1. The sources may have any value, but $I_s = 1$ is convenient. We then calculate the voltages at all the nodes, and this gives the port voltages from which one column of the port Z matrix can be determined.

Suppose, for example, that port 1 uses nodes 1 and 2, and node 1 is the positive terminal of the port. We excite node 1 by $I_{s1} = 1\text{A}$ and node 2 by

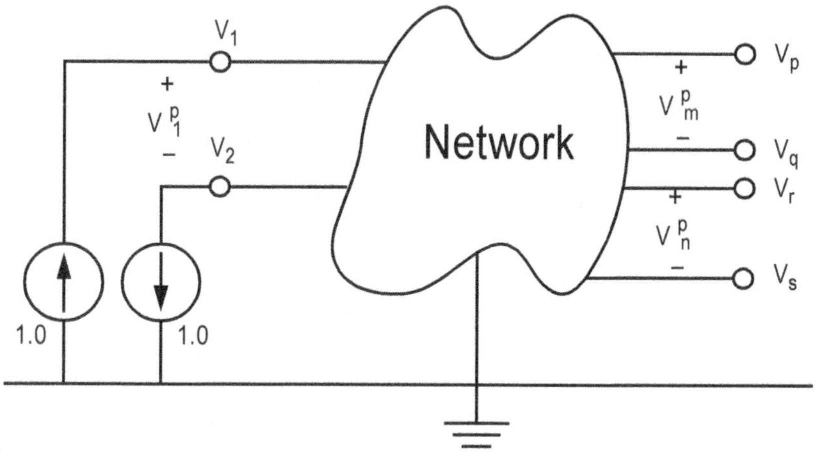

Figure 5.16 Conversion of a nodal matrix to a port matrix. The pairs of nodes representing a port are excited by a pair of current sources and the voltages at all the ports provide a column of the network's Z matrix. The current sources are then moved to the next port and the process is repeated until the port-format Z matrix is filled.

$I_{s2} = -1$A; this is equivalent to a single 1A current source connected from node 2 to node 1. We now have

$$\begin{bmatrix} V_1 \\ V_2 \\ \ldots \\ \ldots \\ V_N \end{bmatrix} = \begin{bmatrix} Z_{11} & Z_{12} & \ldots & \ldots & Z_{N1} \\ Z_{21} & Z_{22} & \ldots & \ldots & Z_{2N} \\ \ldots & \ldots & \ldots & \ldots & \ldots \\ \ldots & \ldots & \ldots & \ldots & \ldots \\ Z_{N1} & Z_{N2} & \ldots & \ldots & Z_{NN} \end{bmatrix} \begin{bmatrix} 1 \\ -1 \\ 0 \\ \ldots \\ 0 \end{bmatrix} \tag{5.48}$$

The voltage at port 1 is

$$V_1^p = V_1 - V_2 = Z_{11} - Z_{12} - (Z_{21} - Z_{22}) \tag{5.49}$$

So

$$Z_{11}^{p} = \frac{V_1^p}{I_s} = Z_{11} - Z_{12} - (Z_{21} - Z_{22}) \tag{5.50}$$

where the superscript p indicates a port-format quantity. All other port voltages are found similarly, and this gives the first column of the port-format Z matrix. The current source is then moved to the nodes that constitute port 2, and the second column of the port Z matrix is found. The process is repeated until all columns of the port-format matrix are filled in. Finally, the resulting Z matrix can be converted to a Y matrix by inversion or to an S matrix per (3.20).

A little thought will show that the indefinite admittance matrix of a large circuit is both large and *sparse*; that is, it contains mostly zeros. The size of the matrix is equal to the number of nodes in the circuit, which can be very large. Each position off the matrix's main diagonal represents a connection, so a position is nonzero only when an element is connected between the nodes corresponding to that position. Since most node pairs in any circuit are not connected by an element, most of the positions are zero. Conventional matrix manipulations, such as Gaussian reduction, are inefficient, as most operations involve multiplying zero by zero and adding it to zero. Furthermore, it makes no sense to waste computer memory by storing a large number of zeros.

Sparse matrix techniques address these problems, providing much greater speed and computational economy than conventional ones. One method, for example, employed in the widely used Spice circuit simulator, stores nonzero matrix elements in a linked list and uses Gaussian reduction combined with various heuristics to minimize "fill-ins" (zero locations converted to nonzero) and computation. Other methods use iterative solutions that converge rapidly to a solution. For more information on such methods, see [5.4–5.6].

References

[5.1] Ruston, H., and J. Bordogna, *Electric Networks: Functions, Filters, Analysis*, New York: McGraw-Hill, 1966.

[5.2] Van Valkenburg, M. E., *Network Analysis* (3rd ed.), Englewood Cliffs, NJ: Prentice-Hall, 1974.

[5.3] Vlach, J., and K. Singhal, *Computer Methods for Circuit Analysis and Design*, (2nd ed.), New York: Chapman and Hall, 1994.

[5.4] Stoer, J., and R. Bulirsch, *Introduction to Numerical Analysis* (3rd ed.), New York: Springer-Verlag, 2002.

[5.5] Tewarson, R., *Sparse Matrices*, New York: Academic Press, 1973.

[5.6] Lau, H. T., *A Numerical Library in C for Scientists and Engineers,* Boca Raton, FL: CRC Press, 1995.

Chapter 6

Circuit and Element Modeling

This chapter addresses the problem of modeling circuit elements in such a way as to create a valid, accurate circuit model. This process superficially seems straightforward: obtain characterizations of the various circuit elements, simulate them when necessary (say, with an electromagnetic simulator) drop them into a circuit simulator with the appropriate connections, click on the *analyze* button, and stand back and watch the fun. Unfortunately, simulation of high-frequency circuits involves a degree of subtlety that is often unrecognized, and failing to appreciate this subtlety results in analyses having errors or, at best, mediocre accuracy. We discuss some of those considerations in this chapter.

6.1 Circuit Characterization

Inevitably we must use S parameters to represent a device, frequently a transistor, a passive element in chip form, or a strip transmission-line structure analyzed by an electromagnetic (EM) simulator. Often there is a temptation to use such data as if it represented a lumped element: stitch the parts together with direct connections or, when necessary, transmission lines, and dump it into the circuit simulator. Although this seems to be the most common procedure for analyzing a high-frequency circuit, it ignores a number of important constraints and characteristics that are inherent in S-parameter data. We examine those in this section and suggest some ways to deal with them.

6.1.1 Wave and I/V Characterization

A circuit model is based on nodes and conductors. Our transmission line models, both single lines (Chapter 1) and coupled lines (Chapter 2), treat the wave as a voltage or current on a conductor; a transmission line, however, supports a guided wave, and a specific voltage or current often cannot be identified in any meaningful way. For example, a rectangular waveguide can be viewed by a transmission-line model, but it is impossible to identify a specific voltage or current that defines the waves in the guide. Even in strip transmission lines, the current is distributed over the conductors, and at the ends there is no single point or node where a unique line current is observed. Additionally, S parameters often represent distributed structures that have no clear input or output nodes, so S parameters are strictly valid only for the type of connection used to measure them. How do we reconcile these characteristics in a circuit?

The choice of viewing a wave on a transmission line as a voltage, current, power, or electromagnetic wave actually doesn't matter very much until another line or a circuit element is connected. Then, the current and field distributions on the two sides of the interface do not match. The resulting discontinuity must be modeled in some way, invariably by a model consisting of lumped or, often, distributed circuit elements.

A discontinuity model accounts for the differences in the structures at the interface. For example, consider a step junction in a microstrip line; that is, a line that connects to a second, narrower one. The "squeezing" of the current into the narrower line is inductive, while the fringing electric fields at the corners of the step add capacitance. Thus, an appropriate model for such a junction is a series inductor and shunt capacitor. The values of those elements depend on the dimensions of the lines and the substrate dielectric constant. In general, they are not constant with frequency.

More complicated interconnections require more complex models. An example of the latter is a transistor chip connected by bond wires to the end of a microstrip line. We shall examine these in more detail as we proceed.

6.1.2 Characterization of Discrete Components

6.1.2.1 Measurement and Application

As an example of the considerations in using high-frequency components, we begin with a chip capacitor. Many kinds of chip capacitors exist, in various forms and sizes, for use in circuits over a broad range of frequencies. At high frequencies, single-layer chip capacitors are used most commonly; these consist of two metal layers with an intervening dielectric. They are normally mounted directly on the circuit metallization with the metal layers parallel to the substrate; one layer is directly bonded to the metal, and the other is connected by ribbon or wire bonds. At lower frequencies, multilayer chip capacitors are used. These consist of a number of interleaved metal layers, embedded in a dielectric, with metal end caps. The end caps are bonded to the circuit, usually by soldering. Multilayer chip capacitors are available in standard sizes, often remarkably small: "0201" size chips, for example, are 0.020 × 0.010 × 0.010 inches, or 0.5 × 0.25 × 0.25 mm. Such small chips are used in applications where small size is essential, such as cellular handsets and other portable electronic components.

Figure 6.1(a) shows a chip capacitor mounted on a substrate and Figure 6.1(b) shows a lumped-element, equivalent-circuit model. C_s is the chip's capacitance and L_s is the series inductance, which exists because of the displacement current in the capacitor. The inductance depends largely on the overall shape and size of the chip, and less on its value; for an 0402-size chip, L_s is approximately 0.2 nH. The series resistance, R_s, models the loss mechanisms in the chip, dielectric losses, and the power loss in the resistance of the capacitor's plates. Finally, C_p models fringing capacitance between the end caps. C_p is strongly affected by the substrate on which the

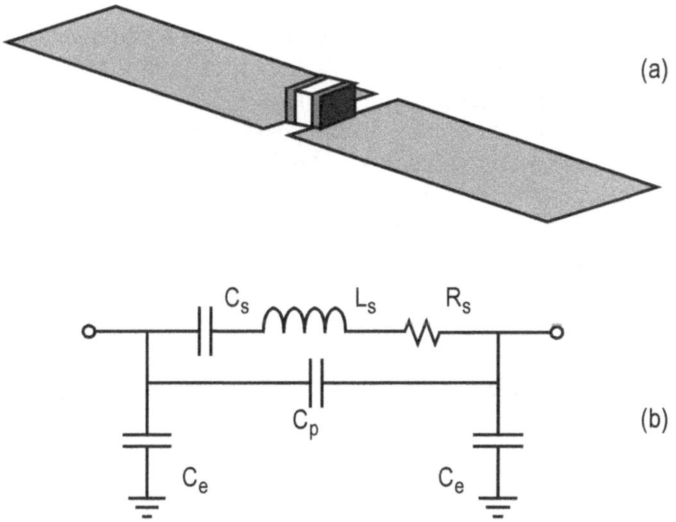

Figure 6.1 A chip capacitor mounted in a microstrip circuit (a); its lumped-element equivalent circuit (b). The equivalent circuit implicitly assumes that its reference planes are at the terminals of the chip.

capacitor is mounted, as well as the fringing fields around the ends of the connecting lines. C_e is largely from the fringing capacitance between the end caps and the ground plane. C_e inevitably is affected by the width of the microstrip line where it connects to the chip.

In addition to the chip's parasitics, the connection between the microstrip line and the chip creates a discontinuity. That discontinuity generates spurious modes in the vicinity of the connection, and those depend on the way that the chip is connected into the circuit. The modal structure is different for parallel and series interconnections and for different types of substrates and line dimensions. That discontinuity generally includes a series inductance, caused by the concentration of the current distribution near the connection, and a shunt capacitance, resulting from fringing fields at the open end of the microstrip line. The series inductance is often small, especially if the line width is not much greater than the chip width, so it is often successfully ignored. The open-end capacitance then becomes part of C_e. Capacitive coupling between the microstrips' open ends may also increase C_p.

C_s, R_s, and L_s are largely intrinsic to the chip; that is, they depend little on the way that the chip is mounted or used. C_p and C_e depend more strongly on extrinsic factors such as the substrate's dielectric constant, its thickness, and the width of the connecting lines. An important effect of C_p is the creation of a parallel resonance; C_p resonates with L_s at high frequencies, where the reactance of C_s is negligibly small. This resonance creates an upper limit to the usefulness of the chip, and since C_p depends strongly on extrinsic factors, so does that parallel resonance.

Now, consider the problem of creating a lumped-element model or a set of S parameters for a chip capacitor. The chip is mounted in a gap between the ends of microstrip lines in a test fixture; the test fixture is usually a "split block" fixture, which can be disassembled so that calibration standards can be placed in the gap where the chip will be mounted. These standards are measured with a vector network analyzer, and the effects of the fixture, its connectors, and connecting cables, as well as errors in the network analyzer, are removed mathematically. Additionally, the calibration process places the reference plane at the end of the microstrip lines, where the chip will be connected. The calibrated fixture then presents virtually perfect, quasi-TEM microstrip source and load to the chip.

The chip, mounted on a test carrier, is then placed in the fixture and the S parameters are measured. When this is done, those S parameters include all the interconnection discontinuities, as well as the chip's substrate-dependent parasitics. Thus, those S parameters, and any equivalent circuit derived from them, are strictly valid *only when the chip is used precisely in the manner in which it was measured.* Stating this differently, we could say that the normalizing impedance of the S parameters is not simply some standard value, but instead is the impedance of the microstrip line used in the measurement. Using the chip on other substrates, with lines having other dimensions, or in a different configuration (say, connected in shunt with the microstrip line when the S parameters were measured in series) will give results having some degree of error. This is true even if those lines have the same characteristic impedance, and thus the same normalizing impedance, as those used in the measurement.

Thus, before using a chip component in a high-frequency circuit, the designer must know something about the environment in which it was measured; an equivalent-circuit model or a set of S parameters alone is not sufficient. The location of the reference planes likewise should be confirmed; the author has occasionally seen these located at the center of the chip, instead of the chip's edges, as one would expect.

Some advanced models have attempted to include the effects of the interconnecting line width and the substrate dielectric constant and thickness. Such models eliminate much of the uncertainty in using chip components in a wide variety of applications.

Most chip S parameters are measured in a series microstrip configuration. Using those S parameters when the chip is connected in shunt requires care. It is common to treat the shunt connection as a tee junction plus the chip capacitor, but the chip usually is not separated from the tee junction by enough microstrip to prevent the coupling of their discontinuity fields and thus an invalid model. The best way to avoid this problem is to leave a length of microstrip between the tee junction and the chip. Where this is not possible, one must either rethink the entire design approach, obtain an appropriate model, or suffer some degree of inaccuracy.

6.1.2.2 Lumped-Element Model

The lumped-element, equivalent-circuit model of the chip in Figure 6.1(b) can be found in a straightforward manner. We note that any pi circuit can be expressed in terms of its Y parameters as shown in Figure 6.2(a), and the parts of the model in Figure 6.1(b) can be readily identified. Since the circuit is reciprocal, $Y_{12} = Y_{21}$, so Y_{12} does not appear explicitly. The Y parameters are best determined by measuring S parameters in a well-calibrated fixture and converting them to Y parameters per (3.25).

We begin with the determination of the upper branch, whose admittance is Y_{21}. R_s is found at series resonance, where $\text{Re}\{Y_{21}\} = -1 / R_s$. C_s is presumed to be known, as it is the low-frequency chip capacitance, so L_s can be found readily from the resonant frequency. If C_s is unknown, it can be found from the Q of the series resonance or from low-frequency mea-

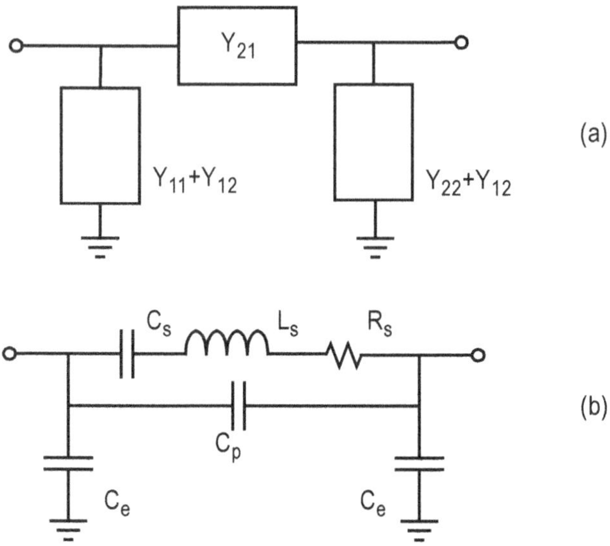

Figure 6.2 A set of Y parameters can be expressed as a pi circuit (a) and equated to the elements of the lumped-element model (b).

surements. C_p is found from the parallel-resonant frequency or from $\mathrm{Im}\{Y_{21}\}$ at series resonance. Finally, the end capacitances are determined from $\omega C_e = Y_{11} + Y_{21}$.

Similar procedures can be used to determine models of many kinds of passive structures that are well modeled by a pi circuit. If the structure is more naturally described by a tee circuit, an analogous procedure using Z parameters is appropriate.

More general methods for creating RLC equivalent circuits from measured S parameters of passive structures have been developed [6.1-6.5]. Some of these are intended for the creation of compact models; others are designed for allowing those structures to be used in time-domain simulators. In general, the goal is to approximate a network function $F(s)$ by the form

$$F(s) = \frac{N(s)}{D(s))} \tag{6.1}$$

where $N(s)$ and $D(s)$ are polynomials in the LaPlace transform variable, s. This can be done by various kinds of rational-function approximations; a Pade approximation [5.4] is a common choice. The function can then be expanded in the form

$$F(s) = A + Bs + \sum_{k=1}^{P} \left(\frac{C_k}{s - p_k} + \frac{C_k^*}{s - p_k^*} \right) \tag{6.2}$$

and lumped-element equivalents can be associated with each term. The idea can be expanded to apply to matrix quantities; then, all terms of the matrix are treated similarly.

While this idea seems simple, it entails many numerical difficulties in the form of ill conditioning and inability to scale easily to large or active circuits. Most of these problems have been addressed in the literature.

6.1.3 EM-Simulated Circuit Elements

6.1.3.1 EM Simulators

Many types of EM simulators are in common use. They can be roughly divided into two classes: three dimensional, predominantly planar simulators, sometimes called "2 1/2 D" or simply *planar* simulators, and full three-dimensional (*full 3D*) simulators.

Planar simulators generally use moment methods, which originally were limited to zero-thickness metal structures on a planar substrate of one or more layers. Improvements to their analytical methods, however, developed over the years, have allowed much greater versatility, so certain structures having nonzero dimensions perpendicular to the plane can be accommodated. Among these are nonzero conductor thickness, ground vias, and "dielectric bricks," small isolated pieces of dielectric. Modern planar simulators are fast and accurate and do not require disturbing amounts of computer memory.

Full 3D simulators treat all three dimensions equally. They can analyze such problems as coaxial transmission lines, coaxial connectors, and waveguide structures. Their memory use is greater and computational speed slower than planar simulators for problems of comparable magnitude, and for predominantly planar structures, planar simulators usually provide better accuracy. For this reason, problems involving strip transmission lines are best analyzed by planar simulators.

Planar simulators can use a closed or open formulation, which can be gridded or ungridded. *Closed formulation* means that the structure is enclosed in a metal box; *open formulation* means that the substrate dimensions are infinite in the horizontal plane. In a gridded formulation, the metal edges must align with a predefined grid; in an ungridded one, no such grid exists. The requirement of a closed structure is inconvenient, as is the need for a grid, but these characteristics generally result in greater accuracy than in the open and ungridded formulations. The latter also require more operator skill, as they are subject to errors that are often pitfalls for an inexperienced user. Closed, gridded formulations are generally more robust and require less skill on the part of the user.

6.1.3.2 De-Embedding

All planar EM simulators analyze the entire structure presented to them, including the lines from the ports to the structure of interest and the parasitics associated with the ports themselves. Thus, all such simulators include provisions to de-embed each port from the structure of interest. Although many types of ports and de-embedding techniques exist, their purposes are the same: the de-embedding process removes the effect of the port and an interconnect, typically a microstrip line, from the analysis, leaving only the structure whose S parameters are desired.

As an example, consider the structure shown in Figure 6.3(a), a bent stub with a ground via. The structure of interest is connected to two ports through lengths of interconnecting microstrip line. We want the S parameters at the reference planes shown in the figure, the edges where the interconnecting lines meet the stub.

Each "port" consists of an infinitesimal gap between the end of the strip and a short connection to ground. The gap contains a voltage source, which can be zero. Currents throughout the structure are computed; in effect, Y parameters are determined. The ground connections and the gaps represent significant discontinuities, which generate evanescent (nonpropagating) waves in their vicinities. We want the S parameters of the stub alone, not including the effects of the ports, their discontinuities, or the lines from the ports to the stub, so the ports and intervening lines must be de-embedded.

Figure 6.3(b) shows the de-embedding standard for the input line (port 2). The standard consists of that port and line segment cascaded with a reversed version of the same structure. The S parameters of that standard are computed by exciting it in an even and an odd mode. When the ports are excited by an even mode, the center of the standard is a virtual open circuit; with an odd mode, it is a virtual short circuit. From these two cases, the S

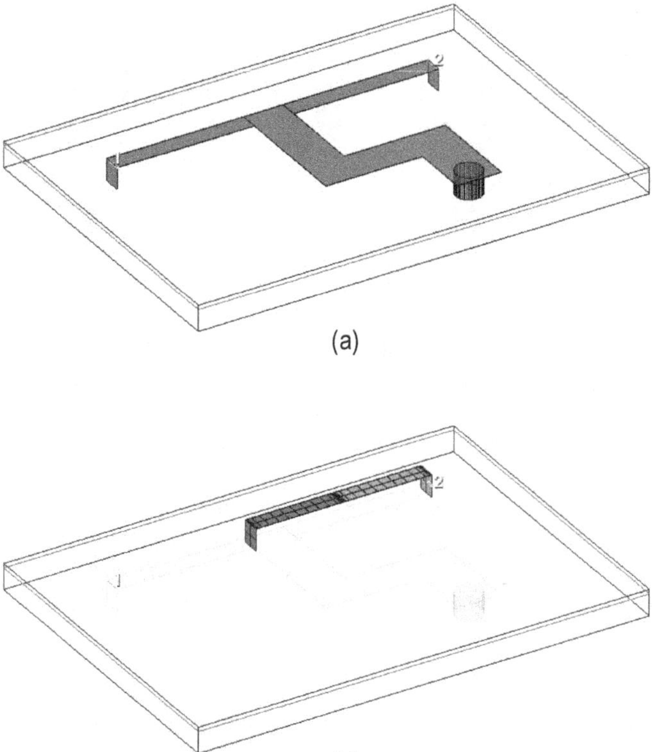

(a)

(b)

Figure 6.3 EM-simulated stub (a); de-embedding model (b). The borders around the structures are intended to show the substrate layers and their relative thicknesses; they do not mark the edge of the substrate, which is infinite.

parameters of the port and the interconnecting line can be determined. When the port and interconnect have been characterized, it is then a simple matter to remove its effects from the EM simulation. For example, we could convert the S matrix into a T matrix, invert the T matrix, and cascade it with the T matrix of the entire structure.

It is important to recognize that the de-embedded structure includes any discontinuity parasitics at the reference plane. For this reason, the resulting S parameters of the de-embedded structure are strictly valid only when it is excited by a line of the same width and long enough so that the evanescent fields near the reference plane do not couple to those of other discontinuities.

The discontinuity effects at the reference plane are part of the structure, so it is essential that they be part of the model; that is, after all, the only way that they will be included when the structure is used in a circuit simulation. Because they are included, however, the same considerations exist in using an EM-simulated circuit element as in using a chip component. The EM analysis of an EM-simulated structure is valid only for the connecting-line widths used in the analysis. Furthermore, it is valid only for the environment in which it was simulated. For example, a step junction analyzed as a simple, series microstrip structure is not, in general, valid when used in a coupled-line filter, where other microstrips are close by. It is also invalid when another junction is connected close enough so that the fields from the two junctions might couple.

The de-embedding process requires care. If the circuit element is to be connected to a microstrip line, the lines between its reference plane and the port in the EM simulation must be long enough so that the evanescent modes near the port do not couple to those near the structure. Otherwise, the "de-embedded" S parameters include something of the effect of the port, which is not likely to be what is desired. In most simulators, the user can check for a clean quasi-TEM mode by displaying the current or field distribution on the line from the port.

It is also important to maintain the substrate dimensions so that, at the highest frequency, additional high-order propagating modes are not generated. Although the EM simulator can deal with these additional modes, the circuit simulator does not, as all elements are described by their dominant, quasi-TEM characteristics. Thus, the effect of any additional propagating modes in the EM-simulated structure are not evaluated elsewhere in the circuit. Use of an "overmoded" substrate is always a bad practice, not just from a circuit-simulation viewpoint. At high frequencies, such substrates are subject to surface waves and radiation, whose effects are often difficult to control and invariably deleterious.

6.1.3.3 EM Database Elements

Although modern EM simulators are remarkably powerful, evaluating a structure by EM analysis has a high computational cost. At the same time, high-frequency circuits often use large numbers of the same types of elements, such as tees, step junctions, open-ended stubs, and so on. It makes sense, then, to analyze a large number of such elements, store the results in a database, and interpolate those results when an element is needed. This process speeds the analysis significantly, as an interpolation of a database is much faster than an EM simulation.

Database elements have a number of limitations. First, the elements are analyzed in isolation, so one must be careful to avoid locating them close to other elements, where the evanescent fields at their junctions might couple. Second, the results are valid only for the dominant TEM mode. For the reasons given at the end of Section 6.1.3.2, this limitation is unavoidable. Third, the database is likely to include data for common substrate materials and dimensions; if the element is significantly different from those in the database, the interpolation may not give accurate results. When any of these restrictions cannot be met, the simplest solution is to perform an EM analysis of the structure in question.

6.1.3.4 Use of EM Results in Nonlinear Analysis

Nonlinear analysis requires special considerations for using EM analysis and S parameters. Most fundamental is the need for the S parameters to cover the frequency set used in the EM analysis. The mixing frequencies, f_{mix}, generated in a nonlinear circuit are given by

$$f_{mix} = \sum_{m,\, n,\, p,\, \ldots} mf_1 + nf_2 + pf_3 + \ldots \tag{6.3}$$

where the f_k are the excitation frequencies and $m, n, p\ldots$ are integers, limited in practice to some maximum value that can be set by the nonlinear simulator's user. The EM simulation must cover all these mixing frequencies, including zero. The best accuracy results when minimal frequency interpolation is used. The frequencies in (6.3) are not uniform, so the frequency steps in the EM simulation also should be nonuniform, and should be concentrated in the ranges where most of the mixing frequencies occur. Finally, some of the excitation frequencies are swept over a range, so the frequency set includes all the frequencies of (6.3) evaluated at each step. This is likely to be a large set of frequencies, requiring a long, costly analysis.

A little judgment can avoid the need for an unnecessarily long EM analysis. The excitation frequencies, lowest-order mixing products, and output mixing frequencies of interest (e.g., intermodulation products) are likely to be the most important, so one should make sure frequency points close to these are included in the EM simulation. Some EM-simulated circuit elements (for example, ones in an input low-pass filter) might be exposed to very little energy at most mixing frequencies, so only minimal analysis near these frequencies may be enough. Finally, the devices' parasitic capacitance tends to short-circuit the device at high frequencies, so most high-frequency currents remain in the device and do not excite other circuit elements. In this case, analysis of passive circuit elements at high mixing frequencies may not be needed.

It is especially important to note that dc is part of the frequency set in (6.3). EM simulators are not designed for dc analysis, however, so they usually substitute a low frequency for dc, and that practice usually is acceptable. Conversely, at the highest mixing frequencies in (6.3), the simulation results may be meaningless, as the substrate is severely overmoded. This is not much of a problem, however; because of their high order, there is little energy in the circuit at the highest frequencies.

If an EM simulation is not performed at a particular mixing frequency, the circuit simulator must interpolate the S parameters at frequencies where the structure is analyzed. As long as an analysis at a nearby frequency exists, this is usually a simple matter. However, if the simulator must extrapolate from a modest frequency to a very high one, or to dc, strange and troublesome values can result. It is not unusual, for example, to obtain active-circuit S parameters from such an extrapolation. To avoid such difficulties, one should simply include dc data and at least one high-frequency point in all EM simulations and S-parameter sets. Even if the S parameters at extreme frequencies are inaccurate, the inclusion of S parameters at extreme frequencies prevents the use of wildly incorrect data resulting from large extrapolations.

Oscillator analysis is especially sensitive to EM simulation results and to S-parameter data sets. In interpolating S-parameter sets, the simulator may create discontinuities in the data. Oscillator analysis, for many reasons, is a delicate process; the "graininess" of S-parameter data can upset it. A classic example is a 3D EM analysis of a high-Q resonator, where phase and magnitude steps are large, even if made very close together in frequency.

One solution is to fit the S parameters to an equivalent circuit. This can often be done if the structure is relatively simple, such as a resonator or a ground via hole. When S parameters must be used, the type of interpolation can have a large effect on the success of the analysis, and many circuit simulators allow the user a choice of several interpolation methods. Linear interpolation is probably the worst option. Better options are cubic spline or rational-function interpolation, which minimizes those discontinuities.

6.1.4 Correction of Reference-Plane Locations

Manufacturers of chip capacitors, inductors, and other such components often provide S-parameter characterizations of their products. Unfortunately, those characterizations often do not include information as to the way the chips are measured. Sometimes even the location of reference planes is not specified.

The logical location for the reference plane of a chip component is at the outer edges of its terminals. If the reference plane is located in the center of the chip, as is occasionally the case, the chip effectively has a negative-length transmission-line segment connected to its terminals. Moving a reference plane to the edge of the chip's terminal requires connecting a positive-length transmission-line segment. That transmission line must appear in the circuit's schematic but must not be included in its layout. This is a confusing situation, one that could easily lead to layout errors.

Moving the reference plane of a two-port can be accomplished by modifying its S parameters. Figure 6.4 illustrates the problem. The two-port's input and output reference planes are located at R_1 and R_2, respectively, and we wish to move them to $R_1'$and R_2', as shown in the figure. We assume that the two-port includes transmission-line segments at its input and output, having propagation constants γ_1 and γ_2 and characteristic impedances equal to the ports' normalizing impedances. When the reference planes are moved in a direction opposite to that of the *a* wave, the waves change as follows:

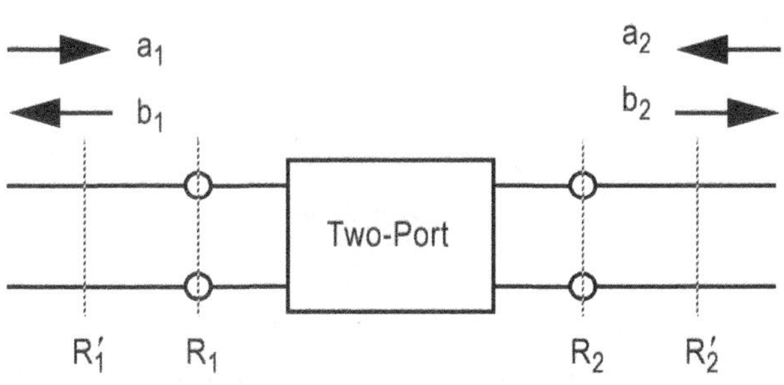

Figure 6.4 Shifting the reference plane from one location to another is equivalent to adding a length of transmission line whose characteristic impedance is equal to the normalizing impedance of the port.

$$\begin{aligned}
a_1' &= a_1 \exp(j\gamma_1 l_1) \\
b_1' &= b_1 \exp(-j\gamma_1 l_1) \\
a_2' &= a_2 \exp(j\gamma_2 l_2) \\
b_2' &= b_2 \exp(-j\gamma_2 l_2)
\end{aligned} \tag{6.4}$$

where

$$\begin{aligned}
l_1 &= R_1' - R_1 \\
l_2 &= R_2' - R_2
\end{aligned} \tag{6.5}$$

The primed quantities are those of the new reference plane (i.e., $l_{1,2}$ are positive when the reference plane is extended outward from the original port). We find directly that

$$S_{11}' = \left.\frac{b_1'}{a_1'}\right|_{a_2' = 0} = \left.\frac{b_1 \exp(-j\gamma_1 l_1)}{a_1 \exp(j\gamma_1 l_1)}\right|_{a_2 = 0} = S_{11} \exp(-2j\gamma_1 l_1) \tag{6.6}$$

and

$$\begin{aligned}
S_{21}' &= \left.\frac{b_2'}{a_1'}\right|_{a_2' = 0} = \left.\frac{b_2 \exp(-j\gamma l_2)}{a_1 \exp(j\gamma l_1)}\right|_{a_2 = 0} \\
&= S_{21} \exp(-j(\gamma_1 l_1 + \gamma_2 l_2))
\end{aligned} \tag{6.7}$$

Similarly,

$$\begin{aligned}
S_{12}' &= S_{12} \exp(-j(\gamma_1 l_1 + \gamma_2 l_2)) \\
S_{22}' &= S_{22} \exp(-2j\gamma_2 l_2)
\end{aligned} \tag{6.8}$$

It would be convenient to express this as a product of matrices, but for S parameters, that is not possible. In the case of T matrices, however, it is simple. The T matrix of the extensions at the input and output are just those of transmission lines. The T matrix of the input line, T_1, is

$$T_1 = \begin{bmatrix} \exp(-j\gamma_1 l_1) & 0 \\ 0 & \exp(j\gamma_1 l_1) \end{bmatrix} \tag{6.9}$$

and the output line, T_2, is

$$T_2 = \begin{bmatrix} \exp(-j\gamma_2 l_2) & 0 \\ 0 & \exp(j\gamma_2 l_2) \end{bmatrix} \tag{6.10}$$

The T matrix with the extensions is given by pre- and postmultiplying the T matrix by these:

$$\begin{aligned} T' &= T_1 T T_2 \\ &= \begin{bmatrix} T_{11}\exp(-j(\gamma_1 l_1 + \gamma_2 l_2)) & T_{12}\exp(-j(\gamma_1 l_1 - \gamma_2 l_2)) \\ T_{21}\exp(j(\gamma_1 l_1 - \gamma_2 l_2)) & T_{22}\exp(j(\gamma_1 l_1 + \gamma_2 l_2)) \end{bmatrix} \end{aligned} \tag{6.11}$$

If the input and output lines are ideal, $\gamma_1 l_1 = \theta_1$ and $\gamma_2 l_2 = \theta_2$, where $\theta_{1,2}$ are the reference plane extensions in degrees of phase. Shifting either reference plane in the opposite direction simply requires changing the sign of $l_{1,2}$ or $\theta_{1,2}$.

6.1.5 De-Embedding by Negative Images

Occasionally it is necessary to remove the effects of elements at the input or output of a two-port. Often these are device parasitics that are not desired in the two-port's characterization, or must be removed as part of a modeling process. For example, the input structure of a discrete chip transistor measured in a test fixture includes several series and parallel reactances. Such a structure is shown in Figure 6.5(a). A model of the input consists of a parallel capacitance associated with the end of the microstrip line and the series inductance of the bond wire. The bond wire also has a small capacitance to ground, which can be modeled by capacitors at each end; the one at the microstrip end is absorbed into the microstrip end capacitance. The resulting model is shown in Figure 6.5(b). This is not a perfect model, but it should be adequate if the frequency is not too high.

The S parameters are measured with the reference plane at the point where the bond wires are connected to the microstrip, and S parameters of the bond wires are calculated by EM simulation or, possibly, measurement. The model in Figure 6.5(b) is determined by fitting it to the latter set of S parameters. We now need to remove it from the measurement of the entire structure.

One method, of course, is to calculate the T matrix of the bond wires, invert it, and multiply the T matrix of the entire structure by it. This is fine in principle, but it can be clumsy to implement in a circuit simulator. An alternative is to cascade the equivalent circuit with its negative image circuit, as shown in Figure 6.6. The entire structure is cascaded with the negative image circuit in the circuit simulator, and the input parasitics are cancelled.

If the input parasitic includes a transmission line, a transmission-line segment with negative length can be used. From (6.9) and (6.10), it should be clear that the inverse of a transmission line's T matrix is simply the T matrix of a negative-length line. Likewise, the T matrix of the negative-image circuit is the inverse of the T matrix of the original circuit.

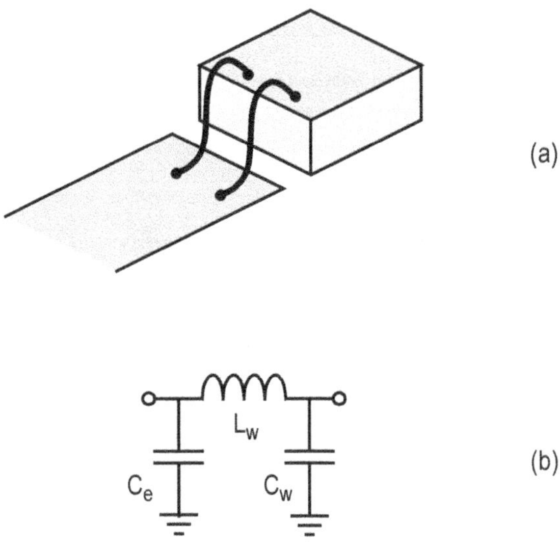

Figure 6.5 A chip connected by bond wires to a microstrip line (a); the equivalent circuit (b). C_e is the microstrip end-effect capacitance, L_w is the inductance of the wire, and C_w is its capacitance. A component of C_w at the microstrip end is absorbed into C_e.

6.2 Some Useful Nonexistent Components

A few kinds of elements are especially useful in high-frequency circuit analysis and modeling, even though they do not exist in a practical form. In some cases, they are useful for circumventing the limitations of circuit simulators; in others, they facilitate analyses that would be difficult or impossible without them. In this section we examine a few of them and show how they can be useful.

6.2.1 Transformer

Although an ideal transformer cannot be realized in high-frequency form, it can be quite useful in modeling circuits and components. A transformer is a floating two-port, shown in Figure 6.7; port 1 is traditionally called the *primary* and port 2 the *secondary* winding. An ideal transformer consists of two windings in which the fields from each winding entirely couple the other; the self-inductances of each winding have large reactances (ideally infinite) relative to the impedances in the circuit and no loss. The requirement for perfect coupling means that

$$M = \sqrt{L_1 L_2} \tag{6.12}$$

where M is the mutual inductance of the coupled coils and L_1, L_2 are their self inductances. An ideal transformer is characterized entirely by its *turns ratio*, n, which we define here as the ratio of the number of turns in the secondary to that of the primary.

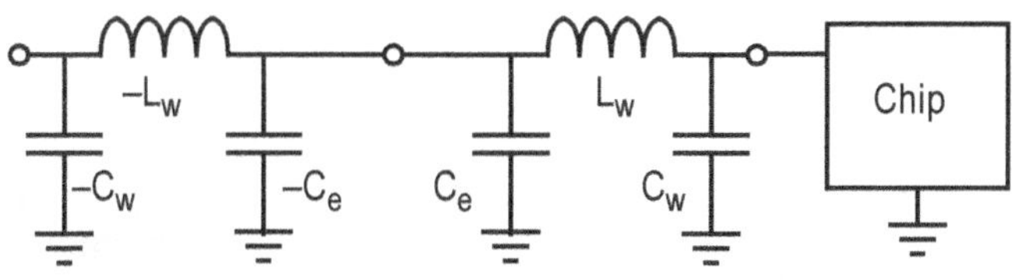

Figure 6.6 Removal of the input parasitics in Figure 6.5 can be accomplished by cascading it with its negative-image circuit, then analyzing the combination.

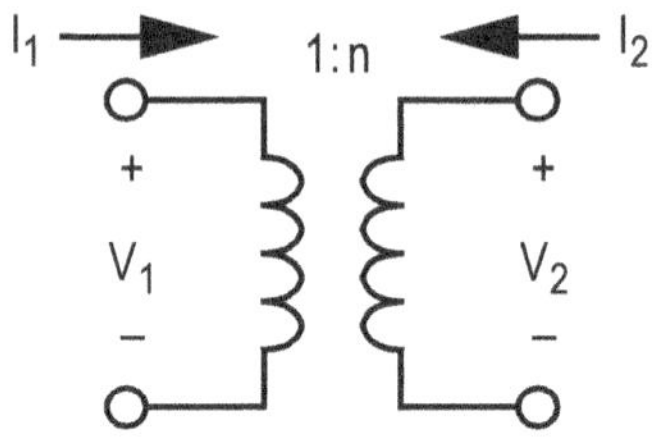

Figure 6.7 An ideal transformer with currents and voltages indicated. The turns ratio is from winding 2 to winding 1.

The voltages and currents in an ideal transformer are related:

$$V_2 = nV_1$$
$$I_2 = \frac{-I_1}{n} \tag{6.13}$$

and it is easy to show that if port 2 is terminated in some impedance Z_L, the input impedance at port 1, Z_{in}, is

$$Z_{\text{in}} = \frac{Z_L}{n^2} \tag{6.14}$$

One obvious use of a transformer is to shift impedance levels in a circuit. This capability was especially useful in the past, when circuit simulators were designed so that all ports used a standard 50Ω impedance, and provisions for changing that impedance were clumsy. Even today, the most common format for transistor S parameters assumes that both ports use the same normalizing impedance. If different normalizing impedances are desired, a transformer can be used to shift the input or output impedance to a different value.

A transformer can also create a two-terminal port when only grounded ones are available in the simulator. One side of the transformer is connected to the circuit nodes that represent the two-terminal port; the other side is connected to the grounded port, with one terminal grounded. This capability must be used with care, however, as it is easy to leave a floating circuit on the ungrounded side; then the nodal matrix is singular. This problem was discussed in Section 5.2.

Figure 6.8 shows a tapped transformer. In this case, the secondary is divided into two parts with a connection between them. As a result, the transformer has two turns ratios, from the primary winding to each of the two secondaries. Then,

$$\begin{aligned} V_2 &= n_2 V_1 \\ V_3 &= n_3 V_1 \\ I_1 &= -(I_2 n_2 + I_3 n_3) \end{aligned} \tag{6.15}$$

With loads Z_{L2} and Z_{L3} on ports 2 and 3, respectively, we easily obtain

$$Z_{in} = \frac{1}{\dfrac{Z_{L2}}{n_2^2} + \dfrac{Z_{L3}}{n_3^2}} \tag{6.16}$$

which is just the parallel combination of the input impedances from the individual secondaries. Equations (6.15) and (6.16) indicate that the tapped transformer is equivalent to two simple transformers, in which the primaries are in parallel and the secondaries are in series.

A tapped transformer can separate odd and even modes. This capability is especially useful for testing baluns and other imperfect even/odd mode structures. Figure 6.9 shows how this can be accomplished. The circuit in the block is a parallel strip balun, which we found in Section 3.2.4 to be notoriously mediocre. The balun is one-quarter wavelength long at 10 GHz

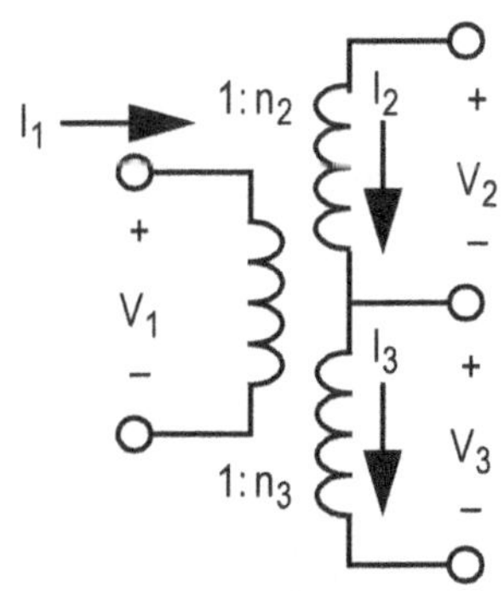

Figure 6.8 An ideal tapped transformer is a useful component for circuit modeling. Usually it is most useful when $n_2 = n_3$, but we do not limit our consideration to that case alone.

and has $Z_{0e} = 250$ and $Z_{0o} = 25\Omega$. It is therefore intended for a 50Ω load across its output terminals, or 25Ω from each terminal to ground. The transformer's turns ratios are therefore chosen to present 25Ω to each side of the balun's output.

The transformer couples the odd mode from the balun to port 1 of the circuit and the even mode to port 2. The value of the impedance at port 2 is more or less arbitrary; it is probably best chosen to represent the even-mode input impedance of the circuit that terminates the balun. Then, the loss can be used to estimate the effect of the even mode in the circuit; for example, in a balanced mixer it could be used to determine one or more

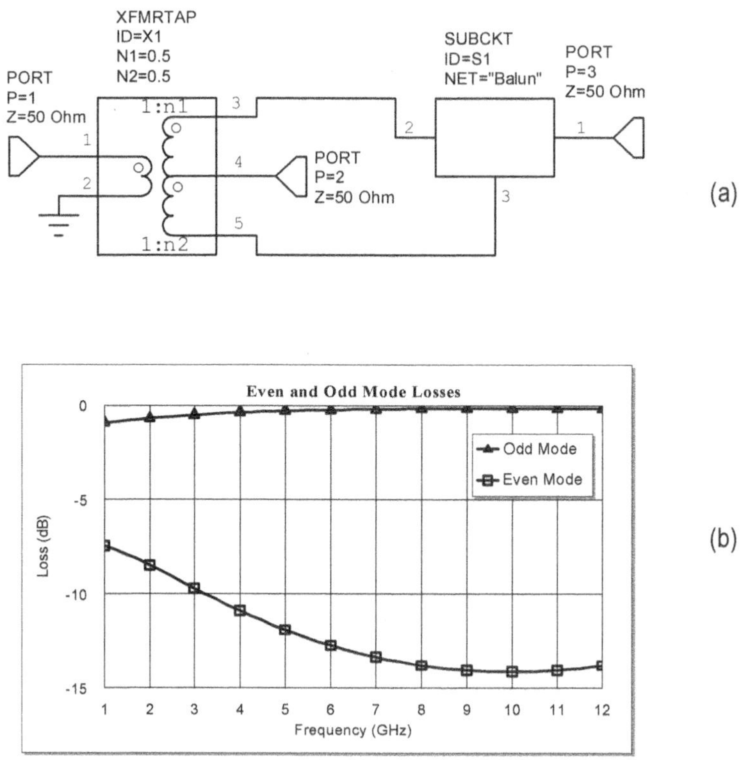

Figure 6.9 A transformer can be used to separate odd and even modes in a circuit (a). It is especially useful for evaluating baluns, as shown here. The balun is excited at port 3; the even mode is sensed at port 2 and the odd at port 1. The even- and odd-mode losses are shown in a plot (b).

port isolations. It could also be set to 125Ω, half the even-mode impedance (since the transformer effectively connects the two even-mode outputs in parallel) and one could obtain the levels of the even and odd modes in an optimally terminated balun.

In Section 3.2.5 and Figure 3.5, we saw that a transformer could model an ideal 180-degree hybrid when $n_2 = n_3 = 1/\sqrt{2}$. We noted that the port connected to the center tap, port 3 in the figure, has an impedance half that of the other three, which are equal. This port could be converted to the impedance of the other three through the use of a second transformer, of course.

6.2.2 Gyrator

A *gyrator* is a lossless, passive, nonreciprocal two-port. While not a practical component, a gyrator is useful for modeling various kinds of microwave circuit elements.

A gyrator is described by the Y matrix

$$Y = \begin{bmatrix} 0 & G \\ -G & 0 \end{bmatrix} \tag{6.17}$$

or the Z matrix

$$Z = Y^{-1} = \begin{bmatrix} 0 & -R \\ R & 0 \end{bmatrix} \tag{6.18}$$

$R = 1/G$ is called the *gyrational resistance*. If both ports are normalized to R, the S matrix is

$$S = \begin{bmatrix} 0 & -1 \\ 1 & 0 \end{bmatrix} \tag{6.19}$$

The symbol for a gyrator is shown in Figure 6.10.

A gyrator reverses voltage and current at its ports; that is, the voltage at port 2 equals the current at port 1, and the current at port 2 equals the negative of the voltage at port 1. The constant of proportionality is R or G, as appropriate. The input impedance of a gyrator, Z_{in}, terminated in some impedance Z_L, is easily shown to be

$$Z_{in} = \frac{R^2}{Z_L} \tag{6.20}$$

This shows that a gyrator inverts impedance; a capacitance at one port, for example, generates an inductive reactance at its opposite port.

A gyrator is not, intrinsically, a useful microwave element; gyrators are most useful for modeling many kinds of components, including ones that have no Y matrix representation and therefore cannot be entered directly into a nodal matrix. In other cases, their use is not essential, but it results in better matrix conditioning than other kinds of elements. This is especially true of their use as current sensors in nonlinear analysis.

6.2.2.1 Transformers Modeled by Gyrators

An ideal transformer has no Y matrix representation (although it has an S matrix), so it cannot be used directly in nodal analysis. A simple, ideal transformer can be realized by cascaded gyrators, however, and that representation can be entered into a nodal matrix. Notwithstanding a couple of caveats, the representation is well conditioned.

Figure 6.11 shows a simple, ideal transformer consisting of two cascaded gyrators. From (6.18),

$$V_1 = -R_1 I_2 \tag{6.21}$$

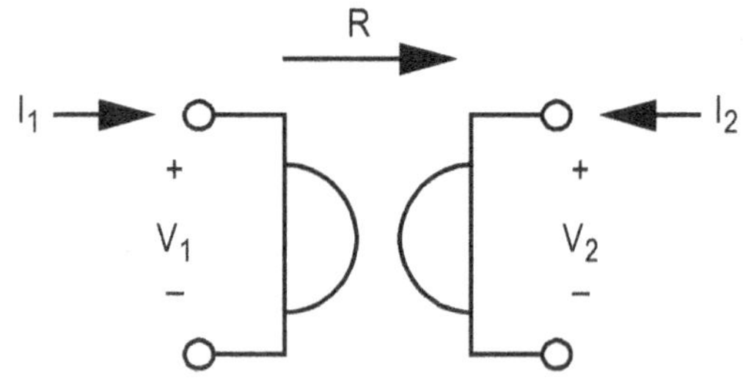

Figure 6.10 Symbol for a gyrator. R is the gyrational resistance; its direction is clearly essential, as the component is nonreciprocal.

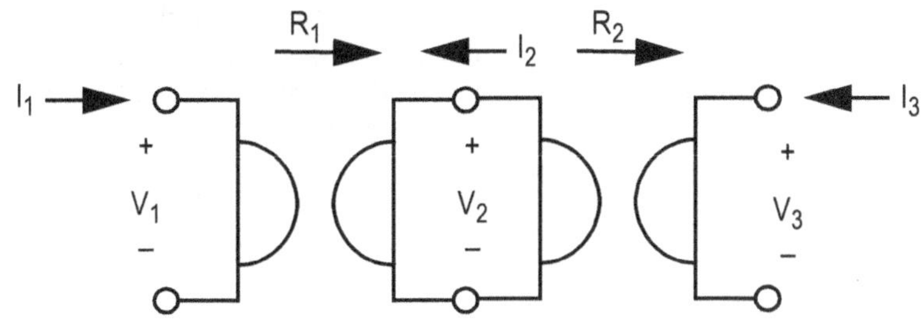

Figure 6.11 Gyrator realization of a transformer. The turns ratio, as defined in Figure 6.7, is $n = R_2 / R_1$. The absolute values of R_1 and R_2 are not important, as long as their ratio is correct and, for good matrix conditioning in nodal analysis, extreme values are avoided.

$$V_2 = R_1 I_1 \tag{6.22}$$

$$V_2 = -R_2 I_3 \tag{6.23}$$

$$V_3 = -R_2 I_2 \tag{6.24}$$

From (6.21) and (6.24),

$$V_3 = \frac{R_2}{R_1} V_1 \tag{6.25}$$

and from (6.22) and (6.23),

$$I_3 = \frac{-R_1}{R_2} I_1 \tag{6.26}$$

showing that the gyrator cascade behaves as a transformer with

$$n = \frac{R_2}{R_1} \tag{6.27}$$

We showed earlier that a tapped transformer can be modeled as two transformers with the primaries in parallel and the secondaries in series. That transformer can be modeled by gyrators in the same way. A simpler model, however, using one less gyrator, is shown in Figure 6.12. In that realization,

$$n_2 = \frac{R_2}{R_1}$$
$$n_3 = \frac{R_3}{R_1} \tag{6.28}$$

Two important cautions must be observed in using these models. First, the important quantities are the ratios in (6.27) and (6.28), not the absolute values of $R_{1,2,3}$. Nevertheless, extreme values should be avoided or matrix conditioning may suffer. Second, it is extremely easy to create isolated nodes (nodes with no direct path to ground) in those models, which would cause the nodal matrix to be singular. This is easy to avoid; a high-value resistor from an isolated node to ground is usually sufficient.

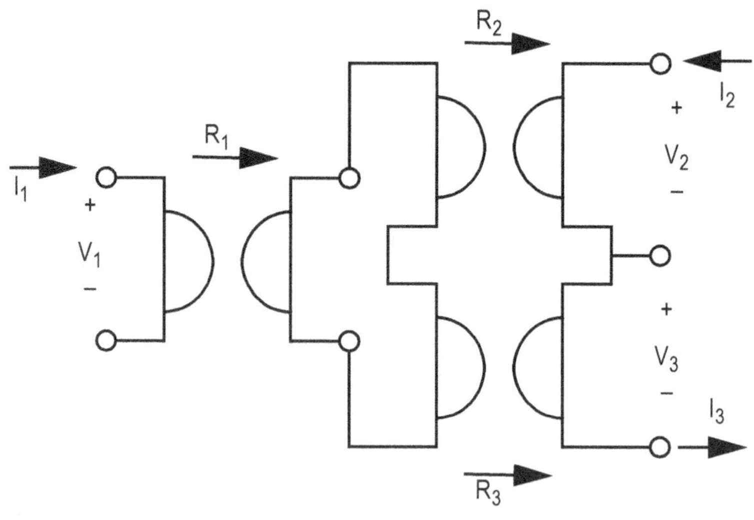

Figure 6.12 Gyrator model of an ideal tapped transformer.

6.2.2.2 Circulator Model

In Section 3.2.3, we noted that a circulator can be modeled by a gyrator as long as the ports have the same normalizing impedances (which is invariably the case in practical circulators), and the gyrational resistance is equal to that normalizing impedance. Figure 3.3 shows how this is done. Since an isolator is simply a circulator with a terminated port, the gyrator realization can be used similarly to model an isolator.

With our understanding of the nodal (indefinite admittance) matrix from Chapter 5, we can show that the gyrator circuit of Figure 3.3 is equivalent to a circulator. We begin by treating (6.17) as a representation of a port matrix with grounded negative nodes at each port, then "float" the grounded node, creating an indefinite admittance matrix. This is accomplished by adding a third row and column so that all rows and columns sum to zero. The result is

$$Y = \begin{bmatrix} 0 & G & -G \\ -G & 0 & G \\ G & -G & 0 \end{bmatrix} \tag{6.29}$$

Normalizing this to $Z_0 = 1 / G$ and substituting into (3.23) gives

$$\begin{aligned} S &= \begin{bmatrix} 1 & 1 & -1 \\ -1 & 1 & 1 \\ 1 & -1 & 1 \end{bmatrix}^{-1} \begin{bmatrix} 1 & -1 & 1 \\ 1 & 1 & -1 \\ -1 & 1 & 1 \end{bmatrix} = \frac{1}{2} \begin{bmatrix} 1 & 0 & 1 \\ 1 & 1 & 0 \\ 0 & 1 & 1 \end{bmatrix} \begin{bmatrix} 1 & -1 & 1 \\ 1 & 1 & -1 \\ -1 & 1 & 1 \end{bmatrix} \\ &= \begin{bmatrix} 0 & 0 & 1 \\ 1 & 0 & 0 \\ 0 & 1 & 0 \end{bmatrix} \end{aligned} \tag{6.30}$$

which is the S matrix of the circulator in Figure 3.2.

6.2.2.3 Current Sensor

In nonlinear circuit analysis, it is sometimes necessary to use current as an independent variable. This is sometimes accomplished by using a low-value resistor to sense current. Unfortunately, since current in high-frequency circuits is often small (say, on the order of microamps) and the low-value resistor must also be small (perhaps $10^{-6}\,\Omega$), the resulting voltage is also quite small. This leads to numerical problems in the simulation.

Two better ways to accommodate current variables are (1) to use modified nodal analysis [6.6], and (2) to convert currents to voltages using a gyrator, as in Figure 6.13. Using a gyrator is simpler if the number of such variables is not large. The gyrational resistance can be selected to create a scaled current, which has the same approximate numerical magnitude as the voltages in the circuit, allowing good matrix conditioning.

If port 2 of the gyrator, where the voltage is sensed, is open circuited, port 1 has zero input impedance, so it does not affect the circuit. In practice, the need to prevent floating nodes necessitates at least one resistor between a node of port 2 and ground, or elsewhere in the circuit, but still the effect on the impedance at port 1 is minimal.

6.2.2.4 Controlled Sources

A voltage-controlled current source (VCCS) is the only controlled source that can be entered directly into a nodal matrix. Other types of controlled sources can be realized by connecting gyrators to the input, output, or both of a VCCS. The possibilities are shown in Figure 6.14 along with expressions for their transfer quantities.

6.3 Some Problematical Circuit Elements

Many circuit elements are difficult to model at high frequencies, yet they must be used in practical circuits. Modeling these often can be successful,

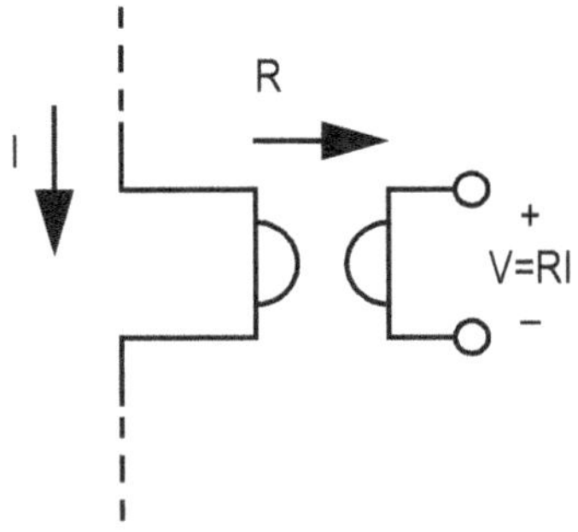

Figure 6.13 A gyrator can be used as a current sensor. This is especially helpful in nonlinear circuit analysis.

however, as long as reasonable precautions are taken, based on an understanding of the places where the difficulties are likely to be encountered and a clear understanding of the circumstances where accuracy is needed and where it isn't. With these items in mind, one can model such components with success.

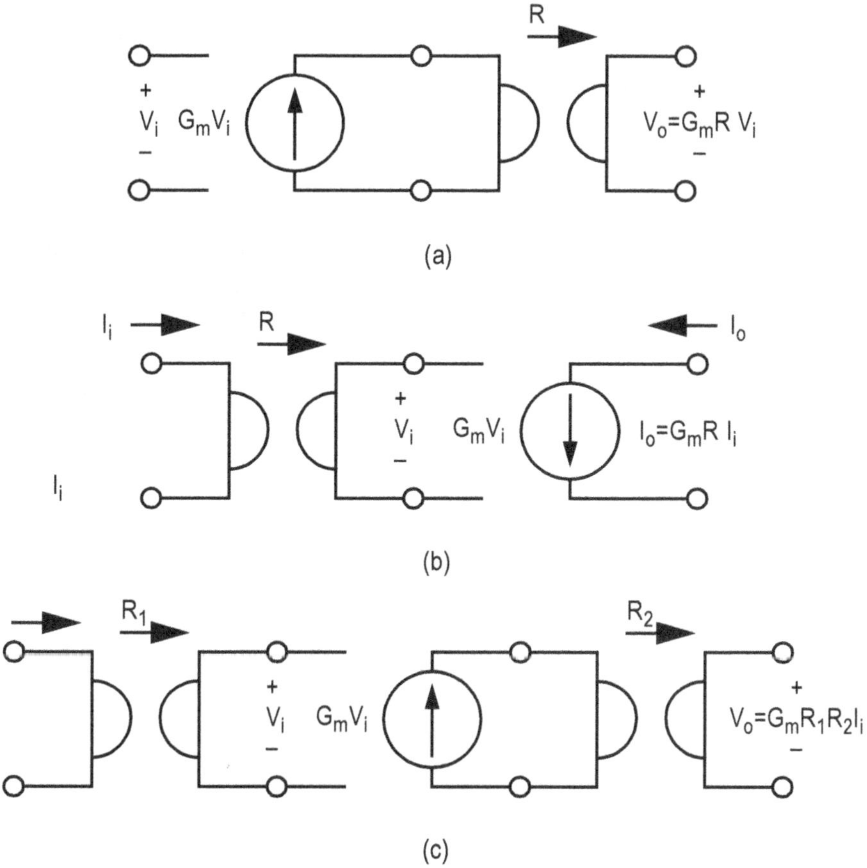

Figure 6.14 Controlled source realizations using gyrators: (a) voltage-controlled voltage source; (b) current-controlled current source; (c) current-controlled voltage source. The user of these equivalents must be careful to avoid the creation of floating nodes.

6.3.1 Bond Wires

A bond wire over a ground plane has series inductance and a small distributed shunt capacitance. Distributed structures often can be modeled by lumped-element circuits, as long as they are short relative to a wavelength. A bond wire usually meets this criterion, so it can be modeled by a tee or pi circuit, as shown in Figure 6.15. The pi circuit is probably most common, as it does not break the wire inductance into two parts. The tee circuit may be more useful in cases where the pi circuit cannot be used; for example, where elements at the ends of the wire short-circuit the capacitances.

The problem, as always, is to determine the values of the equivalent-circuit elements. Even when they are known, however, bond wires, in fabrication, are subject to large variations in their lengths and heights. In an environment where fabrication is not uniform, one might well question the need for an accurate model. Thus, good approximations are probably adequate.

Empirical results show that an arched bond wire can be modeled with reasonable accuracy by a straight conductor plus vertical connections as long as the total length of the wire in the model is the same as that of the real bond wire. Figure 6.16 shows an EM simulation of a pair of bond wires modeled by straight sections and a comparison to a pi model of the wires. The pi model is obtained as in Section 6.1.2.2 by fitting the values of the inductance and capacitance to plots of $1 / Y_{11}$. Explicit fitting to Y_{21} is not needed, as the effects of capacitance and inductance are easily separated in

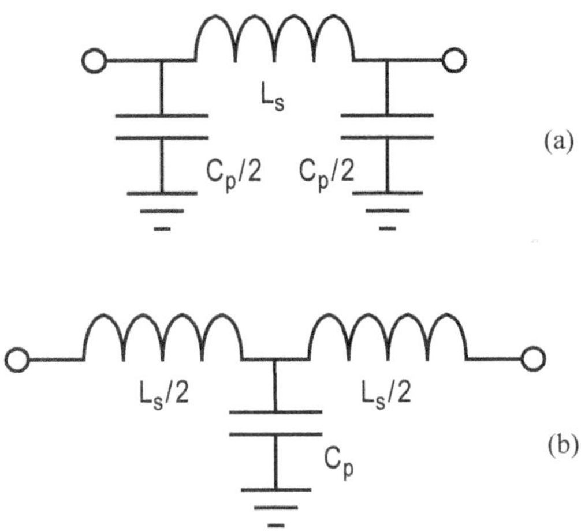

Figure 6.15 Two bond-wire models: (a) pi model; (b) tee model.

Y_{11}, and fitting to $1 / Y_{22}$ is likewise unnecessary, as the structure is symmetrical. It is clear that the simple pi model agrees quite well with the EM results.

In the EM simulation, the reference planes are located at the outer edges of the vertical posts. As such, the model includes the end-effect capacitances of the lines, and it is likely that most of the capacitance in the pi is actually from those microstrip ends. As explained earlier in this chapter, the model includes the parasitics of the interconnection, as well as the wire inductance and capacitance itself. The entire structure between the quasi-

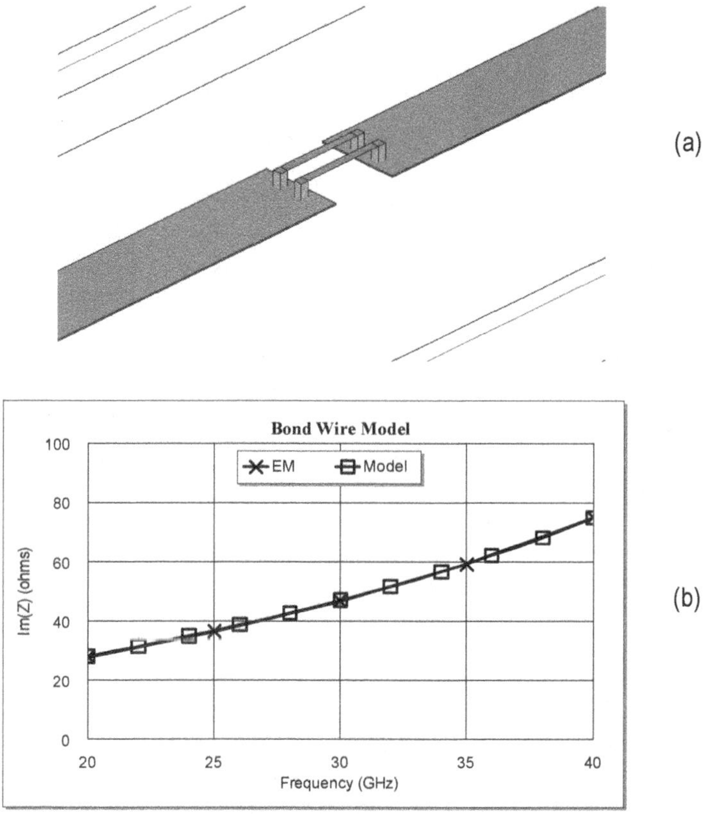

Figure 6.16 A pair of bond wires (a) and their EM simulation (b) compared to a pi circuit model, 20-40 GHz. The bond wires are 250 μm long and 50 μm high. $L_s = 0.205$ nH, $C_p / 2 = 0.024$ pF. Most of the capacitance is probably from the microstrip ends, not the wire itself.

TEM transmission lines is characterized; in circuit design, this is usually what we need.

It is important to recognize that the closeness of the two bond wires allows their magnetic fields to couple. As a result, the inductance of the two wires is greater than half that of a single wire in isolation. One cannot assume that multiple bond wires behave like simple, isolated inductors in parallel.

6.3.2 Bond Wires to Chips

A bond wire is an inductive parasitic, so minimizing its length is often necessary. For this reason, chips are preferentially mounted with their surfaces level with the top of the substrate. Occasionally, however, that is not possible and the chip must be mounted on the surface of the substrate, so the wire ends are at two different levels. Since planar EM simulators usually do not allow for truncated dielectrics, this kind of structure cannot be simulated as easily as that of Figure 6.16.

One obvious option is simply to use a 3D EM simulator, although that may entail higher computational cost than can be justified for the needed accuracy. A second is to use a planar EM simulator with an approximate model, as shown in Figure 6.17(a), in which a pair of bond wires is modeled by an L-shaped combination of strips and posts. The EM simulation uses two layers, the substrate and an air layer above it equal to the chip thickness; the terminal at the chip end is on the top of the air layer and the opposite end is on the substrate. This does not model the discontinuity at the chip end perfectly, but it should result in a good estimate of the bond-wire inductance as long as the length of the strips in the EM model equals that of the real wires. Because it is asymmetrical, a circuit model, Figure 6.17(b), has unequal capacitances at the ends so both Y_{11} and Y_{22} must be used. Figure 6.17 shows the agreement between the circuit and EM models, which is quite good up to at least 40 GHz.

6.3.3 Cell Interconnections in Large Devices

Small transistors consist of a single cell or occasionally a small number of cells interconnected in a simple manner. Large devices are often realized as parallel interconnections of those smaller devices or cells. It is common, in models provided by integrated-circuit foundries, to model those large devices by using a model of the smaller devices and interconnecting them with standard microstrip elements, such as step and tee junctions.

From our previous discussions, it should be clear that such models are not likely to be accurate, as the microstrip discontinuities are too close to-

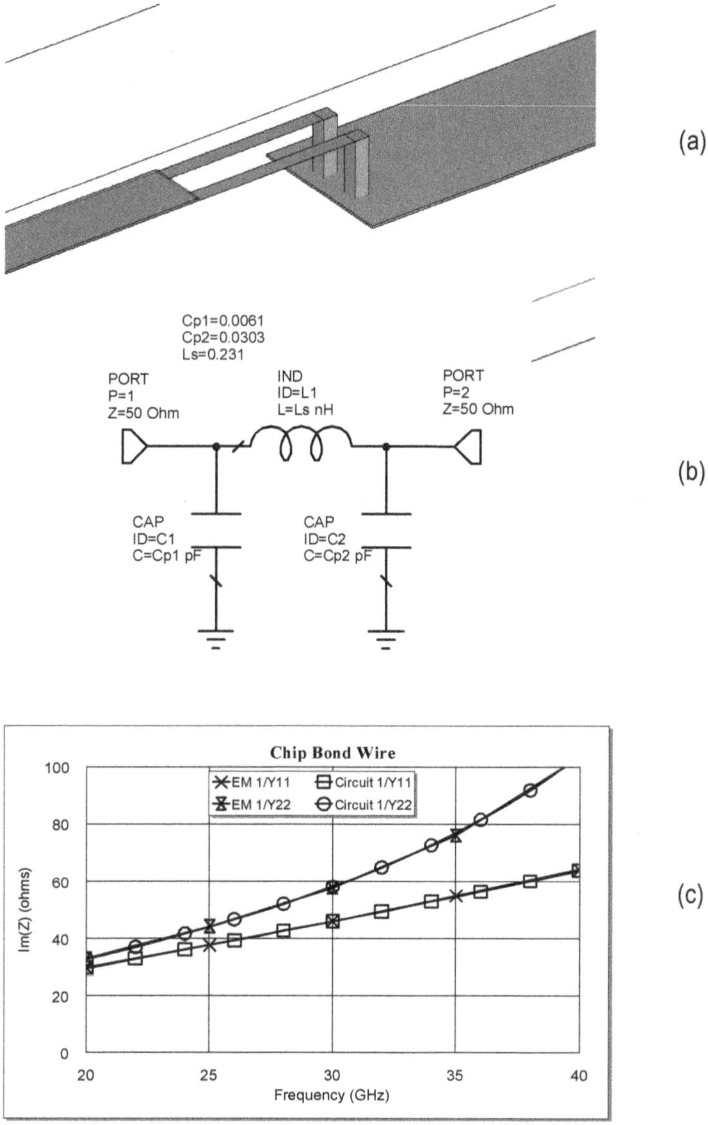

Figure 6.17 A pair of chip bond wires is modeled by posts and horizontal strips (a). The post height is 100 μm and the strips are 275 μm long. The circuit model is shown in (b), and a comparison of the circuit and EM results is shown in (c). Port 2 is the microstrip end.

gether and their fields are certain to couple. Additionally, treating the large device as a set of smaller, individually modeled devices ensures that even a simple circuit has a high computational cost; this is especially true in power devices, which are likely to be quite large and require nonlinear analysis. The only good reason for treating the cells or small devices as individuals is to account for differences in device temperatures, voltages, and currents, which are likely to be small in a well-designed power transistor. Thus, the justification for this complexity is weak.

It is usually best to treat the large power device as a single, scaled version of the smaller devices, and to model the interconnections at the input and output as single two-ports. Analyzing the interconnections with a planar EM simulator, and expressing them as two-ports, is usually a relatively simple problem. If the device must be modeled as an interconnection of cells, the only accurate way to characterize the cell interconnections is by EM simulation.

6.3.4 Housing Effects

The housing in which a high-frequency circuit is mounted can have a strong effect on its operation. Unfortunately, however, the design of the housing and the location of the circuit within it are often unknown at the outset of the design process, so it can be difficult, at design time, to avoid housing-related difficulties.

A rectangular metal housing is, in effect, a kind of waveguide resonator that is lightly loaded by a dielectric (the substrate). The circuit's stray fields couple to this resonator in a largely unpredictable manner, and as a result, resonances sometimes appear in the circuit's passband. The most effective way to avoid such resonances is to make the housing small enough so that all resonances occur outside the circuit's frequency range.

A closed, rectangular resonator can be viewed as a truncated rectangular waveguide; the lowest-frequency resonance requires the dominant TE_{10} mode and a housing length equal to one-half guide wavelength. The guide wavelength, λ_g, is

$$\lambda_g = \frac{\lambda_0}{\sqrt{1 - \left(\frac{\lambda_0}{2a}\right)^2}} \tag{6.31}$$

where λ_0 is the free-space wavelength and a is the width of the housing. We assume that its height is less than $\lambda_0 / 2$. A resonance occurs when its length, d, is $\lambda_g / 2$. Solving this for the resonant frequency gives

$$f = \frac{v_c \sqrt{a^2 + d^2}}{2ad} \tag{6.32}$$

where v_c is the free-space velocity of light. The loading by the dielectric substrate reduces the resonant frequency, but because the loading is light, the reduction is usually only a few percent.

Keeping the height of the housing's interior and one horizontal dimension less than $\lambda_0 / 2$ is the best way to minimize radiation and unwanted resonances. When that is not possible, loading the housing's interior with RF-absorptive material is usually effective in minimizing resonances.

Even when the housing is small enough to avoid waveguide-like resonances, it can affect the circuit's performance. Microstrip circuits do not contain their fields perfectly, and those fields can couple to adjacent structures. Two phenomena are important to keep in mind. The first is surface waves, which are waves guided by the discontinuity between the substrate and the air above it. The second is simply radiation from the circuit itself. Surface waves occur when the substrate has a high dielectric constant (e.g., alumina or GaAs) and its thickness is an appreciable fraction of a wavelength. Radiation is more of a problem when the dielectric constant is low, as in most inexpensive composite substrate materials. The best way to minimize radiation and surface wave generation is to minimize the substrate thickness. For example, as a general rule, alumina substrates in the common 635-μm thickness probably should not be used above 18 GHz; for higher frequencies, a 250-μm substrate is likely to be more satisfactory.

6.3.5 Transmission-Line Loss

Conventional, closed-form equations for most types of transmission lines do not model loss well. EM simulators are much better, of course, but one cannot use an EM simulator for every microstrip line in a circuit. Even with accurate loss models, however, the characterization of the metal conductivity remains a problem. Textbook values of metal conductivity are reliable at dc or low frequencies, but at high frequencies, where skin effect dominates, they are not as reliable. The metallizations in high-frequency circuits are microscopically rough, and the grain size of the metals is an appreciable fraction of a skin depth, so the effective conductivity is generally greater than the textbook value. In microstrip models, for example, the author usually sets the conductivity to at least double its textbook value; often that is not high enough. Loss measurements can be used to calibrate this value.

6.3.6 Thick Metal in EM Simulations

Originally, planar EM simulators based on moment methods required zero-thickness metallization. Many modern planar EM simulators allow for nonzero metal thickness. In single strips, a zero-thickness assumption is often acceptable in all but high-precision applications, such as filters. In coupled lines, it often is not acceptable. This is especially the case in integrated circuits, where metal thickness is not small relative to the gap width.

Two zero-thickness lines, one above the other, separated by an amount equal to the metal thickness, approximate a nonzero line surprisingly well. The lines should be pinned together with via posts at appropriate intervals; it is not necessary to fill the entire gap with via material. This model is accurate for coupled-line analysis when the line spacing is on the order of the metal thickness.

See Section 2.2.4 for more information on the problem of coupled lines and thick metal.

6.3.7 Poorly Modeled Circuit Elements

Some circuit elements are inherently much easier to model accurately than others. The closed-form microstrip models used in today's circuit simulators are accurate to a few percent, even for rather extreme dimensions and dielectric constants, and are almost perfect for conventional substrates and ordinary line dimensions. Other kinds of components often are not well modeled by closed-form expressions, and even when well modeled, have a narrow range of validity. Among these are such elements as microstrip radial stubs and circular resonators. The analysis of those elements is a complex electromagnetic problem, one that does not lend itself easily to simplification. While models of such components may be adequate in limited ranges of frequency, dimensions, and dielectric constants, it is unlikely that they will ever have good, wide-range closed-form models.

The best solution to this dilemma is simply to avoid such elements. Limiting a circuit design to accurately modeled circuit elements is often viewed as an undesirable constraint, but probably no worse than such ubiquitous constraints as size, weight, and dc power. When those elements cannot be avoided, EM simulation is an option, as long as the circuit does not contain too many of them. In any case, their use involves higher costs, either in design time or in troubleshooting and tuning after the circuit is constructed.

References

[6.1] Rautio, J. C., "Synthesis of Compact Lumped Models from Electromagnetic Analysis Results," *IEEE Trans. Microwave Theory Tech.*, Vol. MTT-55, 2007, p. 2548.

[6.2] Beyene, W. T., and J. E. Schutt-Aine, "Transient Analysis of Diode Switching Circuits Using Asymptotic Waveform Evaluation," *IEEE Trans. Computer-Aided Design of Circuits and Systems*, Vol. 16, 1997, p. 1447.

[6.3] Kuznetsov, D. B., and J. E. Schutt-Aine, "Optimal Transient Simulation of Transmission Lines," *IEEE Trans. Circuits Syst.,* Part 1, Vol. 43, 1996, p. 110.

[6.4] Kim, T., X. Li, and D. J. Allstot, "Compact Model Generation for On-Chip Transmission Lines," *IEEE Trans. Circuits Syst.*, Part 1, Vol. 51, 2004, p. 459.

[6.5] Brazil, T. J., "Accurate and Efficient Incorporation of Frequency-Domain Data within Linear and Non-linear Time-Domain Transient Simulation," *IEEE MTT-S International Microwave Symposium Digest*, 2005.

[6.6] Vlach, J., and K. Singhal, *Computer Methods for Circuit Analysis and Design*, New York: Chapman and Hall, 1994.

Chapter 7

Active Two-Ports

By *active two-ports* we mean, primarily, amplifiers. In previous chapters, especially Chapters 3 and 5, we developed a number of important expressions for gain, reflection coefficients, and stability, and we developed analytical techniques for dealing with multiports described by S matrices. In this chapter we consolidate those, derive a few more useful quantities, and, especially, develop the underlying logic for amplifier design.

7.1 Amplifier Theory

7.1.1 Summary of Previous Results

In the previous chapters, we derived a number of useful relations for input/output reflection coefficients, gain, and stability. In this chapter, we make frequent use of them, so it is helpful to summarize them here.

7.1.1.1 Gain

In Section 3.3.2 we saw that an amplifier's gain, in terms of power, can be defined three ways:

1. *Transducer gain*: When we loosely speak of gain, we usually mean transducer gain. Transducer gain, G_t, is defined as

$$G_t = \frac{P_d}{P_{av}} \tag{7.1}$$

where P_d is the power delivered to the load and P_{av} is the available power from the source. Specifically,

$$G_t = \frac{|S_{21}|^2(1 - |\Gamma_s|^2)(1 - |\Gamma_L|^2)}{|(1 - S_{11}\Gamma_s)(1 - S_{22}\Gamma_L) - S_{12}S_{21}\Gamma_s\Gamma_L|^2} \tag{7.2}$$

where Γ_s and Γ_L are the source and load reflection coefficients, respectively.

2. *Power gain*: Power gain is defined as

$$G_p = \frac{P_d}{P_{in}} \tag{7.3}$$

where P_{in} is the input power; that is, the power dissipated in port 1 of the two-port. An expression for the power gain of a terminated two-port is

$$G_p = \frac{|S_{21}|^2(1 - |\Gamma_L|^2)}{|1 - S_{22}\Gamma_L|^2 - |S_{11} - (S_{11}S_{22} - S_{12}S_{21})\Gamma_L|^2} \tag{7.4}$$

which can also be expressed as

$$G_p = \frac{|S_{21}|^2(1 - |\Gamma_L|^2)}{(1 - |\Gamma_{\text{in}}|^2)|1 - S_{22}\Gamma_L|^2} \tag{7.5}$$

where Γ_{in} is the input reflection coefficient. The source reflection coefficient, Γ_s, does not appear in the expression, as its definition implies that the input port is conjugate matched.

3. *Available gain*: Available gain, G_a, is

$$G_a = \frac{P_{av2}}{P_{av}} \tag{7.6}$$

where P_{av2} is the available power from the output port, port 2 of the two-port. It is given by

$$G_a = \frac{|S_{21}|^2(1 - |\Gamma_s|^2)}{|1 - S_{11}\Gamma_s|^2 - |S_{22} - (S_{11}S_{22} - S_{12}S_{21})\Gamma_s|^2} \quad (7.7)$$

or

$$G_a = \frac{|S_{21}|^2(1 - |\Gamma_s|^2)}{(1 - |\Gamma_{\text{out}}|^2)|1 - S_{11}\Gamma_s|^2} \quad (7.8)$$

where Γ_{out} is the output reflection coefficient. This is the dual of power gain, as it implies a conjugate-matched output.

7.1.1.2 Input and Output Reflection Coefficients

The input and output reflection coefficients of a two-port terminated at its input or output, respectively, were derived algebraically in Section 3.3.1.2 and with network graphs in Section 5.1.4.1. We obtained

$$\Gamma_{\text{in}} = S_{11} + \frac{S_{12}S_{21}\Gamma_L}{1 - S_{22}\Gamma_L} \quad (7.9)$$

and

$$\Gamma_{\text{out}} = S_{22} + \frac{S_{12}S_{21}\Gamma_s}{1 - S_{11}\Gamma_s} \quad (7.10)$$

7.1.1.3 External Stability

In amplifier design, we are usually concerned with *external* stability; that is, stability as a function of the source and load terminations. Section 3.4.1 contains a more complete discussion of stability, which is worthwhile to review. We showed that the terminated two-port is stable, in the sense that neither the input nor the output impedance has a negative real part, for any passive termination, when

$$K = \frac{1 - |S_{11}|^2 - |S_{22}|^2 + |\Delta|^2}{2|S_{21}S_{12}|} > 1 \quad (7.11)$$

and

$$|\Delta| = |S_{11}S_{22} - S_{21}S_{12}| < 1 \tag{7.12}$$

Δ is the determinant of the S matrix. Most well-designed small-signal transistors meet the condition (7.12), so it is often ignored and only (7.11) is evaluated. Unfortunately, many conditions can cause (7.12) to be violated, so in all cases it is wise to evaluate both. Equations (7.11) and (7.12) are sometimes called the *K-factor* criterion and when they are met, the device or circuit is said to be *unconditionally stable*. When they are not met, the device or circuit is *conditionally stable*. It is worth noting that K is not changed by reactive matching, nor is the condition in (7.12). Thus, as long as there are no lossy elements in the matching circuits, a stability evaluation of a solid-state device is valid for the amplifier in which it is used.

A second approach is to find the region of Γ_L for which $|\Gamma_{in}| < 1$. We do this by finding the locus of Γ_L values for which $|\Gamma_{in}| = 1$; the result is a circle, as circles in Γ_L map into circles in $|\Gamma_{in}|$. We obtain

$$\Gamma_L = \frac{(S_{22} - \Delta S_{11}^*)^*}{|S_{22}|^2 - |\Delta|^2} + \frac{|S_{21}S_{12}| \exp(j\theta)}{|S_{22}|^2 - |\Delta|^2} \tag{7.13}$$

where θ varies from 0 to 2π. This is called an *output stability circle*. Similarly, we can find the region of Γ_s for which $|\Gamma_{out}| < 1$. This is called an *input stability circle*; symmetry considerations immediately provide the result

$$\Gamma_s = \frac{(S_{11} - \Delta S_{22}^*)^*}{|S_{11}|^2 - |\Delta|^2} + \frac{|S_{21}S_{12}| \exp(j\theta)}{|S_{11}|^2 - |\Delta|^2} \tag{7.14}$$

and the radius and center of the circle can be found by inspection.

The location of the *stable regions*, the region of Γ_L that maps into $|\Gamma_{in}| < 1$ and the region of Γ_s for which $|\Gamma_{out}| < 1$, still must be determined, as the stable region could be either the inside or the outside of the circle. The simplest way is to select a point in the Γ_L or Γ_s plane and determine the resulting $|\Gamma_{in}|$ or $|\Gamma_{out}|$, respectively. An obvious choice is $\Gamma_s = 0$ or $\Gamma_L = 0$, as these result in $\Gamma_{out} = S_{22}$ and $\Gamma_{in} = S_{11}$, respectively. Then, if $|S_{22}| < 1$, $\Gamma_s = 0$ is in the stable region; similarly, if $|S_{11}| < 1$, $\Gamma_L = 0$ is in the stable region.

7.1.2 Gain Circles

Loci of values in the Γ_L or Γ_s plane that provide a specific value of gain are useful for amplifier design. For the gain to be a function of only one of these quantities, we must make some kind of assumption about the termination on the opposite port. Assuming that the opposite port is conjugate matched is the simplest and most useful; then, loci of Γ_L provide values of power gain and those of Γ_s provide available gain. By now, it should come as no surprise that those loci, for a specific value of gain, are circles in their respective reflection-coefficient planes.

Equations for constant-gain circles are found in a straightforward but algebraically tedious manner, in which (7.4) (power gain) or (7.7) (available gain) are equated to a constant and are solved for Γ_L or Γ_s, respectively. We find the radius of the power-gain circle, R_p, to be

$$R_p = \frac{(1 - 2Kg_p|S_{12}S_{21}| + g_p^2|S_{12}S_{21}|^2)^{1/2}}{\left|1 + g_p(|S_{22}|^2 - |\Delta|^2)\right|} \tag{7.15}$$

and its center, C_p, to be

$$C_p = \frac{g_p C_2^*}{1 + g_p(|S_{22}|^2 - |\Delta|^2)} \tag{7.16}$$

where Δ is again the determinant of the S matrix and

$$g_p = \frac{G_p}{|S_{21}|^2} \tag{7.17}$$

G_p is the value of power gain for which the circle is drawn. Finally,

$$C_2 = S_{22} - \Delta S_{11}^* \tag{7.18}$$

Similar expressions provide circles of constant available gain. The radius of the available-gain circle, R_a, is

$$R_a = \frac{(1 - 2Kg_a|S_{12}S_{21}| + g_a^2|S_{12}S_{21}|^2)^{1/2}}{\left|1 + g_a(|S_{11}|^2 - |\Delta|^2)\right|} \tag{7.19}$$

and the center, C_a, is

$$C_a = \frac{g_a C_1^*}{1 + g_a(|S_{11}|^2 - |\Delta|^2)} \tag{7.20}$$

where

$$g_a = \frac{G_a}{|S_{21}|^2} \tag{7.21}$$

$$C_1 = S_{11} - \Delta S_{22}^* \tag{7.22}$$

and G_a is the value of available gain for which the circle is drawn.

In the past, before computer circuit-analysis software was ubiquitous, gain circles were an effective graphical tool for broadband amplifier design. The designer plotted gain circles for the desired value of gain at points across the passband, and values for the reflection coefficient were selected to provide an easily realizable termination. This process required a good intuitive sense of what constituted a realizable termination, usually born of many years of staring at Smith charts. The impedance of the opposite port was then found by (7.9) or (7.10), as appropriate.

Today, gain circles are still valuable, as they provide a quick visual assessment of the options for matching a particular solid-state device. The circles also provide a good visual indication of difficulties that might arise in using that device; for example, large-diameter circles usually imply that significant input-output interaction may occur or the desired value of gain is lower than optimum. They can also be combined with stability circles to identify regions of terminating impedance that might best be avoided.

7.1.3 Simultaneous Conjugate Match

The discussion of gain circles indicates that there is no unique set of (Γ_S, Γ_L) values for any particular value of gain. Any point on, say, an available-gain circle requires a unique corresponding value of $\Gamma_L = \Gamma_{\text{out}}^*$, so an infinite number of (Γ_S, Γ_L) pairs provide any achievable value of gain. The one exception to this rule occurs when the available gain is maximized; then the gain "circle" becomes a point (which can be viewed as a degenerate circle) and the input and output are both conjugate matched.

This *simultaneous conjugate match* condition exists only if the device is unconditionally stable.

Simultaneous conjugate match conditions can be found in the way one might expect: in (7.9) and (7.10), set

$$\begin{aligned} \Gamma_{in} &= \Gamma_s^* \\ \Gamma_{out} &= \Gamma_L^* \end{aligned} \tag{7.23}$$

and slog through the algebra. The result is a little complicated,

$$\Gamma_s = \frac{B_1 \pm (B_1^2 - 4|C_1|^2)^{1/2}}{2C_1} \tag{7.24}$$

$$\Gamma_L = \frac{B_2 \pm (B_2^2 - 4|C_2|^2)^{1/2}}{2C_2} \tag{7.25}$$

where

$$B_1 = 1 + |S_{11}|^2 - |S_{22}|^2 - |\Delta|^2 \tag{7.26}$$

$$B_2 = 1 + |S_{22}|^2 - |S_{11}|^2 - |\Delta|^2 \tag{7.27}$$

and C_1 and C_2 are given in (7.22) and (7.18). Equations (7.24) and (7.25) give two roots each; invariably, only one has a magnitude less than unity. The gain achieved under simultaneous conjugate match conditions is called the *maximum available gain* (MAG) or occasionally G_{max}. It is an important figure of merit for solid-state devices.

Substituting Γ_s or Γ_L into any of our previous gain expressions gives a result that can be reduced to

$$\text{MAG} = \left|\frac{S_{21}}{S_{12}}\right|(K - (K^2 - 1)^{1/2}) \tag{7.28}$$

where K is the K factor, given by (7.11).

7.1.4 Figures of Merit for Solid-State Devices

7.1.4.1 Maximum Available Gain and Maximum Stable Gain

We have already mentioned that the maximum available gain (MAG) is an important figure of merit for a solid-state device. It is defined, however, only at frequencies where the device is unconditionally stable. Most well-designed FETs and BJTs are conditionally stable below some particular frequency and are unconditionally stable at higher frequencies, but any device may well be used at frequencies where it is only conditionally stable. In that case, it helps to have a figure of merit for the conditionally stable case.

One figure of merit is called the *maximum stable gain*, or MSG. MSG is defined as the maximum available gain at the frequency where $K = 1$. From (7.28), it is

$$\mathrm{MSG} = \left|\frac{S_{21}}{S_{12}}\right| \tag{7.29}$$

$K = 1$ occurs at only one frequency; however, MSG as defined by (7.29) is traditionally evaluated over a range of lower frequencies, where $K < 1$, although it does not have much applicability to amplifier design in that frequency range. Still, it is a ratio of quantities that relate to the performance of a device in amplifier applications, and thus is a valid figure of merit. Designing for gain greater than MSG is generally viewed as a bad practice. Figure 7.1 shows the MSG and MAG of a small-signal microwave FET; the behavior of the quantities is typical of such devices.

7.1.4.2 f_{max} and f_t

Figure 7.1 shows that the MAG rolls off with frequency at an approximate rate of 6 dB per octave. Along such a curve, the product of the gain and frequency is a constant; if we know the gain-bandwidth product, we can estimate the MAG at any frequency. The quantity f_{max} is that gain-bandwidth product; f_{max} is sometimes called the *maximum frequency of oscillation*, as f_{max} is the maximum frequency at which the device has gain, a fundamental requirement for a simple feedback oscillator.

f_{max} is determined by fitting a line having a slope of 6 dB per octave to the MAG curve and extrapolating to the horizontal axis. The axis intercept is f_{max}. As such, it is an extrapolated quantity, not one that is measured directly, and the maximum available gain of the device at that frequency is likely to be lower than unity.

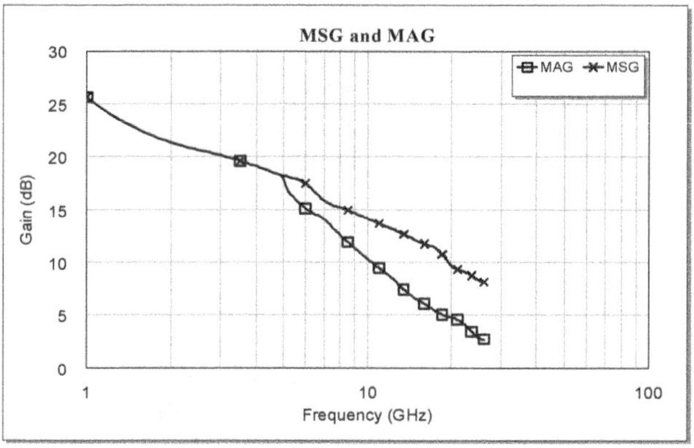

Figure 7.1 MAG and MSG of a small-signal FET from 1 to 26 GHz. The point where $K = 1$ is 5.0 GHz. MSG is plotted at frequencies where $K > 1$, as well as the range where $K < 1$, although it has little meaning in that frequency range.

Another quantity of interest is f_t, the current gain-bandwidth product. This quantity arises in the analysis of the simple equivalent circuit in Figure 7.2, which approximates both FET and bipolar devices. An analysis of the circuit shows that

$$H_{21} = \frac{I_2}{I_1} = \frac{G_m}{Cj\omega} \tag{7.30}$$

where G_m is a transconductance. According to this model, $|H_{21}|$ rolls off at a rate of 6 dB per octave. The point where $|H_{21}| = 1$ is

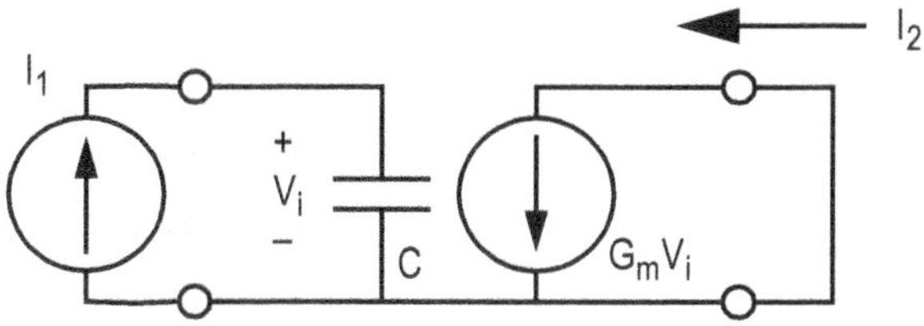

Figure 7.2 The idea of f_t is based on this simple equivalent circuit of a solid-state device.

$$f = f_t = \frac{G_m}{2\pi C} \tag{7.31}$$

Indeed, Figure 7.3, which shows a plot of $|H_{21}|$ vs. frequency for the small-signal FET whose gains are plotted in Figure 7.1, indicates that, at high frequencies, this real device does indeed roll off approximately at the rate of 6 dB per octave. Again, we fit a curve having a slope of 6 dB per octave to the high-frequency part of the curve, and extrapolate to the horizontal axis. The axis intercept is f_t.

Although beloved of analog-IC designers, f_t is not a very useful figure of merit for RF and microwave devices. The performance of such devices is strongly affected by quantities that do not appear in (7.31) or Figure 7.2, such as the device's terminal resistances, especially gate resistance in FETs and base resistance in BJTs. Other resistances and parasitic reactances also affect f_{max} but not f_t. Its value lies in the fact that device characteristics and dimensions affect G_m and the input capacitance approximately proportionally, so f_t is most useful as a measure of how well those quantities are optimized. Although f_t does not predict much about the gain or bandwidth achievable with a particular device, optimizing it can go a long way toward optimizing performance. For example, in bipolar devices, f_t is a strong function of collector bias current, so it provides a clear criterion for selecting an amplifying device's bias.

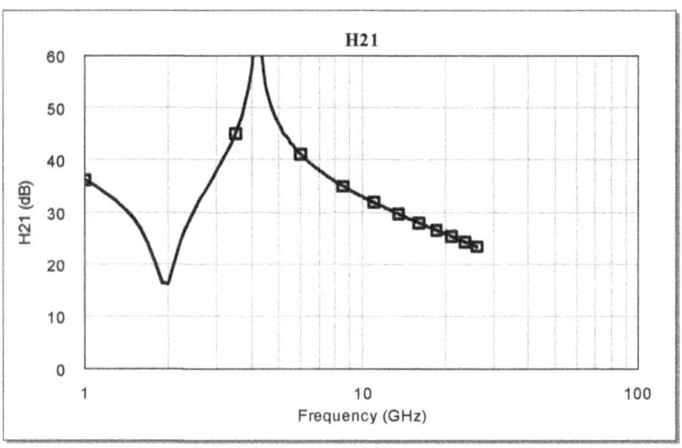

Figure 7.3 H_{21} in decibels of the device whose gains are shown in Figure 7.1.

7.1.5 Power Considerations

Most small-signal linear amplifiers are not designed to provide a particular level of output power. In some cases, however, even in small-signal amplifiers, output power is an important consideration. Such applications include drivers for frequency multipliers and mixer local oscillators. In that case, we must use a power-amplifier design approach, even though we are not producing a power amplifier in the usual sense.

A conjugate-matched output provides maximum output power only when the output voltage and current are much lower than the dc bias values, so the output power is determined solely by the gain and input level. In a power amplifier, however, the drain or collector bias voltage and current limit the output power, and the load impedance must be optimized to account for them. In that case, the output is no longer conjugate matched, and the amplifier's small-signal gain is somewhat lower than could be achieved with a conjugate-matched output. If noise performance is not a concern, as is usually the case in such applications, the input can be conjugate matched to optimize gain.

In the following, we consider the case of a linear amplifier in which the load is selected for maximum output power, not for gain or other criteria. The amplifier is a linear, "class-A" design, which means that the device conducts drain or collector current throughout the entire period of the excitation waveform. Other modes of operation are possible, but we will not discuss them here. A complete discussion of power amplifiers is beyond the scope of this chapter, but they are well covered in [7.1–7.3].

We consider the rather idealized FET circuit shown in Figure 7.4, in which the FET's drain is loaded by a resistor, R_L, and a resonant circuit. R_L represents the output load. We assume that the FET's drain-to-source resistance is much greater than R_L, so we can ignore it, and that its drain-to-source capacitance is absorbed into the resonator capacitance. In practice, a matching circuit would be used, and within the band of interest it would have the same fundamental characteristics as the resonator. That is, it would present the desired load resistance to the drain, its limited bandwidth would guarantee that harmonic distortion components (which always exist, despite our best efforts to avoid them) are rejected from the output, and the matching circuit, like the ideal parallel-LC resonator, would present a short circuit to the drain at all harmonics. Harmonic-distortion components of current simply idle in the drain and LC resonator or matching circuit, so they do not affect the drain voltage waveform.

The FET's drain current, $I_d(t)$, is

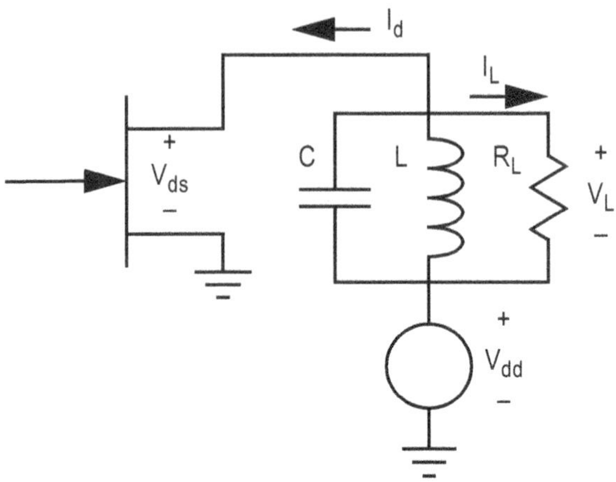

Figure 7.4 A FET whose output is matched to maximize output power can be modeled as shown. The resonator guarantees that the load voltage will be sinusoidal and provides a path to the drain for the bias current.

$$I_d(t) = I_{dd} + \frac{V_{dd} - V_{ds}(t)}{R_L} \tag{7.32}$$

where V_{dd} and I_{dd} are the bias voltage and current and $V_{ds}(t)$ is the drain-to-source voltage. As there can be no dc voltage across the inductor, the load voltage, $V_L(t)$, must be

$$V_L(t) = V_{ds}(t) - V_{dd} \tag{7.33}$$

The inductor carries the dc component of the drain current, so the load current, $I_L(t)$, is

$$I_L(t) = -(I_{ds}(t) - I_{dd}) \tag{7.34}$$

Equation (7.32) presents a set of conditions on $V_{ds}(t)$ and $I_d(t)$ that must be satisfied at the FET's drain. They describe a straight line in a plot of $I_d(t)$ as a function of $V_{ds}(t)$. The FET's drain I/V characteristic provides a second set of conditions, so if we draw (7.32) on a plot of that characteristic, the points of intersection must be those that describe the $V_{ds}(t)$ and $I_d(t)$ waveforms. Figure 7.5 shows this; the straight line is called a *load line*.

The output power is maximized when $V_{ds}(t)$ and $I_d(t)$ both have their greatest variations. According to this plot, $I_d(t)$ can vary from zero to some minimum determined by the knee of the uppermost I/V curve; in practice, however, it is limited to some low value by the softness of the gate-to-drain I/V characteristic near pinchoff and the need to minimize distortion. $V_{ds}(t)$ can vary from some low voltage, determined by the same knee that limited $I_d(t)$, to the voltage corresponding to minimum current. Since $V_L(t)$ and $I_L(t)$ are simply the RF components of $V_{ds}(t)$ and $I_d(t)$, maximizing the variation of $V_{ds}(t)$ and $I_d(t)$ maximizes $V_L(t)$ and $I_L(t)$.

To maximize output power, it should be clear that V_{dd} and I_{dd} should be located halfway between the extremes of $V_{ds}(t)$ and $I_d(t)$. The power delivered to the load, P_d, under these conditions is

$$P_d = \frac{1}{2}|V_{ds}(t)||I_d(t)| = \frac{1}{2}\left(\frac{V_{max} - V_{min}}{2}\right)\left(\frac{I_{max} - I_{min}}{2}\right) \tag{7.35}$$

where V_{max}, V_{min}, I_{max}, and I_{min} are the respective maximum and minimum values of $V_{ds}(t)$ and $I_d(t)$. The load resistance that provides this output power is

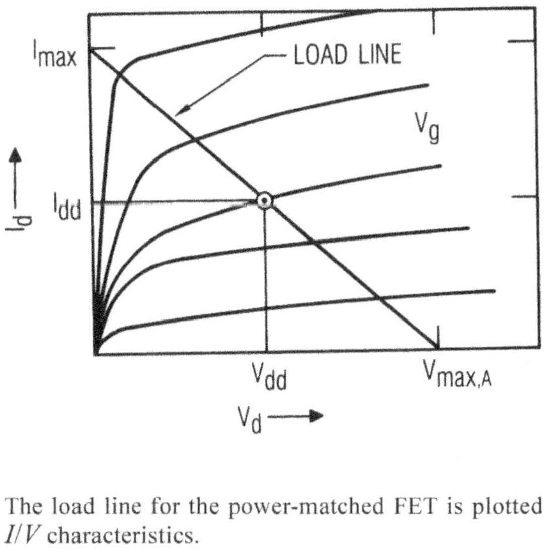

Figure 7.5 The load line for the power-matched FET is plotted on a curve of drain I/V characteristics.

$$R_L = \frac{\dfrac{V_{max} - V_{min}}{2}}{\dfrac{I_{max} - I_{min}}{2}} = \frac{V_{max} - V_{min}}{I_{max} - I_{min}} \approx \frac{V_{dd}}{I_{dd}} \tag{7.36}$$

If it were possible for the $V_{ds}(t)$ and $I_d(t)$ waveforms to reach the axes, V_{dd} and I_{dd} would be centered between those extremes, the approximation in (7.36) would become an equality, and

$$P_d = \frac{1}{2} V_{dd} I_{dd} \tag{7.37}$$

The amplifier's output power would then equal half the dc input power; the *dc-to-RF efficiency*, at maximum output power, would then be 50%. This is the theoretical limit to the efficiency of a class-A amplifier. In practice, the maximum amplifier efficiency is usually significantly lower, typically in the range of 35% to 40%.

Since a class-A amplifier is a linear component, the equations derived earlier for gain, input reflection coefficient, and output reflection coefficient are valid. It should be clear that, since the load impedance is selected for reasons that do not involve matching, the output of a power amplifier is likely to be poorly matched. For example, consider a FET operating at $V_{dd} = 3$V with a maximum current of 100 mA, so $I_{dd} = 0.05$. From (7.36) $R_L \approx 60\Omega$, a quantity well below the value of a few hundred ohms that is likely to be needed for a matched output. This clearly has an effect on the amplifier's gain. It also shows why a power amplifier's gain is invariably lower than that of a similar small-signal amplifier.

7.1.6 Distortion

When an amplifier is operated near the limit of its output power capability, it invariably distorts the signal. Even at low signal levels, however, weak nonlinearities in the device cause low levels of distortion, which can be surprisingly troublesome. The practical upper limit of an amplifier's signal-handling capability is most often established by that distortion, not by its maximum sinusoidal output power.

The usual, textbook treatment of distortion is fairly simple. We assume that the amplifier has an input-output characteristic given by

$$V_o = a_1 V_i + a_2 V_i^2 + a_3 V_i^3 + \dots \tag{7.38}$$

where V_i is the input voltage and V_o the output voltage. We apply a two-tone input signal given by

$$V_i = V_1 \cos(\omega_1 t) + V_2 \cos(\omega_2 t) \tag{7.39}$$

where V_1 and V_2 are the magnitudes of the inputs, usually assumed to be equal, and ω_1 and ω_2 are their respective closely spaced frequencies. These are substituted into (7.38), and the result contains the following terms:

$$\begin{aligned} V_o = {} & \left(a_1 V + \frac{3}{2} a_3 V^3\right)[\cos(\omega_1 t) + \cos(\omega_2 t)] \\ & + \frac{3}{4} a_3 V^3 [\cos((2\omega_2 - \omega_1)t) + \cos((2\omega_1 - \omega_2)t)] + \dots \end{aligned} \tag{7.40}$$

where we have assumed that $V_1 = V_2 = V$. The first term in (7.40) shows gain saturation of both tones, as a_1 and a_3 invariably have opposite signs. The second term gives the intermodulation distortion (IM) products, new tones at $2\omega_2 - \omega_1$ and $2\omega_1 - \omega_2$. These are third order, as they involve three frequency terms, $\omega_2 + \omega_2 - \omega_1$ and $\omega_1 + \omega_1 - \omega_2$, respectively. Equation (7.38) produces other IM products as well; for example, second-order products at $\omega_2 - \omega_1$ and $\omega_1 - \omega_2$. Since ω_1 and ω_2 are closely spaced, the third-order distortion products also lie close to the excitation frequencies, so they cannot be removed by filtering. Conversely, most even-order products lie well away from the excitation frequencies, so they are usually easy to reject by filtering. When more complex excitation waveforms are used, intermodulation distortion causes new frequencies to be generated immediately outside the waveform's spectrum, broadening its bandwidth. Frequency components can even be generated within the waveform's spectrum.

Most interesting is the cubic relationship between the tone voltages seen in (7.40); this implies that the levels of the distortion products vary 3 dB for each decibel change in the input tone magnitudes.

While the foregoing description seems hopelessly naive, it does, in fact, capture the behavior of real systems and components. A more sophisticated model using Volterra methods, as described in [7.1], shows that all kinds of weakly nonlinear components, not just amplifiers, exhibit the same type of behavior; that is, nth-order IM products vary in amplitude n dB per 1-dB change in the input level, as long as the input level is kept well below gain saturation. In either case, we find that the levels of the linear tones and distortion products vary as shown in Figure 7.6. Because the curves are straight and have different slopes, they inevitably intersect; the intersection point, which can be specified in terms of either input or output power, is called the *nth-order intermodulation intercept point*, usually abbreviated IP_n and always expressed in decibels. The difference between the input and output intercept points is simply the gain.

Given the intercept point and the slope, we can easily determine the ratio of the linear to IM power at the output. The IM level in dBm is given by

$$P_{IMn} = nP_1 - (n-1)IP_n) \tag{7.41}$$

where P_{IMn} is the level of the nth-order IM product, n is the order, and IP_n is the nth-order intercept point for that particular product. Thus, the intercept point defines the IM level at any excitation level; this is a simple and accurate way to predict IM levels in weakly nonlinear components or systems.

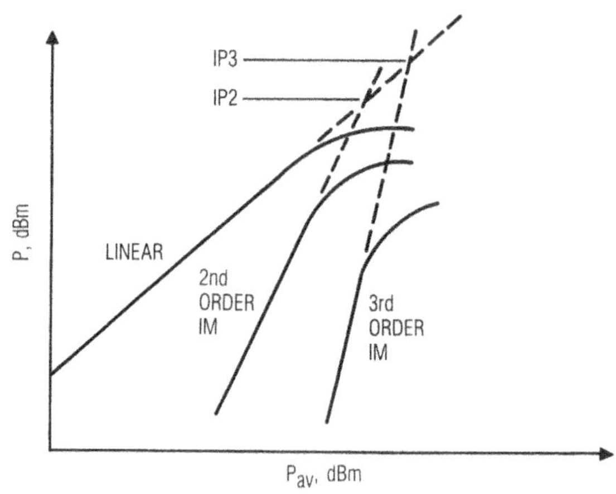

Figure 7.6 Input and output power curves showing linear, second- and third-order two-tone products. By tradition, tone levels are identical and the power levels are those of a single tone.

7.2 Noise

7.2.1 Noise Temperature and Noise Figure

Noise is a randomly varying voltage or current that exists in all kinds of electrical and electronic circuits. As such, a noise voltage or current is not predictable, but its statistics usually can be determined. Thus, we characterize it by its statistics, especially its time-averaged value and standard deviation. Electrical noise usually is well modeled as a Gaussian random process; that is, its voltage or current, at any instant, has a Gaussian probability distribution. In that case, the noise power equals the square of its standard deviation, and its average simply equals any dc voltage or current component on which the noise is superimposed, which is usually either zero or of no interest.

In the frequency domain, electrical noise has a continuous (but not necessarily constant) power spectrum in watts per hertz. Usually the spectral density can be treated as a constant over some restricted bandwidth centered on a frequency of interest; if the noise spectrum is not flat, we can narrow our observation of the noise to a segment of the spectrum small enough so that the assumption is valid. The noise is then viewed as a number of such spectral components. This practice allows the "white Gaussian noise" model to be used extensively in linear electronic systems.

Noise is generated by a variety of mechanisms. Many of these are associated with the irregular way in which electrons move through solid-state devices, and, as such, are well beyond the scope of this fundamental treatment. See [7.4-7.6] for more complete information on noise in electronic devices. One mechanism must be discussed, however: all lossy devices at a temperature above absolute zero generate noise. This type of noise is related to blackbody radiation and is called *thermal noise*.

Any lossy element, regardless of its type, has an available thermal noise power spectrum given by

$$S = KT \tag{7.42}$$

where S is the power spectral density in watts per hertz, K is Boltzmann's constant in joules per kelvin, and T is absolute temperature in kelvins. Since the spectrum is white, the available power, P_{av}, in any bandwidth B is

$$P_{av} = KTB \tag{7.43}$$

From (7.42), that power spectral density can be characterized by the temperature of the element, so temperature can be used as an alias for noise spectral density. Furthermore, in noise analysis, the bandwidth is an incremental one in the vicinity of the frequency of interest, so the noise temperature can be used as a power quantity with the bandwidth implicit.

Figure 7.7 shows circuit noise models of a resistor. The noise can be modeled in two ways, either as a series voltage source or shunt current source. Although we usually characterize voltages and currents by their magnitudes, noise is characterized, by convention, as an RMS quantity. A simple analysis readily shows that the noise power delivered to a matched source is as given in (7.43).

7.2.1.1 Noise Temperature

Any two-port can be viewed as a noise source, as a noise voltage is present at its output port. That noise is the result of noise sources inside the two-port and of noise power applied to the input. In characterizing the noise of the two-port, however, we are concerned only with the components of the available output noise power that arises within the two-port; input noise, while often a significant (even dominant) part of the noise output power, is not part of the two-port's noise characterization. As we shall see, however, it is useful to compare that noise to the noise of a standard termination, defined by noise temperature.

For a number of reasons, it is convenient to view the output noise by a kind of "system-level" view, in which the output noise arises entirely from

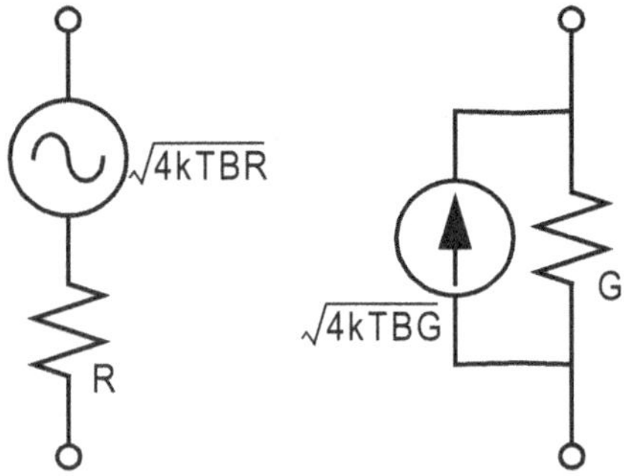

Figure 7.7 Thermal noise in a resistor can be modeled by a series voltage source or a shunt current source. The direction of the current or polarity of the voltage source has no significance for single sources.

a thermal noise source at the input,. The two-port itself, then, is noiseless. The idea is shown in Figure 7.8. The noise temperature of the source, T_n, called the *two-port's equivalent input noise temperature*, or simply its *noise temperature*, is

$$T_n = \frac{P_{d,n}}{KBG_t} \tag{7.44}$$

where $P_{d,n}$ is the component of the noise output power due to noise generated in the two-port that is delivered to a standard load. G_t, as before, is the two-port's transducer gain. Inevitably, the input source is noisy, and has a temperature T_s. The total output noise power delivered to the load then is

$$T_L = G_t(T_n + T_s) \tag{7.45}$$

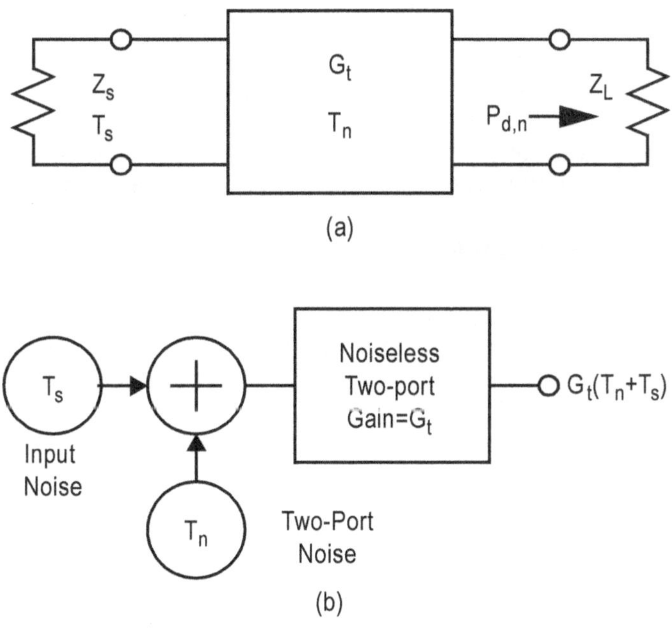

Figure 7.8 A noisy two-port (a) can be described by a model (b) in which the two-port is noiseless and its noise arises from an input source.

7.2.1.2 Noise Figure

Amplifiers are usually characterized by their noise figures at a number of in-band frequencies. Noise figure, F, is defined as the total output noise divided by the component of the noise from the source alone. The source and load are purely resistive and have the same values for which G_t is defined. Thus,

$$F = \frac{G_t(T_n + T_s)}{G_t T_n} = \frac{T_n + T_s}{T_s} \tag{7.46}$$

This quantity is unique only if the input noise temperature has a standard value. By convention, that value is 290K, traditionally designated T_0. Then (7.46) becomes

$$F = \frac{T_n + T_0}{T_0} = 1 + \frac{T_n}{T_0} \tag{7.47}$$

Clearly, one could specify the noise of the two-port either by noise temperature or by noise figure. The use of noise figure has become customary in the communications industry; in other fields, especially radiometry, the use of noise temperature is probably more common. Noise figure is invariably stated in decibels while noise temperature is always stated in kelvins.

7.2.2 Noise Figure Optimization

Optimizing the noise figure of an amplifier generally requires *mismatching* the input. This surprising result is a consequence of the number and types of noise sources in the device. A detailed explanation can be found in [7.4].

The noise figure of a particular two-port is a function solely of the source impedance or admittance. It is given by the following expression:

$$F = F_{\min} + \frac{R_n}{\mathrm{Re}\{Y_s\}}|Y_s - Y_{s0}|^2 \tag{7.48}$$

where Y_s is the source admittance; Y_{s0} is the value of source admittance that provides minimum noise figure, $F_{\min}$; and R_n is the noise resistance.

R_n can be viewed as a constant, having units of resistance, that establishes the sensitivity of the noise figure to source admittances differing from Y_{s0}.

Equation (7.48) can be expressed in terms of reflection coefficient. Substituting

$$Y_s = \frac{1}{Z_0}\left(\frac{1-\Gamma_s}{1+\Gamma_s}\right) \tag{7.49}$$

for Y_s and a similar expression for Y_{s0} gives

$$F = F_{\min} + 4\frac{R_n}{Z_0(1-|\Gamma_s|^2)|1+\Gamma_{s0}|^2}|\Gamma_s-\Gamma_{s0}|^2 \tag{7.50}$$

where Γ_{s0} is the reflection coefficient corresponding to Y_{s0}. Equations (7.49) and (7.50) define the four noise parameters of any two-port; they are the source admittance or reflection coefficient (real and imaginary parts), the noise resistance, and the minimum noise figure. R_n is often provided in normalized form, R_n / Z_0. In general, these quantities are not constant with frequency or, in the case of a solid-state device, dc bias. $F_{\min}$ is not affected by lossless matching, but the other noise parameters are.

Loci of constant noise figure are circles in the input reflection-coefficient plane. For a specified value of noise figure, F, the center of the circle, C_F, is

$$C_F = \frac{\Gamma_{s0}}{1+N} \tag{7.51}$$

and the radius, R_F, is

$$R_F = \frac{1}{1+N}(N^2 + N(1-|\Gamma_{s0}|^2))^{1/2} \tag{7.52}$$

where

$$N = \frac{F-F_{\min}}{4(R_n/Z_0)}|1+\Gamma_{s0}|^2 \tag{7.53}$$

These are useful for assessing the sensitivity of a device's noise figure to its source reflection coefficient, as well as for the initial design of a low-noise amplifier.

7.2.3 Noise Figure of an Attenuator

Any matched, passive, lossy two-port can be viewed, for noise purposes, as an attenuator, so the results in this section apply to lossy transmission lines and similar lossy elements. We assume that the loss results entirely from power dissipation, not reflection, so the noise is thermal. With these results, and those of the next section, we shall see that even very small input losses can have a severe effect on the noise figure of an amplifier.

Suppose the two-port in Figure 7.8 is a matched attenuator in thermal equilibrium with its source and load terminations at some temperature T. At its output port, the attenuator plus the input termination is indistinguishable from a simple two-terminal resistive load, so the noise power delivered to the output termination, in terms of noise temperature, is simply T. Some of this noise arises in the source termination, and some in the attenuator itself. The noise at the load that comes from the source is simply $G_t T$, so the noise from the attenuator must be $T - G_t T$. Dividing this by G_t gives us the equivalent input noise temperature of the attenuator, T_a:

$$T_a = T\left(\frac{1}{G_t} - 1\right) \tag{7.54}$$

This is usually written

$$T_a = T(L - 1) \tag{7.55}$$

where L is the loss factor; $L = 1 / G_t$.

The noise figure of the attenuator is found simply by substituting (7.55) into (7.47). We obtain

$$F = 1 + \frac{T(L-1)}{T_0} \tag{7.56}$$

which gives the interesting result that $F = L$ when $T = T_0$.

7.2.4 Cascaded Stages

Figure 7.9 shows a series of M cascaded stages and its noise model. Its output noise is given by

$$T_L = T_1 G_1 G_2 \ldots G_M + T_2 G_2 G_3 \ldots G_M + \ldots + T_M G_M \tag{7.57}$$

where T_m and G_m are the noise temperature and gain, respectively, of the mth stage. The input equivalent noise temperature is, as usual, the output noise divided by the gain, so

$$T_n = \frac{T_L}{G_1 G_2 \ldots G_M} = T_1 + \frac{T_2}{G_1} + \frac{T_3}{G_1 G_2} + \ldots + \frac{T_M}{G_1 G_2 \ldots G_{M-1}} \tag{7.58}$$

That is, the contribution of each stage to the cascade noise temperature is simply its noise temperature divided by the gain ahead of it.

The noise figure of the cascade is found by substituting (7.58) into (7.47). We obtain

$$F = F_1 + \frac{F_2 - 1}{G_1} + \frac{F_3 - 1}{G_1 G_2} + \ldots + \frac{F_M - 1}{G_1 G_2 \ldots G_{M-1}} \tag{7.59}$$

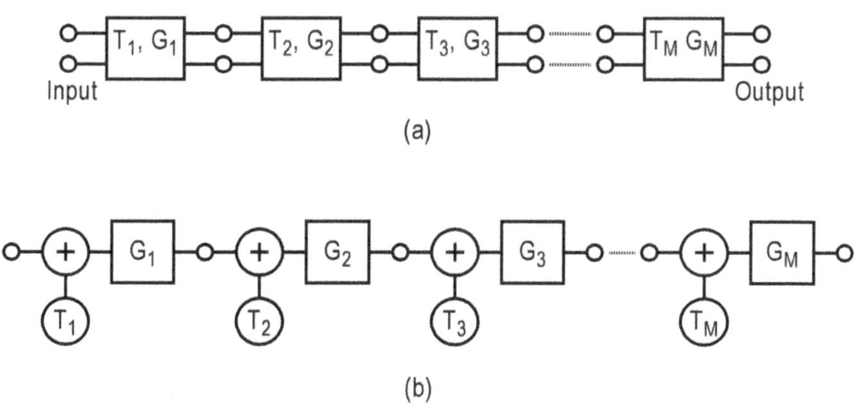

Figure 7.9 A series of cascaded stages (a) and its noise model (b).

Similarly, the contribution of each stage after the first to the cascade noise figure is its noise figure, minus one, divided by the gain ahead of it.

A clear implication of (7.58) and (7.59) is that the early stages of a cascade contribute the most to its noise temperature or noise figure; thus, the noise of those stages is most important in establishing a low system noise figure. At this point, it's easy to see the effect of input attenuation on the noise figure or noise temperature. From (7.59) we see that the first-stage (now the attenuator) noise figure is approximately equal to its loss, and the contribution of the second stage (the amplifier) is now divided by the attenuator's gain, a number less than 1. If the attenuator's temperature is T_0, we find that the increase in noise figure, in decibels, is equal to the attenuator's loss in decibels. In view of the fact that the minimum noise figure of many microwave FETs can be as low as a few tenths of one decibel, it is clear that even small losses can be significant, even dominant, in establishing a system's noise figure.

7.3 Amplifier Design

7.3.1 Device Bias in Amplifier Design

The bias voltages applied to terminals of a bipolar transistor or a FET, and the resulting collector/drain current, have important effects on the performance of the device. We noted in Section 7.1.4.2 that f_t and f_{max} are strong functions of bias current. Minimum noise figure is also strongly dependent on collector/drain current, as is output power.

7.3.1.1 Bipolar Devices

The f_t of a bipolar transistor increases monotonically with collector current until it reaches a maximum; it then drops rapidly at currents above that value. The f_{max} behaves similarly, and its peak is usually close to the f_t maximum. To achieve maximum gain and bandwidth, a bipolar transistor should be operated at or near that f_t peak. If that results in more current than is desired, a device having a smaller emitter area can be used, as the current at maximum f_t scales in proportion to emitter area. A smaller device, correctly biased, usually provides better performance than a larger device with suboptimal collector current.

The minimum noise figure of a bipolar transistor decreases with collector current until it reaches a minimum; above that minimum, it usually increases gradually. The current at minimum noise figure is typically well below the f_t maximum, but since the noise figure is only modestly sensitive to collector current, a bipolar transistor generally can be biased near the f_t maximum without a great noise-figure penalty. As long as the collector voltage is above the knee of the collector voltage-current characteristic, the minimum noise figure is only a weak function of collector voltage.

Although Section 7.1.5 uses FET characteristics to illustrate the optimization of output power, the same principles apply to bipolar transistors. From (7.37) it should be clear that the output power depends on both dc collector voltage and current. Collector voltage is limited by avalanche breakdown in the device, at peak voltage, and collector current is limited by the device's maximum allowable current density, which is on the order of 10^4 A/cm^2, depending on the technology of the device. The product of those values is also limited by power dissipation and the resulting heating of the device. High output power requires large chips having a large total emitter area, which in turn affects f_t and f_{max}.

7.3.1.2 FETs

Unlike a BJT, a junction FET's drain current is limited to a value slightly above its I_{dss}, the drain current at zero gate bias. In most junction FETs designed for operation in the lower microwave region, I_{dss} is a relatively weak function of drain voltage. In very small, high-frequency devices and in the exotic technologies used at millimeter wavelengths, I_{dss} is more strongly dependent on drain voltage, increasing 30% to 50% over a practical drain-voltage range. Insulated-gate FETs do not have such a strict current limit, but they are limited in allowable current density and power dissipation.

The gain of a FET rises rapidly near the gate's pinch-off or threshold voltage, but increases much more slowly at practical values of drain current. The gate-to-source capacitance varies weakly in this region, so most types of FETs do not have the pronounced f_t maximum observed in bipolars. Their noise figures show a pronounced minimum as a function of drain current, however.

As a general rule, a FET's noise figure is optimum around 15% to 20% of I_{dss}, while it achieves optimum gain and low distortion around 50% of I_{dss} or perhaps a bit more. Thus, there is a distinct trade-off between noise, gain, and distortion in selecting a FET's bias. The noise figure of a FET is also more sensitive to source impedance than that of a BJT; we shall discus this further in the following sections.

Power considerations are essentially the same for FETs as for bipolar devices. For maximum power in a linear, class-A amplifier, a FET must be biased close to 0.5 I_{dss}, as the peak drain current is twice the dc value. Power dissipation and avalanche breakdown, as in other types of amplifiers, also limit the drain voltage and current.

7.3.2 Narrowband Amplifier Design

We first consider a small-signal, linear amplifier. We view the device as a two-port, choose values of Γ_S and Γ_L that provide the desired gain, and realize matching circuits that present those reflection coefficients to the device. The problem is first to find some appropriate criterion for designing matching circuits, then to realize them.

7.3.2.1 Matching Approach

Four possibilities for the design of the matching circuits in a small-signal amplifier present themselves:

1. Conjugate match both the input and output. This simultaneous conjugate match is possible only at frequencies where the device is unconditionally stable. This option is attractive, as it is the only one where Γ_S and Γ_L are unique. However, usually only narrowband matching is possible, and if the stability factor K is only marginally greater than 1.0, as is often the case, the resulting amplifier, while theoretically stable, is still close to oscillation. Thus, its advantage is one of its disadvantages: the designer cannot select Γ_S or Γ_L to increase the margin of stability. The amplifier also cannot be designed to minimize noise figure, and conjugate matching the input often results in a poor noise figure.
2. Match the output but not the input. In this case the source reflection coefficient can be selected to achieve a specific value of available gain and to avoid the unstable region identified by an input stability circle. It also allows noise-figure optimization through the selection of an appropriate value of Γ_S, usually entailing a trade-off between gain and noise figure. In most small-signal amplifier designs, noise figure is an important consideration, so this approach is probably the most common one.
3. Match the input but not the output. This requires the use of a power-gain design criterion. It is most often employed when the device's output is loaded for maximum output power. In such amplifiers, the noise figure is usually not a consideration, so an input conjugate match is acceptable. If the amplifier is broadband, however, the input match is the only degree of freedom for adjusting the gain to achieve a flat

passband, so its design becomes a trade-off between input match, stability, and passband flatness.

4. Mismatch both the input and the output. This allows design for a particular value of gain, but, in most systems, requires the use of some other method to provide good input and output VSWRs. Often this is the only practical approach to the design of a broadband, low-noise amplifier. In that case, the input is matched for low noise, and the output mismatched, with a higher output VSWR at the low-frequency end of the band, to provide a flat passband. Achieving a low VSWR in system use often is not difficult; two options are the use of isolators or connecting two amplifiers by quadrature (Lange) hybrids. We discuss quadrature-coupled amplifiers in Chapter 8.

The latter three options are possible when the device is conditionally stable. In those, the values Γ_s and Γ_L are not unique; that is, there is an unlimited number of (Γ_s, Γ_L) combinations that provide any particular value of gain. The problem is to find some criterion for selecting some specific pair, then to synthesize matching circuits that realize those reflection coefficients over the required bandwidth.

The application of these ideas is best illustrated by examples.

7.3.2.2 Example: Low-Noise Amplifier

We wish to design a low-noise amplifier (traditionally called an *LNA*). To keep the example from becoming unnecessarily complicated, we consider narrowband operation at 8 GHz. Option 2 above, match the output but not the input, seems like the obvious choice. We select a packaged FET whose specifications seem appropriate and for which good S and noise parameters at optimum-noise bias are available.

Figure 7.10 shows a plot of the available gain circles for 12 to 15 dB, the input stability circle, and the noise-figure circle for 0.5 dB above F_{min}. At this frequency, F_{min} = 0.35 dB. The S and noise parameters used in these plots were measured at V_{ds} = 2.0V and I_d = 10 mA. The drain current is the value that provides the minimum noise figure.

The plots show clearly the trade-offs in the selection of Γ_s. Since the value of Γ_s that provides minimum noise figure is near the center of the 0.5 dB circle, selecting that value of Γ_s results in a gain of approximately 15 dB when the output is matched. That might be too much gain, though; although the value is well outside the unstable region, the high gain might still cause oscillation because of feedback resulting from stray coupling. If we can tolerate a little higher noise figure, we could select, for example, Γ_s

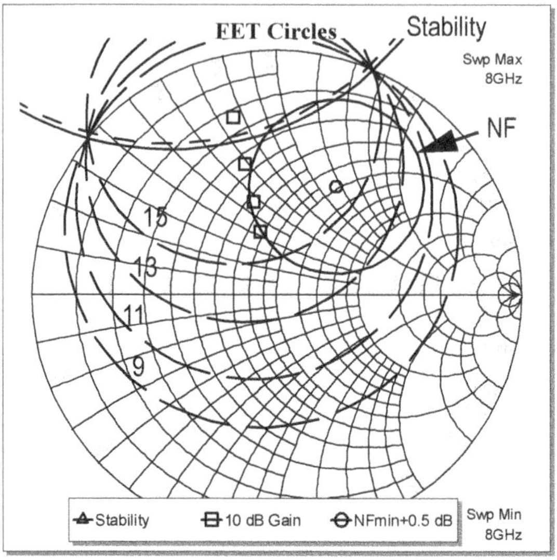

Figure 7.10 Available gain, stability, and noise-figure circles for a FET at 8 GHz. The gain circles are drawn for values of gain from 9 to 15 dB and the noise-figure circle is for $F_{\min}$ + 0.5 DB, which is 0.85 dB.

on the 13-dB gain circle, where it intersects the noise circle. This would also increase the distance between Γ_s and the unstable region. This value of Γ_s corresponds to a source impedance $Z_s = 100 + j24$; that is, 100Ω plus a small reactive part, so it should be easy to realize.

Given the source impedance, we now must find the load impedance. We could simply use (7.10) to determine the output reflection coefficient, and the load reflection coefficient is simply the complex conjugate of that value. A better idea, however, is to design the source matching circuit first, then calculate the output impedance; that way, the output design is based on the realized source reflection coefficient, which may be slightly different from the design's target value.

The required source impedance immediately suggests the use of a quarter-wave transformer to shift the 50Ω source impedance to 100Ω, then follow with a narrow microstrip line that approximates an inductor. On a 635-μm-thick alumina substrate ($\varepsilon_r = 9.8$), we find that a line 0.250 mm wide and 3.8 mm long meets the requirements for the transformer. The inductor is realized with a 100Ω line (50 μm wide) 0.04λ long. Since the design of the inductor is approximate, we have to do a little manual adjustment of the lengths to achieve the desired source impedance. The final length of the transformer is 2.9 mm and the inductive line is 0.38 mm.

With the input matching circuit complete, we can calculate the output impedance. We obtain an output impedance of $46 + j32\Omega$, so the load must be $46 - j32\Omega$. A series-L, shunt-C L-section matching network should make short work of this output impedance. We again use distributed elements, a 50-μm-wide line 1.2 mm long followed by a 0.64-mm-wide open shunt stub, 1.1 mm long. This achieves an output match at 8 GHz.

Figure 7.11 shows the resulting circuit along with its gain and noise figure. The expected result of 0.85-dB noise figure and 13-dB gain has been achieved. Of course, this is not a complete design; it is still necessary

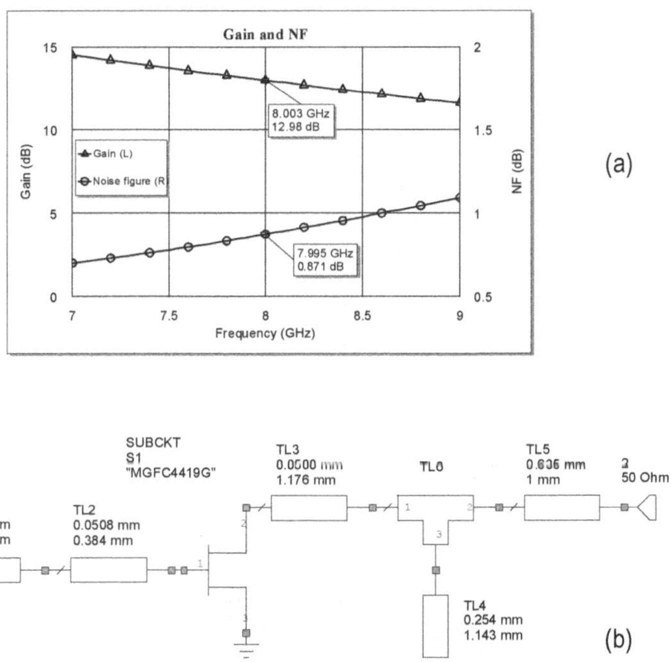

Figure 7.11 Amplifier performance (a) and matching circuits (b). The design goal of 13-dB gain and 0.85-dB noise figure at 8 GHz has been met.

to include discontinuity parasitics for the steps in microstrip width, bias circuitry, and models for substrate via holes, which will be needed for connecting the FET's source terminals to ground. These are best included after the basic design of the matching circuits, as they should perturb the results only minimally, and little adjustment of the matching elements will be needed to compensate for them. The designer should also check both input and output stability circles for the complete amplifier.

Figure 7.11 shows that the amplifier has relatively broad bandwidth, in spite of the fact that it is a narrowband design. Because it has significant gain over a broad band, it is important to check stability outside the band of interest. It is entirely possible for such an amplifier to oscillate at some out-of-band frequency. It is also likely that the amplifier has high gain at low frequencies, as nothing has been done to prevent it.

High out-of-band gain can be a problem when the amplifier is used in a system. For example, suppose that the amplifier is used as a first-stage LNA in a system where out-of-band spurious responses are a problem. Usually a filter is used between the LNA and the mixer to prevent these, but out-of-band gain in the LNA will reduce the effectiveness of that filter. Amplifiers should have gain where it's wanted, and little or none where it isn't.

Thus, it is a good idea to include some elements in the circuit that restrict the out-of-band gain. One possibility is to add quarter-wavelength shunt stubs at the input and output; these can be incorporated into the bias or matching circuits. Another method is to make the matching circuits inherently high-pass or bandpass, or at least to include shunt inductors in the matching circuits to create transmission zeros at $\omega = 0$.

7.3.3 Broadband Design Using Negative-Image Models

While it would be possible to draw gain, noise-figure, and stability circles for a number of points over a broad bandwidth, the previous example should provide a hint of the complexity of such a plot. It would be necessary to plot a number of gain and noise-figure circles, plus stability circles, at a number of points over the amplifier's passband. From this mess, the designer would then have to select Γ_S values at each frequency. Many Γ_S loci would be possible; the selected locus would have to be realizable with a reasonably simple ladder network that minimized the error between the desired and realized Γ_S locus. This would require considerable insight. Then, once the source reflection coefficient was determined, a similar process would be needed to find Γ_L.

Although, in the past, this practice was not uncommon, newer methods using computers are much more practical. One of these is negative-image modeling, an elegant way to make the necessary trades between gain, noise, and stability, while creating source and load networks for the synthesis of matching networks.

7.3.3.1 Negative-Image Modeling

Negative-image modeling is based on the following idea. Suppose we create a matching circuit using negative capacitances or inductances that provides the desired response. This could be a conjugate match, minimum noise figure, a specific value of gain, high stability, or some trade-off between all of these requirements, determined with the help of a circuit simulator and a numerical optimizer. Once the negative-value matching circuit (the "negative-image" circuit) has been finalized, we create a circuit having the same structure but with positive-valued reactances. This "positive image" is our load model. We then synthesize a matching network that achieves, as closely as possible a conjugate match to that positive-value load. If we could achieve a perfect match, that matching circuit would be equivalent to the negative-image circuit. Of course, a perfect match is never possible, but the synthesized matching network is nevertheless guaranteed to be a close equivalent to the ideal, negative-image matching circuit.

This method is effective because most transistors' input and output impedances approximately model simple series or shunt RC networks. A negative image of such a network invariably provides a good match, and it is surprisingly easy to make the necessary trade-offs to achieve the desired performance.

Figure 7.12 illustrates this process. We begin with the circuit in Figure 7.12(a), which uses negative-image networks at the input and output. Since the input of a FET approximates a series RC, we use a series RC for the input matching network; as the output approximates a parallel RC, we use a parallel RC network at the output. If these do not provide the necessary performance, we can experiment with other circuit structures. If we can achieve the desired performance with a series or shunt RC, however, we have a significant advantage: we can use classical matching-network synthesis to realize the matching circuit.

We then synthesize input and output matching circuits using the positive-image networks as load models. This step is illustrated in Figure 7.12(b). It should be clear that, if the matching circuit creates a conjugate match to the positive-image load, it must be equivalent to the negative-image matching circuit. The matching-network synthesis has only a conjugate-match goal; compensation for gain roll-off is accomplished in the optimization of the negative-image model.

Of course, the same broadband matching limitations that apply to any kind of network synthesis also apply to the task of creating a matching circuit from the positive-image model. These were discussed at length in Section 4.4. They do not apply, however, to the negative-image circuit, as it uses negative-valued reactive elements; for example, if the input impedance of the FET in Figure 7.12(a) were $R_S + 1 / j\omega C_S$, the input would have a perfect match over an infinite bandwidth. When the matching circuits in Figure 7.12(b) replace the negative-image circuits, their inevitable imperfection results in some degree of deviation from the amplifier's response with the negative-image circuit. Nevertheless, the performance should be close, and a little numerical optimization should be adequate to perfect it.

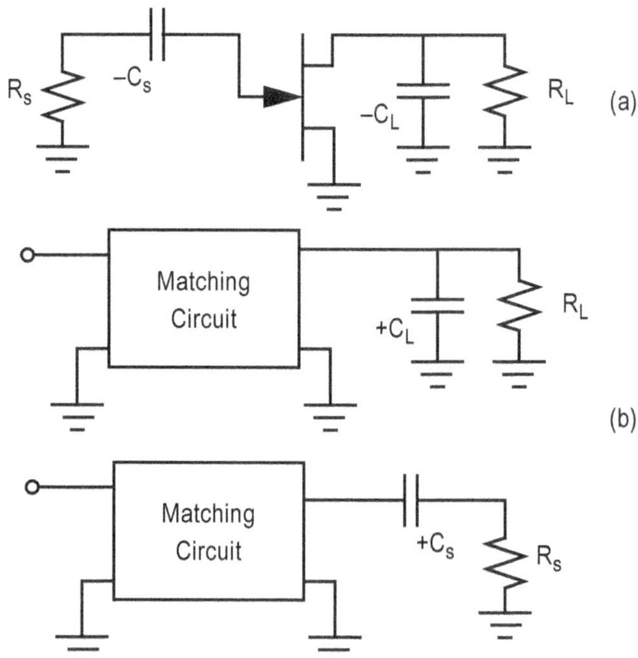

Figure 7.12 In negative-image modeling, we first create matching circuits using negative reactances for the matching elements and optimize the circuit for any desired characteristics (a). The positive-image versions of those networks become load models for synthesizing the real matching circuits (b). If we were to achieve a perfect match to the positive-image models, the matching circuits would then be equivalent to the negative-image matching circuits.

7.3.3.2 Example: LNA Design Using Negative-Image Modeling

While the previous amplifier design example successfully illustrated a design approach for a narrowband amplifier, it was, admittedly, a little naive. The matching circuits used in that design were unable to prevent the existence of high out-of-band gain, which would be a significant problem if that amplifier were used in a microwave system. The example in this section addresses the problem of broadband design while assuring that the low-frequency gain remains tolerably low.

In this example, we design an LNA covering 8 to 12 GHz. The amplifier uses a chip device and is realized on a 0.635-mm-thick alumina substrate. The FET is characterized by S parameters measured at the minimum-noise bias. In the circuit, shown in Figure 7.13(a), we have included inductors to model the bond wires, typically 0.2 nH at the gate and drain. The source bond-wire inductance is considerably lower, as the FET is mounted in a manner that allows several short connections to ground.

We begin by creating negative-image networks for the input and output and analyzing the circuit with them in place. We find quickly that a simple series capacitor is adequate at the input, but the output requires both a capacitor and inductor. The resulting gain and noise figure are shown in Figure 7.13(b). The passband is reasonably flat and the noise figure closely follows the curve of minimum noise figure. In this design, as with most LNAs, the input matching circuit is designed exclusively to optimize the noise figure, while the output is chosen for flat gain. It is impossible to achieve a good output match across the band, as the output must be mismatched at lower frequencies to compensate for the higher gain of the device in that frequency range. The use of the parallel LC network has an additional advantage: the real matching circuit must also be bandpass in nature, so it should effectively prevent high gain at low, out-of-band frequencies.

The next step is to design the input and output matching circuits using the positive-image load models; these are shown in Figures 7.14 and 7.15. Both were derived by applying a little thought and computer optimization; formal synthesis was not needed. The input is a classical series-L, shunt-C arrangement, realized by a series line and stub, followed by a second section for broadbanding, as described in Section 4.2.5. The circuit was designed by using the first series line and stub to achieve a match at band center, then adding the second to broadband it. The initial design was performed with the circuit simulator's tuning capability followed by a brief optimization.

Because of the more complex load network, the output design was a little trickier. The parallel LC part of the load is resonant near 12 GHz; this fact indicates that a second parallel LC resonance connected by a quarter-wave inverter section might be a good approach. The second resonance is realized by a shorted, high-impedance stub; this stub is useful for biasing the device as well. The circuit was adjusted somewhat by tuning, but numerical optimization was needed to perfect it. The relatively low-tech gradient optimizer was adequate for that function.

The third step is to connect the matching circuits to the device and to analyze it. The result is shown in Figure 7.16(a). The gain has a greater slope than that of the negative-image circuit, but the noise figure is fairly close to the optimum. The poor low-frequency noise figure can be attribut-

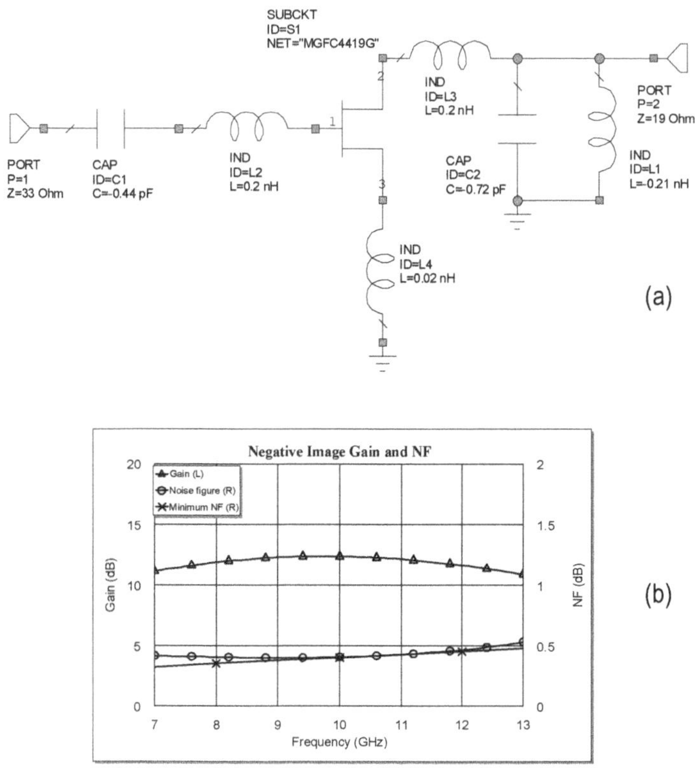

Figure 7.13 Initial design of the amplifier using the negative-image circuits. A simple RC output circuit was not adequate, but the parallel LC network provided flat gain (a). The resulting gain and noise figure compare well to the minimum noise figure of the device (b).

ed to the high sensitivity of the device to small source-impedance variations, a characteristic that is typical of microwave FETs. The result is close enough to the desired performance that simple optimization can perfect it. Figure 7.16(b) shows the performance after an optimization, again making use of the gradient optimizer.

As the purpose of this example is to illustrate the negative-image design approach, we shall leave it at this point. However, we note that a practical design will require the following additional measures:

1. So far, we have not addressed stability. We could have compared the negative-image matching circuits to device stability circles, but we expect the source and load impedances at the device to be different with

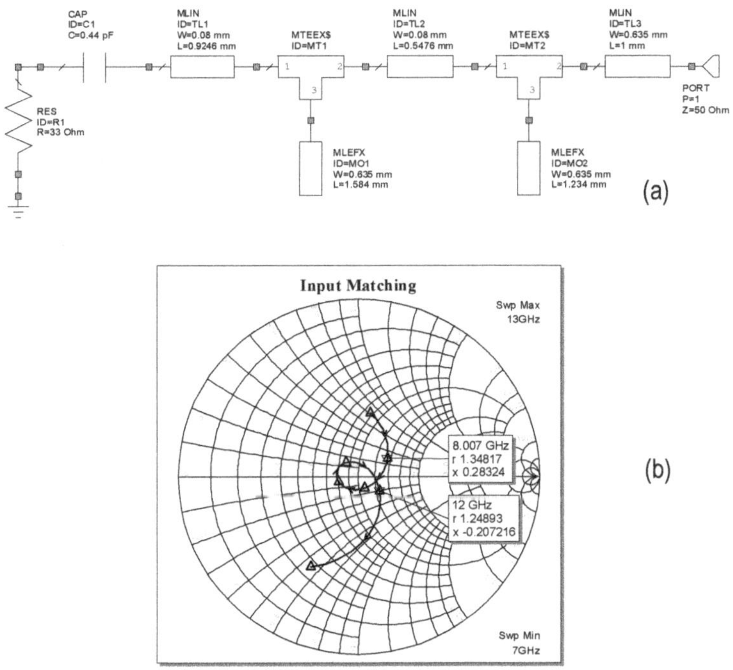

Figure 7.14 Input matching circuit design using the positive image as a load model (a) and its performance (b). This circuit was designed fairly easily, using the circuit simulator's tuning function.

the real matching circuits, so stability is best assessed after those circuits are designed. Since the matching circuits are lossless, we can create stability circles for the complete amplifier and assess stability by the closeness of the circles to the $\Gamma = 0$ point of the input and output reflection-coefficient planes.

As it happens, some of the circles are unacceptably close to the $\Gamma = 0$ point at in-band frequencies. As this is a broadband LNA design, in which Γ_S has a unique value, and Γ_L is constrained by the need for flat, broadband gain, we have limited degrees of freedom for dealing with this state of affairs. One option is to reduce the gain; in this design, the gain was optimized to 12 dB, but a lower value probably would move the value of Γ_L at the device farther from the unstable regions. If that

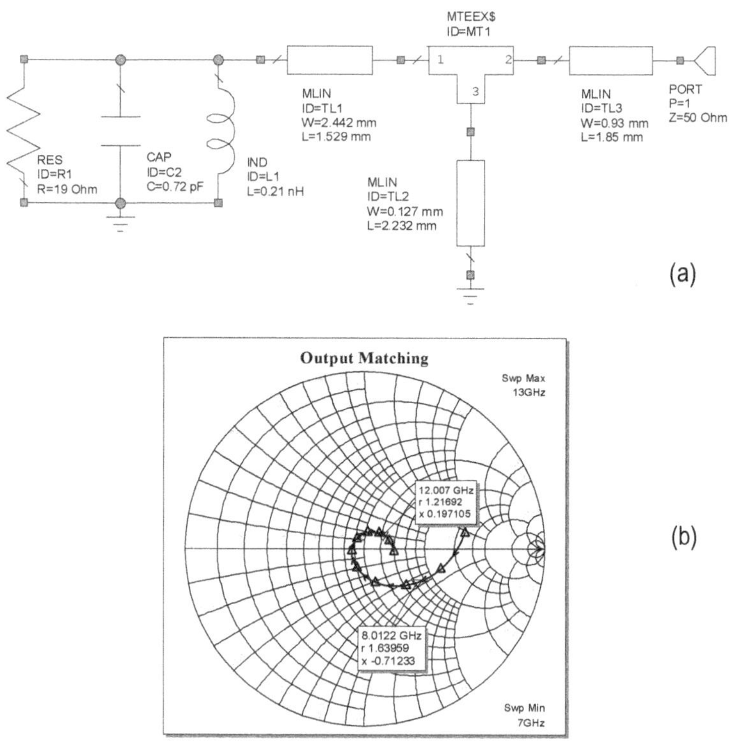

Figure 7.15 Output matching circuit (a) and performance (b).

idea fails, we could add resistive loading to the drain of the device. This might be enough to provide adequate stability margin. If not, resistive loading can be added to the gate, although it will have a strong, deleterious effect on the noise figure. Similarly, the gate tuning could be modified to move the Γ_S points farther from the stability circles, but that would also affect the noise figure and would require retuning of the output circuit. In the end, achieving good stability involves a trade-off with other aspects of performance.

2. The out-of-band gain should also be checked. We find that, on the whole, the out-of-band gain is significantly lower than the in-band,

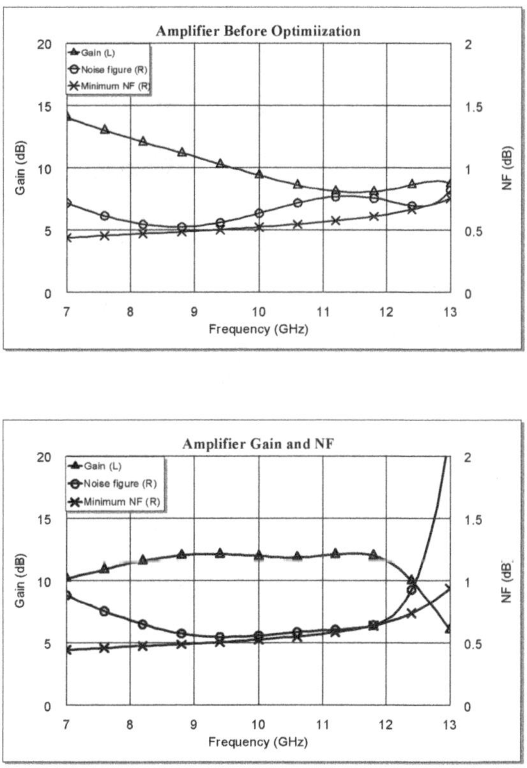

(a)

(b)

Figure 7.16 Gain and noise figure of the amplifier with the input and output matching circuits' original designs (a) and after optimizing the amplifier (b).

but a gain peak (~15 dB) exists near 1 GHz. This is largely caused by the low-pass nature of the input matching circuit. When bias circuits and dc blocking capacitors are included in the design, this peak should disappear.

3. The design needs bias circuits, of course. The output circuit has a shorted stub, which should be ideal for bias. The ground at the shorted end must be lifted and replaced by a bypass capacitor. Some additional resistive loading can be placed in series with the bias line to guarantee low-frequency, out-of-band stability as well, along with a second, high-value bypass capacitor to bypass low-frequency currents. A similar high-impedance, quarter-wave stub must be added at the input to provide gate bias; to minimize its effect on matching, it should be connected to a low-impedance point in the matching circuit. Again, resistive loading can be used in the bias circuit to guarantee good low-frequency stability.

7.3.4 Small-Signal Power Amplifier Design

The term *small-signal power amplifier* may seem self-contradictory. In fact, small-signal amplifiers are often designed like power amplifiers when intended for use as mixer local-oscillator drivers, oscillator buffer amplifiers, and similar applications. For such circuits, we design the output network according to power criteria. As noise figure is rarely a consideration, the input can be designed for flat bandwidth or a conjugate match.

Power amplifiers can be operated in a number of modes, called classes. The mode of operation described in Section 7.1.5 is called *class A*. It provides linear operation at the cost of relatively poor dc-to-RF efficiency. Other classes are possible, generally offering increased efficiency but poorer linearity. In this section we are exclusively concerned with class-A amplifiers, as these are most useful in the applications identified above. See [7.1–7.3] for more about power-amplifier classes. To keep the descriptions concrete, we assume that the amplifier uses a FET. The same considerations apply to BJT amplifiers.

7.3.4.1 Power Amplifier Design

In a power amplifier, the output is loaded to achieve maximum output power under constraints established by dc bias voltage and current. We noted in Section 7.1.5 that the load resistance achieving this is approximately

$$R_L = \frac{V_{dd}}{I_{dd}} \tag{7.60}$$

and the ideal, maximum output power is

$$P_d = \frac{1}{2}V_{dd}I_{dd} \tag{7.61}$$

This implies a dc-to-RF efficiency of 50%, although in reality 35% is more common for microwave class-A amplifiers. The rest of the dc input power must be dissipated in the device, and in small-signal devices, power dissipation, more than the bias voltage or current, usually limits the output power.

Consider, for example, a small-signal FET operating at a drain voltage of 3V and having 100-mA I_{dss}. To accommodate 100-mA peak current, it must be biased around 50 mA. This gives a convenient load resistance of approximately 60Ω and, from (7.61), an output power of 75 mW or almost 19 dBm. The dc power is 150 mW.

These are optimistic numbers, of course. If the efficiency is a more realistic 35%, the output power is closer to 50 mW or 17 dBm, and the power dissipation in the device at full output power is approximately 100 mW. However, the drain current in an ideal class-A amplifier is independent of RF drive level, so removal of the RF excitation causes the device to dissipate the full 150-mW dc power. Thus, the FET must be selected to allow dissipation of that full dc power.

The value of R_L given by (7.60) is only the resistive part of the load. It is, more precisely, the resistance seen at the terminals of the controlled current source that represents the channel in the FET model. For that reason, precise design of a power amplifier requires a nonlinear equivalent-circuit model of the device; S parameters alone are not sufficient to optimize the output power. Even so, class A is a linear mode of operation, so S-parameter design is applicable for determining small-signal gain, bandwidth, and port VSWRs, and with the estimates in (7.60) and (7.61), an acceptably accurate design usually can be produced. The designer must simply load the drain with R_L per (7.60) and add a shunt inductance to resonate the device's output capacitance. A good estimate of that capacitance can be derived from $\mathrm{Im}\{Y_{22}\}$ at the design frequency.

As the drain termination is set by power considerations, the only degree of freedom for adjusting the gain is the selection of Γ_s. Γ_s must be selected not only for flat gain, but to avoid the device's unstable regions and to prevent high out-of-band gain. Since Γ_L is set by power considerations, Γ_{in} at the gate is unique, so synthesizing a matching circuit is straightforward. If the bandwidth is moderate, a simple stub matching circuit may be all that is required; for larger bandwidths, or if Γ_{in} varies with frequency in a complicated manner, real-frequency synthesis may be a good approach. Negative-image modeling is also applicable.

7.3.4.2 Example: Small-Signal, Class-A Amplifier

In this example, we design a small-signal FET amplifier to provide its maximum output power in a linear, class-A mode from 9 to 11 GHz. We select a drain bias current of 50 mA and drain voltage of 3V; these are typical for small-signal devices.

The first step is to design the output matching circuit. From (7.60) we find that we need a load resistance of 60Ω and a reactive part that resonates the device's output capacitance. From the device's Y_{22}, we obtain an estimate of 0.095 pF for that capacitance, –168Ω reactance. This is resonated by a shunt inductance of +168Ω reactance, which is most easily realized by a shunt stub. A 100Ω shorted stub 59 deg in length provides the required reactance. A quarter-wavelength transformer converts the 50Ω load to the desired 60Ω.

The stub is used as a bias-insertion point, so it must be bypassed rather than grounded. Although we have not attended to this detail in other examples, we do so here, as it is helpful in showing how the bias circuit can enhance stability. Stabilizing networks can be used at the input, as well, to enhance stability further. Figure 7.17 shows the output matching circuit and compares Γ_L at the FET's drain to stability circles from 1 to 16 GHz. The drain bias circuit includes a bypass capacitor described by a chip-capacitor model and a series resistance. The capacitor effectively bypasses high frequencies, and for low frequencies, the far end of the loading resistor can be bypassed by a larger capacitor. Partly as a result of this network, the locus of Γ_L values is well removed from the unstable region throughout the 1- to 16-GHz range, and the load impedance of 60Ω is realized accurately.

The input circuit provides a conjugate match to the device, with perhaps a little mismatching around 9 GHz to ensure a flat passband. To prevent high out-of-band gain, the structure of the network is designed to provide some degree of filtering. As such, it uses two shorted shunt stubs and a series line; both shunt stubs are terminated with capacitors and resistors to enhance stability. The capacitors are described by a chip-capacitor model, which includes its Q and both series and parallel resonances. Most important in this application is to include the series resonance. Figure 7.18 shows the input matching circuit, its impedance locus, and input stability circles from 1 to 16 GHz. Without the resistive loading at the ends of the stubs, the Γ_S locus would be very close to the edge of the chart at low frequencies and close to the unstable regions at approximately 7 GHz. The loading moves the locus a bit toward the center of the Γ_S plane, providing a better stability margin with no impact on the in-band performance. Figure 7.18 also shows the input reflection coefficient of the complete amplifier

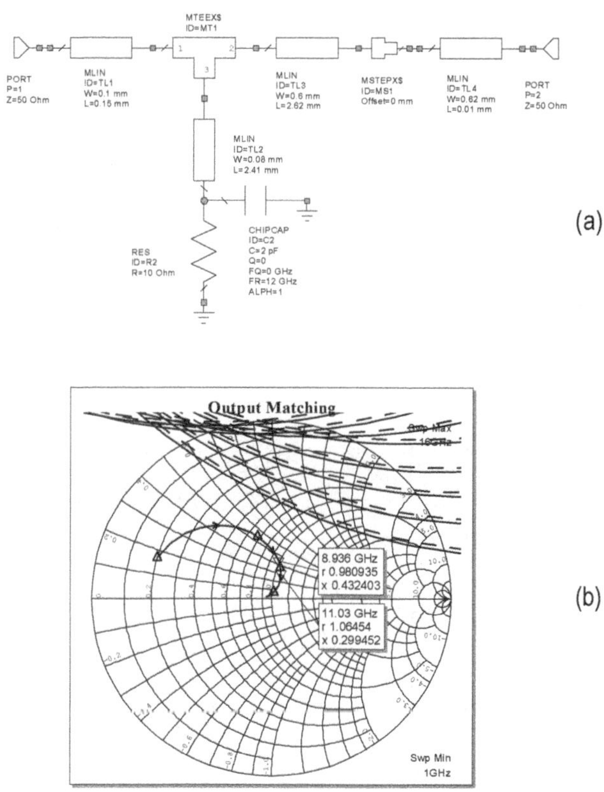

Figure 7.17 Output matching circuit (a) and load reflection coefficient at the device (b) including output stability circles, 1 to 16 GHz. To optimize the matching circuit, the left end is terminated in a shunt 60Ω resistor and 0.095-pF capacitor representing the desired resistive load and the device's output capacitance. The RC network at the end of the stub models the bypassing necessary to use the stub for drain bias.

from 8.5 to 11.5 GHz; the input VSWR is better than 1.5 over the entire 9- to 11-GHz band. Finally, Figure 7.19 shows the amplifier gain. It exhibits a gain slope of less than 1 dB over the 9- to 11-GHz range.

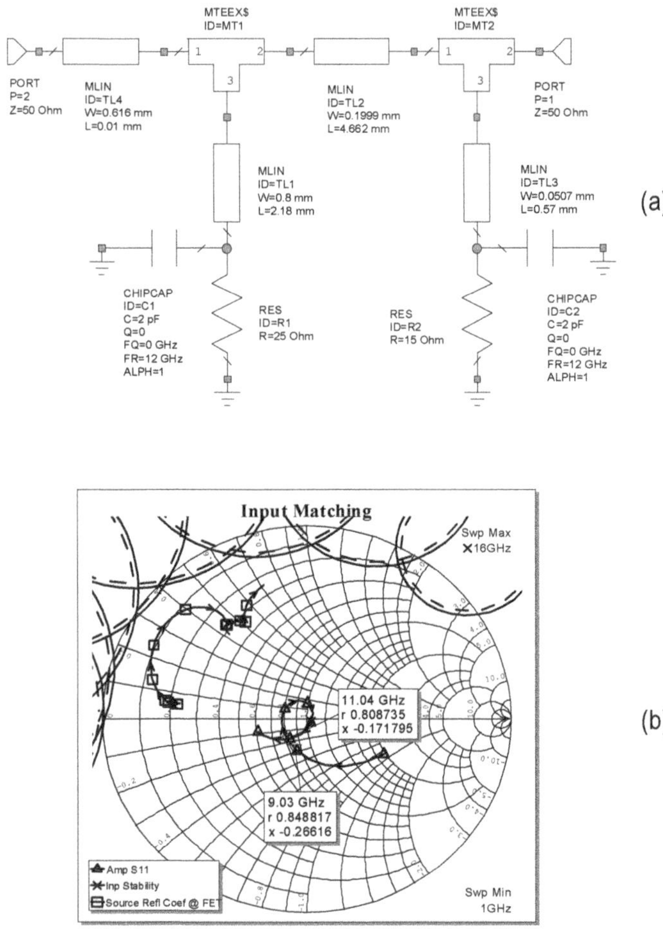

Figure 7.18 Input matching network (a), source reflection coefficient at the device, and input reflection coefficient of the complete amplifier (b). The RC networks are used to increase stability and allow the stubs to be used for gate bias.

7.3.5 Amplifier Design for Dynamic Range

7.3.5.1 Dynamic Range in FET Amplifiers

The dynamic range of a component or system is the range of input power (or, occasionally, output power) over which it can be successfully operated. At the low end it is limited by noise and at the high end by distortion. This is a soft distinction, however; in modern communication systems, the minimum usable signal level is determined by the acceptable bit error rate, which depends in a complex manner on noise level and the type of modulation. Similarly, at the high end, some kinds of systems are more tolerant of distortion than others.

Nevertheless, a useful figure of merit is the input third-order intercept point (IP_3) minus the noise figure, both in decibel quantities. In Section 7.2.1, we discussed ways to optimize the noise figure, and as long as we have all the noise parameters of the device, the criterion for obtaining minimum noise figure is straightforward. Less clear is the criterion for minimizing distortion; although we discussed distortion generally in Section 7.1.6, that theory gives little insight for maximizing the input IP_3 of an amplifier.

Considerable research in nonlinear distortion in FETs has shown that the drain current, I_d, as a function of gate voltage, V_{gs}, when the FET is operated in current saturation, is the dominant nonlinearity. This is sometimes called the device's "nonlinear transconductance," a poorly chosen term, as transconductance is, by definition, a linear quantity. In any case, to opti-

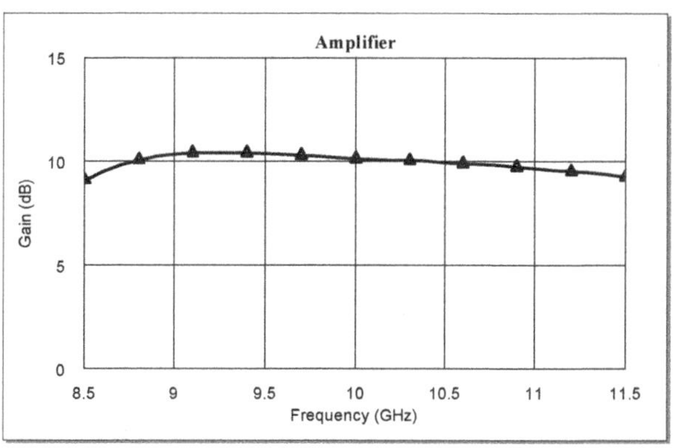

Figure 7.19 Gain of the complete amplifier.

mize the IP_3 we must bias the device in the most linear part of its I_d vs. V_{gs} curve. This is usually around 50% of I_{dss}.

Device size has a strong effect on the IP_3 as well. IP_3 scales in proportion to gate width in FETs or emitter area in bipolar devices. To achieve this scaling, however, the drain or collector current must be increased in proportion to the increase in device size. Additionally, minimum noise figure generally increases moderately with device size.

If we had an appropriate nonlinear model for the device, which included all noise sources, we could optimize the IP_3 numerically. The process is computationally expensive, however, and an appropriate nonlinear model is often unavailable. Fortunately, we can employ the method described in [7.7], which is based on an available-gain design procedure. It allows optimization of the IP_3, although it does not provide quantitative information about the precise value. The procedure is as follows:

1. To minimize IM, the device's drain current bias is set to approximately 0.5 I_{dss}. If the noise figure at this bias current is too high, the drain current can be reduced, but IP_3 is reduced as well. In any case, the device should be biased well away from the knee of the FET's I_d vs. V_{gs} curve.
2. A set of available-gain circles for the device is drawn at a number of points over its frequency range.
3. The values of Γ_s providing maximum output IP_3 are those near the "bottoms" of the curves; that is, closest to the real axis of the Γ_s plane, toward the high-resistance end.
4. Noise circles, measured at the bias point chosen in the first step, can be plotted on the chart as well. This allows a trade-off between noise figure and IP_3.
5. The output IP_3 resulting from this process tends to be fairly constant with gain. Since the input IP_3 is the output IP_3 minus the gain, in decibel quantities, it is best not to make the gain too great.
6. The input matching network is synthesized from the selected Γ_s data; the real-frequency method (Section 4.6.2) is ideal for this process.
7. With the input network in place, the output impedance of the device can be determined and a conjugate-match output network synthesized.

This process is illustrated by the example in Section 7.3.5.3.

7.3.5.2 Wide Dynamic Range Bipolar Transistor Amplifiers

Although their noise figures are generally higher than those of microwave FETs designed for the same frequency range, bipolar transistors achieve IP_3 values that can be surprisingly high. This is especially true of heterojunction bipolar transistors (HBTs), which exhibit even better linearity than silicon BJTs. The high linearity might be surprising, since the curve of collector current I_c as a function of base-to-emitter voltage V_{be} is an exponential function, one of the strongest nonlinearities found in nature. The reason for the good linearity has been traced to an odd cancellation phenomenon between the distortion products generated in the resistive and reactive parts of the device's base-to-emitter junction. This cancellation occurs at frequencies above a cutoff given by

$$f_c = \frac{1}{2\pi R_{be} C_{be}} \tag{7.62}$$

where R_{be} is the base-to-emitter resistance and C_{be} is the base-to-emitter diffusion capacitance. Below this cutoff, cancellation does not occur, and an intercept point several decibels worse than at higher frequencies can be expected. This phenomenon is described in detail in [7.8].

Further conditions for the optimization of BJT and HBT amplifiers have not been studied extensively. Empirically, we find that a conjugate-matched output provides a practical optimum for output IP_3, and the load reflection coefficient should be kept well away from the output stability circle. The noise optimization described in Section 7.2.2 is valid for all types of two-ports, so it is valid for bipolar devices as well as FETs; however, in BJTs and HBTs the noise figure generally is greater than in FETs but not as sensitive to source impedance. For these reasons, an available-gain approach seems appropriate for high-IP_3 bipolar amplifiers. This allows the source impedance to be selected for a reasonably good noise figure and gain, along with a comfortable stability margin.

The IP_3 of a bipolar transistor increases with collector bias current at least up to the maximum f_t point; as all aspects of performance deteriorate rapidly above this value, it makes little sense to bias the device beyond this point. Noise figure also increases with bias current, but not as rapidly as in FETs. Therefore, bipolar devices should be biased to fairly high collector currents when intended for high dynamic range applications.

7.3.5.3 Example: Wide Dynamic Range FET Amplifier

In this section we describe the design of a 9- to 11-GHz wide dynamic range microwave FET amplifier. As the output circuit is straightforward, we consider only the design of the input matching circuit.

We begin by plotting available gain and noise-figure circles using S parameters measured at a drain current that is an appropriate trade-off between noise figure and low distortion. We use a drain-current value that is well above the knee of the I_d vs. V_{gs} characteristic, so the device has good linearity, but not so great as to increase the noise figure severely. We must plot a number of points over the band and select appropriate values of Γ_s at each frequency. Those values must provide a low noise figure but must also be in the low-distortion region. This is a messy plot; to keep things clear, we show only one frequency point in Figure 7.20. The curves for 11 and 13 dB intersect a low-noise region of Γ_s in the right places; the 13-dB curve is a little close to the unstable region, so we choose the 11-dB one.

From the curves, we can select a set of Γ_s values that are located in the desired region. This requires some judgment, and thus is best done manually. We look for points that are inside the 0.5-dB noise circle and are on the 11-dB gain circle; the results are listed in Table 7.1.

Table 7.1

Frequency (GHz)	Normalized Source Impedance
9.0	$1.32+j1.91$
9.5	$1.80+j1.45$
10.0	$1.87+j0.95$
10.5	$1.75+j0.54$
11.0	$1.56+j0.25$

The next step is to synthesize a matching network that realizes this set of source impedances. In this case, we use a real-frequency synthesis to obtain the lumped-element matching circuit shown in Figure 7.21(a). The shunt inductor in that circuit can be realized easily by a transmission-line segment, but the series elements are a greater problem. The series LC suggests realization by an open-circuit stub, which could be converted to a shunt stub as described in Section 4.5.2. The resonant frequency, however, is almost 13 GHz, well outside of our bandwidth, so the accuracy of realizing this as a series stub, based on slope parameters, may not be good, and the conversion to a shunt stub similarly could be inaccurate. The process may still be useful, however, as an initial estimate, which can be optimized numerically.

From slope-parameter considerations (4.46), we find the characteristic impedance of the equivalent open-circuit stub to be 143Ω. From (4.50), we obtain the characteristic impedance of a quarter-wave inverter and a shorted stub. Since there is no impedance transformation, the inverter's characteristic impedance is 50Ω, and the stub is 17.5Ω. These are one-quarter wavelength long at the resonant frequency of the LC combination, 12.9 GHz. The shunt inductor is easily realized by an eighth-wavelength shorted stub of 72Ω impedance. These steps are shown in Figure 7.21(b) and (c).

The matching circuit is then realized in microstrip. Because of the rather severe approximations in the original circuit, optimization is necessary. The result, after a short optimization, is shown in Figure 7.22 along with a plot of the desired and realized values of Γ_S. At this point, it is a simple matter to plot input stability circles and Γ_S on the same graph. These should be evaluated over a wide frequency range, especially frequencies below the

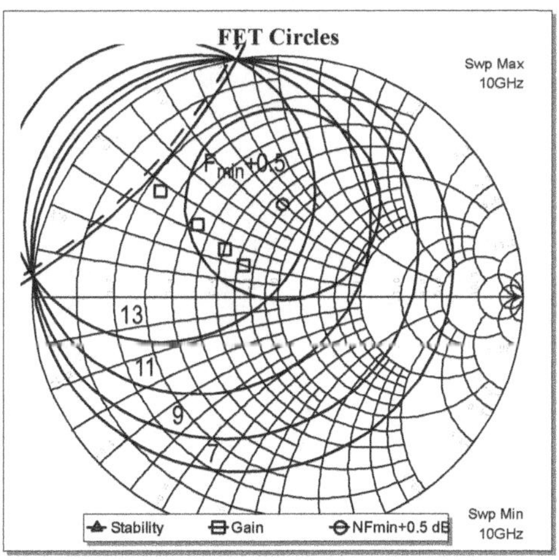

Figure 7.20 Stability, noise, and available gain circles at 10 GHz. The circles represent F_{min} + 0.5 dB and gains of 7, 9, 11 and 13 dB. The FET includes bond wires.

lower edge of the passband. Potential stability problems then should be evident immediately.

The output matching circuit is straightforward. As this is an available-gain design, compensation for the device's gain slope has been provided by the input matching network, so the output is designed for an ordinary con-

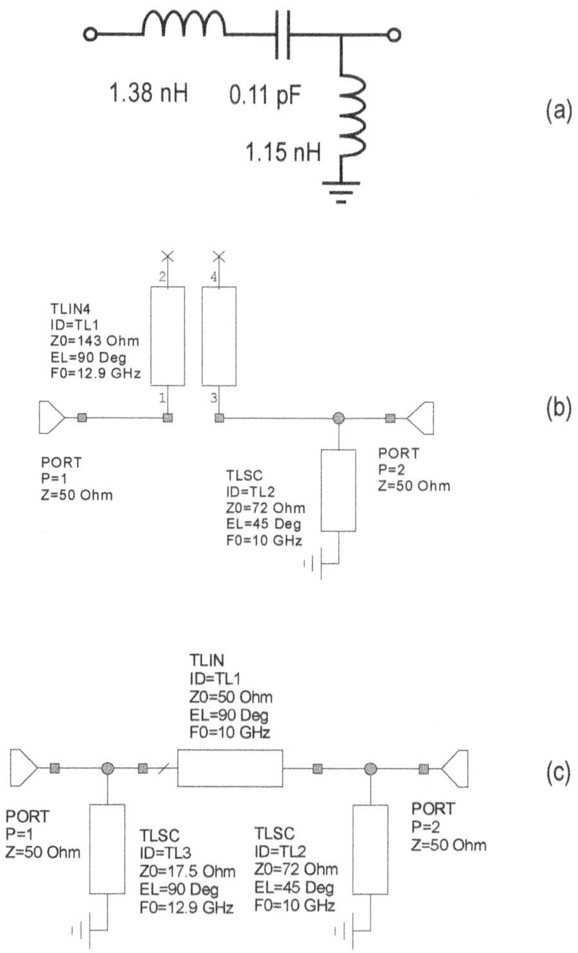

Figure 7.21 Steps in realizing the input matching circuit: (a) lumped-element circuit generated by real-frequency software; (b) approximate distributed equivalent circuit; (c) conversion of the series stub to shunt.

jugate match. The input matching network is connected to the device and the output reflection coefficient is computed. Fortuitously, the entire output impedance locus is close to the center of the Γ_L plane, so simple stub matching at the center frequency should be adequate. As with the LNA of Section 7.3.3.2, appropriate tests of stability and out-of-band gain are necessary to complete and validate the design.

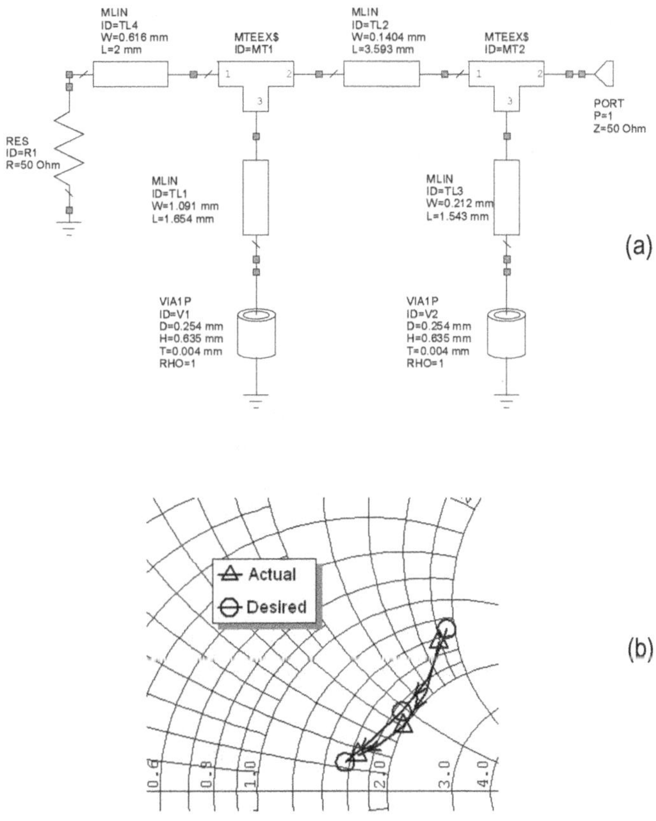

Figure 7.22 Input matching circuit (a) optimized and realized in microstrip, compared to the desired and realized values of Γ_s from 9 to 11 GHz (b).

References

[7.1] Maas, S. A., *Nonlinear Microwave and RF Circuits*, Norwood, MA: Artech House, 2003.

[7.2] Cripps, S., *RF Power Amplifiers for Wireless Communication* (2nd ed.), Norwood, MA: Artech House, 2006.

[7.3] Walker, J., *Handbook of RF and Microwave Power Amplifiers*, Cambridge: Cambridge University Press, 2012.

[7.4] Maas, S. A., *Noise in Linear and Nonlinear Circuits*, Norwood, MA: Artech House, 2005.

[7.5] Yngvesson, S., *Microwave Semiconductor Devices*, Dordrecht: Kluwer, 1991.

[7.6] Sze, S. M., *Physics of Semiconductor Devices* (2nd ed.), New York: Wiley, 1981.

[7.7] Crosmun, A., and S. Maas, "Minimization of Intermodulation Distortion in GaAs MESFET Small-Signal Amplifiers," *IEEE Trans. Microwave Theory Tech.*, Vol. MTT-37, 1989, p. 1411.

[7.8] Maas, S. A., D. Tait, and B. Nelson, "Intermodulation in Heterojunction Bipolar Transistors," *IEEE Trans. Microwave Theory Tech.*, Vol. MTT-40, 1992, p. 442.

Chapter 8

Balanced and Quadrature-Coupled Circuits

A balanced circuit is an interconnection of basic circuit blocks by baluns, transformers, or 180-degree hybrids. Circuits interconnected by 90-degree hybrids are not true balanced structures, but often are called balanced; we discuss them in this chapter but call them *quadrature-coupled* structures. Balanced circuits are used for their ability to reject certain even-order distortion components or simply for power combining. Quadrature-coupled circuits have unique properties, including the ability to provide low port VSWRs when those of the individual circuit blocks are poor. They are especially useful in microwave amplifiers.

8.1 90- and 180-Degree Hybrid Junctions

Hybrid junctions, or more commonly *hybrids*, are matched, passive, four-port devices. In Section 3.2.5 we derived the properties of passive, matched four-ports and determined that only two fundamental types were possible, which we call 180-degree and 90-degree (or *quadrature*) hybrids. The parallel-line directional coupler of Section 2.1.4 is a type of quadrature hybrid when its coupling is 3 dB. It is often realized as a Lange hybrid (Section 2.2.4). Other kinds of structures, such as the branch-line hybrid we examine in Section 8.1.2.2, also can realize a quadrature hybrid.

8.1.1 Characteristics of Hybrids

Hybrids are four-port devices having the following ideal properties:

1. All ports are matched; $S_{11} = S_{22} = S_{33} = S_{44} = 0$. The port-normalizing impedances, however, need not be equal.
2. The four-port is lossless; that is, there is no internal power dissipation.
3. A sinusoidal signal applied to any one port emerges from two other ports with a 3-dB lower level.
4. The phases of two outputs of a quadrature hybrid differ by 90 degrees.
5. In a 180-degree hybrid, the output phase difference is either 180 degrees or 0 degrees, depending on which port is excited.
6. The remaining port is isolated; that is, no signal emerges from it.

In Section 2.3.1 we made the point that a balun is not a hybrid, as its properties are very different. A 180-degree hybrid can replace a balun, however, in some instances.

Hybrids are described schematically as in Figure 3.4. The paths indicate the transmission through the hybrid and the phase shift. The power division in all cases is 3 dB. In the 180-degree hybrid of Figure 3.4(a), the ports are sometimes designated the Σ and Δ ports; if, for example, signals are applied to mutually isolated ports 1 and 2, the signal at port 3 is proportional to their sum, making it the Σ port, and the output at port 4 is the difference; therefore, the Δ port. The designation of Σ and Δ ports depends on the pair of ports that is excited; if ports 3 and 4 are excited, ports 1 and 2 are the Σ and Δ ports, respectively.

No similar terminology exists for the quadrature hybrid in Figure 3.4(b). That hybrid, however, has an intriguing symmetry, in which a signal applied to *any* port is split between the two output ports with a 90-degree phase difference. This property makes the hybrid invaluable for creating many kinds of hybrid-coupled circuits.

While the schematic representations in Figure 3.4 are used regularly, the phase shifts implied by the figure are admittedly somewhat misleading. They seem to imply that the phase shift through the hybrid is 0, 90, or 180 degrees. Per (3.84) and (3.85), this is the case only for an ideal hybrid; real hybrids may include some extra phase shift, which may or may not depend on frequency. As long as that phase shift is identical in each path, so the ideal output phase differences are retained, that additional phase delay is of no consequence in the use of hybrids in practical high-frequency circuits. Real hybrids also have loss, which is normally very small, but in such applications as low-noise amplifiers, it can be significant.

8.1.2 Quadrature Hybrids

8.1.2.1 Coupled-Line Hybrid

The directional coupler of Section 2.1.4, with 3-dB coupling, is a type of quadrature hybrid. We noted in Section 2.2.4 that it was generally not possible to realize edge-coupled directional couplers having 3-dB coupling, although such "tight" coupling can be achieved with broadside-coupled strips. The Lange coupler, Section 2.2.4, achieves 3-dB coupling (and should more correctly be called a hybrid, as we shall in the rest of this chapter) by splitting the parallel strips into a number of interleaved strips, increasing the number of adjacent edges and thus the coupling. On substrates having high dielectric constants, such as alumina ceramic, four strips are usually adequate; on materials having low dielectric constants, such as fused silica, it is not unusual to have six strips.

The S parameters of a hybrid realized by a homogeneous quadrature coupler are found from (2.18) and (2.20) after applying (3.115). We note that the input reflection coefficients of such a coupler are zero at all frequencies, as is its isolation. Then

$$S_{11} = S_{22} = S_{33} = S_{44} \tag{8.1}$$

$$S_{21} = S_{12} = S_{43} = S_{34} = 0 \tag{8.2}$$

$$S_{31} = S_{13} = S_{42} = S_{24} = \frac{\sqrt{1-c^2}}{\sqrt{1-c^2}\cos(\theta) + j\sin(\theta)} \tag{8.3}$$

$$S_{41} = S_{14} = S_{32} = S_{23} = \frac{jc\sin(\theta)}{\sqrt{1-c^2}\cos(\theta) + j\sin(\theta)} \tag{8.4}$$

where c is the voltage coupling factor, which, for an ideal hybrid, must be $\sqrt{2}/2$, and θ is the length in degrees of phase, 90 degrees at the hybrid's center frequency. We have used the port numbering of Section 3.2.5, which differs from that of Chapter 2.[1]

1. The difference in port numbering in the two chapters may be confusing, unfortunately, but it is essential. Numbering the four-port in Section 3.2.5 as in Chapter 2 would destroy the elegant, biconjugate structure of the matrices (3.82), and using the numbering of Section 3.2.5 in Chapter 2 would simply be bizarre. The port numbering throughout this chapter is that of Section 3.2.5 and Figure 3.4.

A remarkable property of a coupled-line directional coupler or hybrid is the frequency independence of its output phase difference. We see, by inspection, that the phase difference between ports 3 and 4, when port 1 is excited, is 90 degrees for all values of θ. This is true only of hybrids realized in homogeneous transmission media, such as stripline; in nonhomogeneous media, such as microstrip, the even- and odd-mode phase velocities differ, and that difference affects the output phase difference. In practical hybrids, however, the difference is small enough that it is usually tolerable.

The amplitude balance of a coupled-line quadrature hybrid is frequency dependent. Figure 2.4 shows the coupling, in decibels, of an ideal hybrid. The 0.5-dB bandwidth is approximately 50%, but it can be extended by overcoupling the hybrid at center frequency; then, the 0.5-dB bandwidth increases to approximately 90%. Such hybrids can be used in quadrature-coupled amplifiers having even wider bandwidths, however, since the imbalance in the output hybrid tends to compensate for that of the input hybrid. This is discussed further in Section 8.2. We shall see, however, that a significant increase in noise figure at the band edges can result.

8.1.2.2 Branch-Line Hybrid

The branch-line hybrid, shown in Figure 8.1(a), is another type of quadrature hybrid. It is a very simple structure; the lengths of all lines are $\lambda / 4$, the series lines' impedances are $Z_0 / \sqrt{2}$, and the shunt lines' impedances are Z_0. All ports have the same normalizing impedance, Z_0.

Analyzing a branch-line hybrid by even- and odd-mode analysis is possible, but the performance of the hybrid hardly justifies the trouble. The loss and isolation bandwidths are shown in Figure 8.1(b); they indicate that the ideal hybrid's bandwidth is only 10% for a 20-dB return loss and isolation. The effect of the tee junctions in practical hybrids is to reduce this bandwidth further. Nevertheless, the simplicity of the structure makes it attractive for applications where broad bandwidth is not needed.

Because of the narrow bandwidths of single-section hybrids, branch-line hybrids are most practical when realized as multisection structures, which have significantly greater bandwidth. Design information for such hybrids can be found in [1.9].

8.1.2.3 Lumped-Element Quadrature Hybrids

Hybrids can be realized by lumped elements as well as distributed ones. Such circuits are especially valuable at low frequencies, where distributed hybrids might be impractically large. Figure 8.2 shows a common realiza-

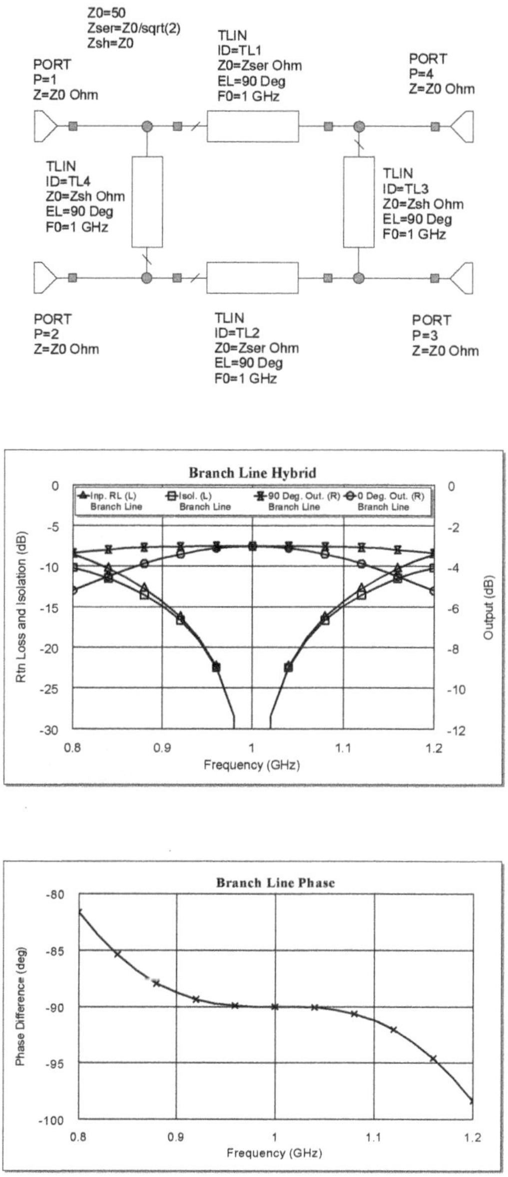

(a)

(b)

(c)

Figure 8.1 A branch-line quadrature hybrid (a), its amplitude performance (b), and phase of port 3 relative to port 4 (c).

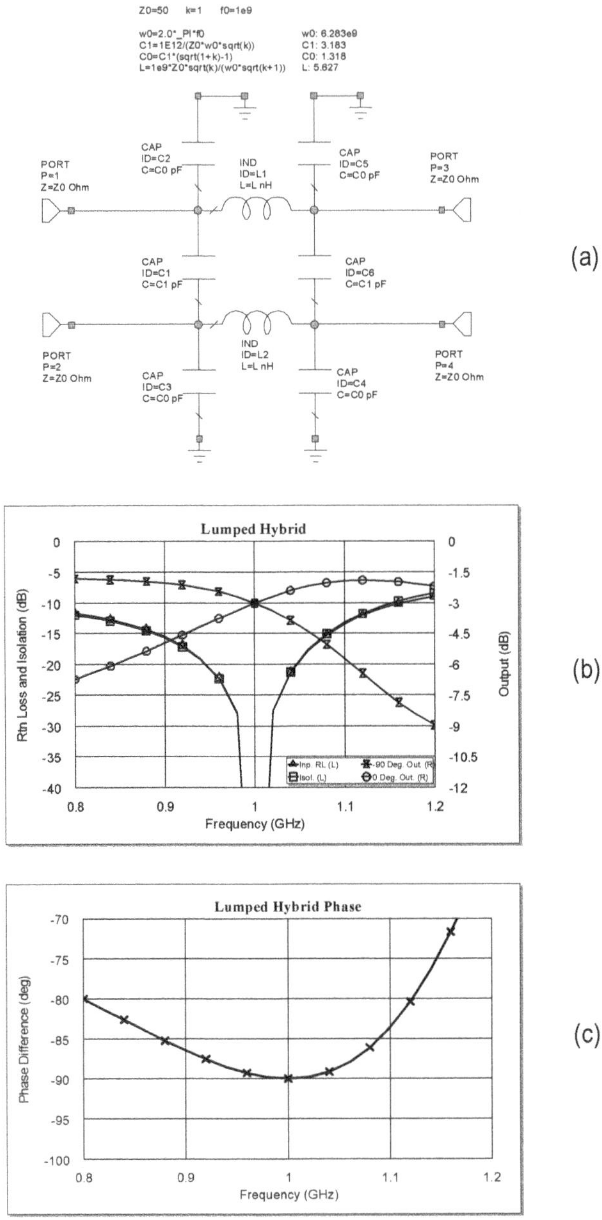

Figure 8.2 Lumped-element quadrature hybrid (a) and performance (b, c). This is a narrowband circuit.

tion of a lumped-element quadrature hybrid along with plots of its performance [8.1]. The bandwidth of the structure is relatively narrow, approximately 5%. Although this is very narrow, it should be adequate for many kinds of circuits.

The design equations for this hybrid are simple:

$$C_1 = \frac{1}{Z_0 \omega_0 \sqrt{k}} \tag{8.5}$$

$$C_0 = C_1(\sqrt{k+1} - 1) \tag{8.6}$$

$$L = \frac{Z_0 \sqrt{k}}{\omega_0 \sqrt{k+1}} \tag{8.7}$$

where ω_0 is the center frequency, C_1 is the value of the four capacitances to ground, C_0 is the value of the two branch capacitances, and L is the inductance. The constant k is the power ratio of the two outputs; for a hybrid, $k = 1$. The port numbering is consistent with Figure 3.4.

Figure 8.3(a) shows another type of lumped-element quadrature hybrid [8.2, 8.3]. It is, in fact, a cascade of two basic hybrids; the circuit can be divided in two to create a usable, but less broadband, hybrid. The single section is limited, however, by a cutoff frequency, f_c, given by

$$f_c = f_0 \frac{4n}{\pi} \tag{8.8}$$

where n is the number of sections and f_0 is the center frequency. When $n = 1$, the cutoff is only 27% above f_0, but for two sections, it is more than double f_0, which is well beyond the bandwidth limits imposed by the LC sections. A disadvantage of this circuit is the need for controlling the coupling factor of the inductors.

The output ports are numbered in Figure 8.3 according to our established convention. Note that the output ports are those of the top right and lower left in the figure; the isolated port is the one in the bottom right corner. The port having the 90-degree phase delay is in the top right.

The design equations are as follows:

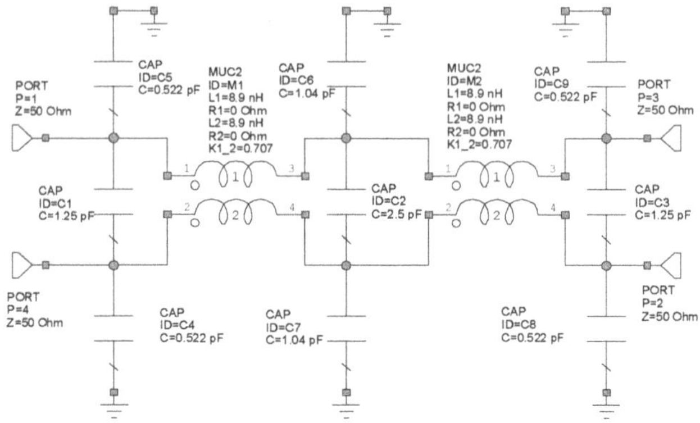

(a)

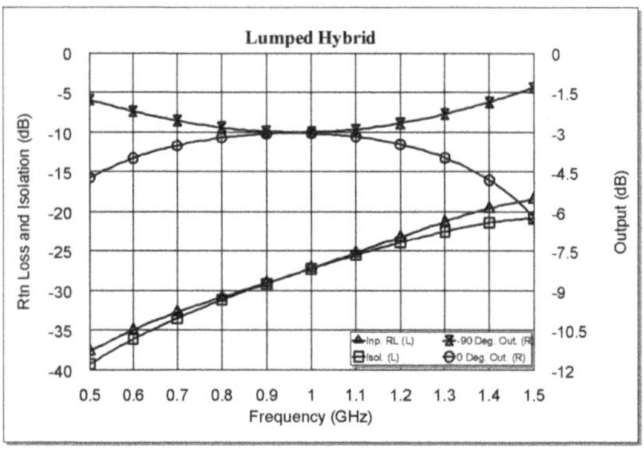

(b)

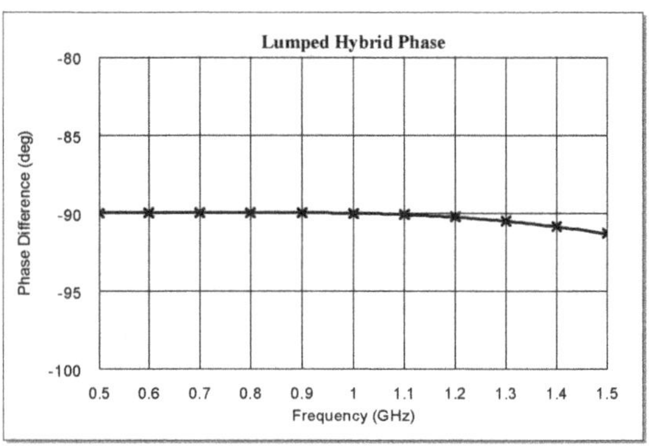

(c)

Figure 8.3 Broadband, lumped-element quadrature hybrid (a); amplitude (b) and phase (c) performance.

$$K = \sqrt{c_p} \tag{8.9}$$

$$M = KL \tag{8.10}$$

$$C_0 = \frac{L - M}{2Z_0^2} \tag{8.11}$$

$$C_1 = \frac{KC_0}{1 - K} \tag{8.12}$$

$$f_0 = \frac{1}{4\pi\sqrt{2(L + M)C_0}} \tag{8.13}$$

where K is the coupling factor of the inductors, c_p is the power coupling factor; for a hybrid, $c_p = 0.5$. C_0 and C_1 are the *single-section* capacitance to ground and the branch capacitance, respectively, as in the hybrid of Figure 8.3. Since this hybrid is a cascade of two single sections, the capacitors in the center of the structure represent parallel combinations of C_0 and C_1, so they must be doubled. Finally, L is the self inductance of the inductors and M the mutual inductance; in any set of coupled inductors L_1 and L_2,

$$K = \frac{M}{\sqrt{L_1 L_2}} \tag{8.14}$$

In this case, the two inductances are equal, so $K = M / L$.

Figure 8.3(b) and (c) show the performance of the hybrid in Figure 8.3(a) centered at 1 GHz. The bandwidth for a 0.5-dB imbalance is more than 25%, and that could be increased by overcoupling slightly at the center frequency. The phase difference is remarkably flat over the 3:1 bandwidth shown, as are the return loss and isolation.

8.1.3 180-Degree Hybrids

8.1.3.1 Rat-Race Hybrid

The rat-race hybrid is probably the most widely used 180-degree hybrid in existence. It is capable of moderately broad bandwidth, 15 to 20%, depending on the needs of the circuit. Variations on the rat-race theme can improve the hybrid's bandwidth or provide unequal power splits.

The rat-race hybrid is shown in Figure 8.4. It consists of a ring 1.5 wavelengths in circumference having a characteristic impedance of $Z_0\sqrt{2}$. The ports are located at intervals of one- and three-quarter wavelengths, as shown in the figure. The path between ports 2 and 4, as in Figure 3.4, is the 180-degree one.

A heuristic development of the hybrid's operation is as useful as any formal, mathematical one. Imagine that port 2 is excited. Because the lengths of the two paths to port 1 differ by one half wavelength, port 1 is a virtual ground, making it the isolated port. Then, because of the quarter-wavelength paths from port 1 to ports 3 and 4, port 1 presents an open circuit to those ports, so it has no effect on them. The hybrid then devolves to a simple circuit in which port 2 is connected to port 3 by a quarter-wave transformer and to port 4 by a three-quarter-wave transformer, thus providing equal power to those ports. The extra half wavelength of line to that port, however, provides an additional 180-degree phase shift. The same kind of reasoning can be applied when other ports are excited.

Figure 8.5 shows the performance of the hybrid when port 2 is excited; the bandwidth for the 0.5-dB imbalance is a little over 20%. The port return loss and isolation over this bandwidth are better than 20 dB.

8.1.3.2 Rat-Race Hybrid with Unequal Power Division

It is possible to design a rat-race hybrid to have unequal port couplings. This could be useful in extending the bandwidth by overcoupling the device at the center of its passband; the designer need only modify the impedances of the lines between the ports. Suppose we excite port 2 and we wish

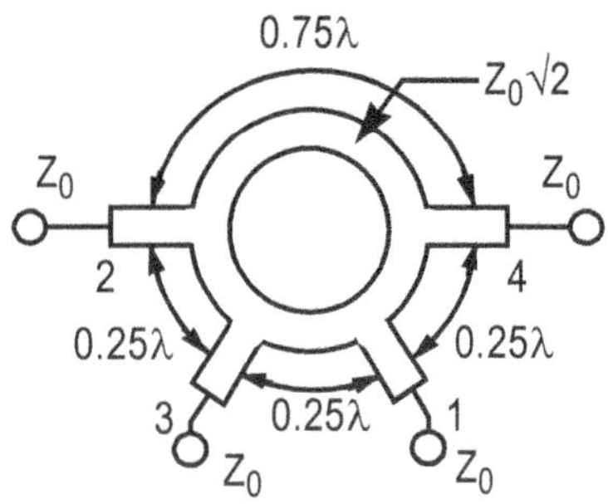

Figure 8.4 Rat-race hybrid.

to have a power difference of 3 dB between ports 3 and 4, with port 3 having the higher power. Then

$$\frac{P_3}{P_4} = c_p \tag{8.15}$$

where P_n is the power at port n. The characteristic impedances of the lines in the ring become

$$Z_{01} = Z_0\sqrt{\frac{1 + c_p}{c_p}} \tag{8.16}$$

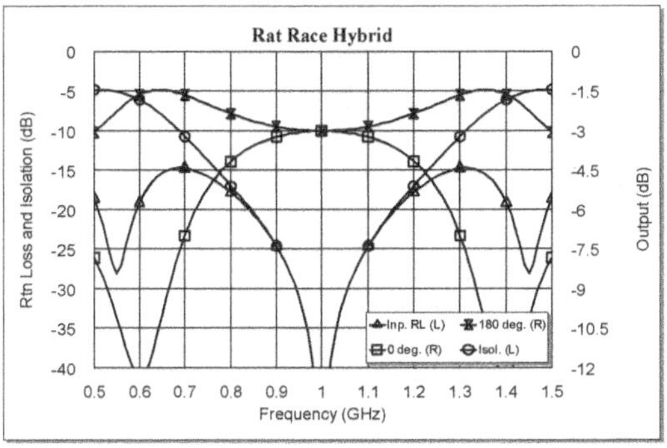

(a)

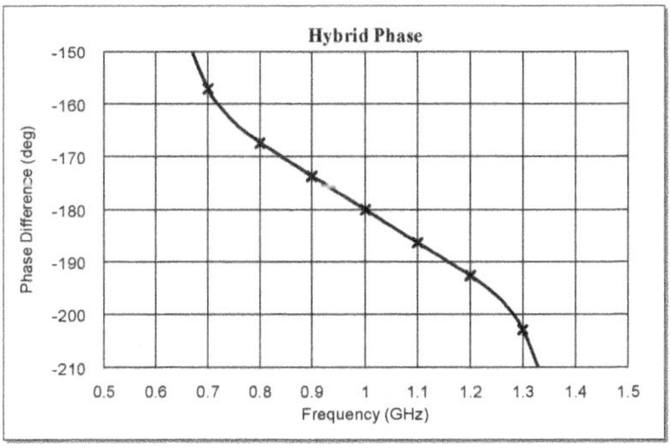

(b)

Figure 8.5 Rat-race hybrid performance when port 2 is excited: power split, isolation, and input return loss (a); phase difference at the output ports (b).

$$Z_{02} = Z_0\sqrt{1 + c_p} \tag{8.17}$$

Z_{01} is the impedance of the lines from port 2 to port 4 and from port 3 to port 1; Z_{02} is used for the remaining lines of the ring. Reversing this pattern would make the power at port 4 greater than port 3 by the same factor. In fact, the design rule is straightforward: Z_{01} and Z_{02} alternate around the ring, and an output port using a Z_{01} path from the excited port is the higher-power port.

8.1.3.3 Broadband Rat-Race Hybrid

In the rat-race hybrid shown in Figure 8.6, the 3λ / 4 path has been replaced by a pair of coupled lines [8.4]. This provides a frequency-independent 180-degree phase shift needed by that path, replacing a half-wavelength of transmission line, whose phase shift is anything but frequency independent. The circuit is capable of 60% bandwidth for 15-dB port return loss; the bandwidth for the 0.5-dB output imbalance is somewhat greater.

A problem in this hybrid is the high values of even- and odd-mode characteristic impedance for the coupled-line section; for a 50Ω port impedance, these are 170.7Ω and 29.3Ω, respectively. These values are too high to be realized in many transmission media, particularly microstrip on

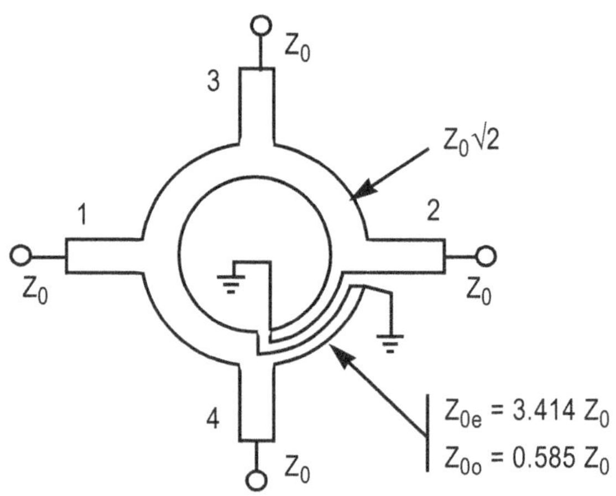

Figure 8.6 A broadband rat-race hybrid is realized by replacing the 3λ / 4 path with a set of coupled lines. This reduces the variation in phase shift of that path with frequency.

an alumina ($\varepsilon_r = 9.8$) substrate. The author has found empirically that more suitable values of these impedances can be used, although the bandwidth is somewhat restricted. Realizing the grounds for the coupled-line section can also be a problem in certain transmission media. They are usually realized by via holes in the substrate, which introduce a parasitic inductance.

8.1.3.4 Marchand Hybrid

A continual difficulty in the design of balanced circuits is the paucity of genuinely broadband 180-degree hybrid designs. The hybrid described in this section is a notable exception. Here we have converted the Marchand balun of Section 2.3.3 into a broadband hybrid. Figure 8.7 shows the circuit and the performance of one possible realization. The conversion is made by adding a pair of quarter-wavelength transmission lines to the output, thus creating a port. Since the even-mode input impedance of the ideal balun is an open circuit, this simply applies in-phase power from the new port to the two original output ports. The characteristic impedances of the new lines in theory should be 70.7Ω for 50Ω port impedances, but reducing this value slightly can improve the bandwidth.

As no theory for the design of such a hybrid exists, it must be designed empirically with the assistance of circuit-analysis software. The even-mode impedance of the coupled lines must be made as high as possible, consistent with the description in Section 2.3.3, and the odd-mode impedance optimized once the even-mode is established. Ideally, the lines to port 1 should not be coupled, but it seems inevitable that they will be located close together, so the coupling must be included in the design. In any case, the even-mode characteristic impedance should be set to 70.7Ω, and the odd-mode as high as possible, with the recognition that it cannot exceed the even-mode impedance. The effect of coupling is to reduce the hybrid's bandwidth somewhat.

The bandwidth of this hybrid ideally can exceed 2:1 with better than an 18-dB return loss at all ports. The bandwidth depends most strongly on the achievable even-mode characteristic impedance of the coupled lines. To maximize this, the usual practices of using multiple coupled lines (as in a Lange hybrid) or a suspended substrate should be employed.

8.1.3.5 Lumped-Element 180-Degree Hybrid

As with the quadrature hybrids, it is possible to design a lumped-element 180-degree hybrid [8.5]. Figure 8.8 shows one realization. Its one design equation is extremely simple:

$$\omega_0 L = \frac{1}{\omega_0 C} = Z_0\sqrt{2} \tag{8.18}$$

As before, ω_0 is the center frequency, Z_0 is the port impedance, and L and C are as shown in Figure 8.8. Compared to the quadrature lumped hybrids, the bandwidth of this hybrid is remarkably good, almost 30% for the 0.5-dB imbalance. The phase balance, better than 3 degrees over a 40% bandwidth, is similarly a very nice feature.

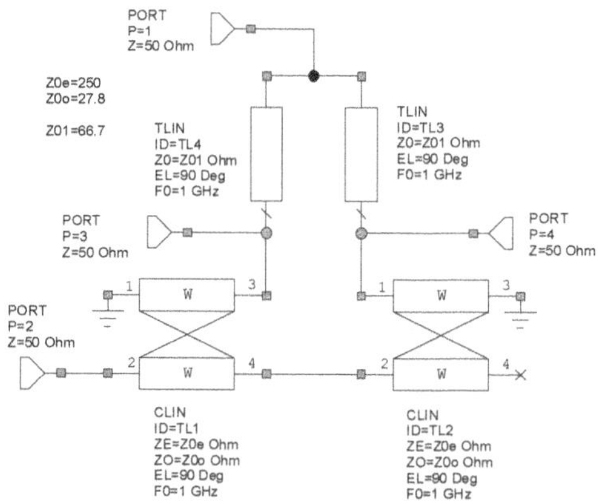

(a)

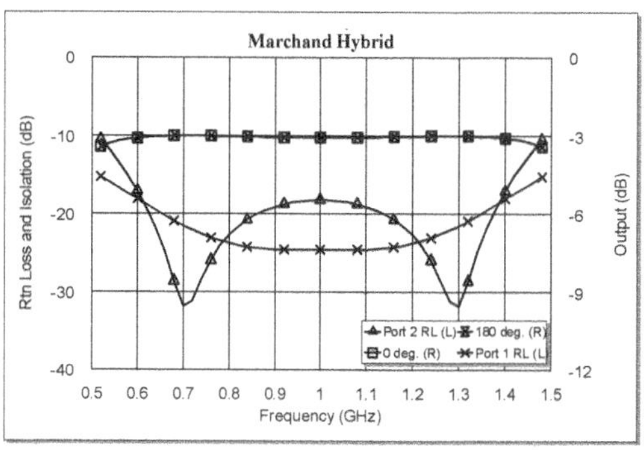

(b)

Figure 8.7 The Marchand balun can be converted into a hybrid with the addition of a port and two single lines (a). The bandwidth ideally can exceed an octave. The isolation is not shown, as it is over 100 dB in the ideal balun. The return losses of ports 3 and 4 are nearly identical to that of port 2, and the power splits of all paths are similarly nearly identical (b).

8.1.4 Practical Considerations

The hybrids described above are all ideal circuits. Real versions of those hybrids inevitably include parasitics that affect all aspects of the hybrid's performance.

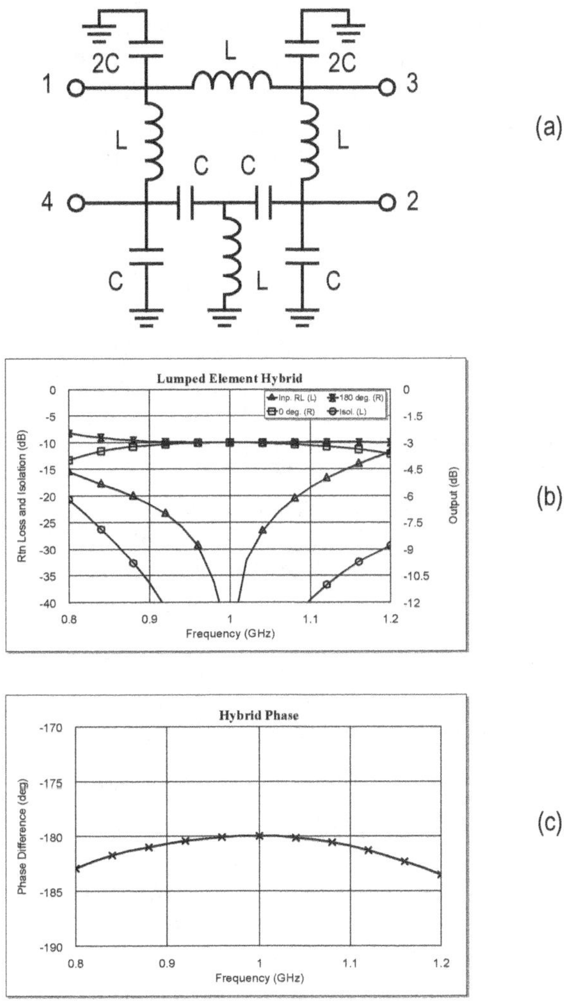

Figure 8.8 180-degree lumped-element hybrid (a); power split, return loss, and isolation (b) when port 2 is excited; phase difference at the output ports (c).

In the rat-race and branch-line hybrids, the tee junctions are the most important discontinuities. At lower frequencies, where the dimensions of the tee are small relative to the line lengths, their effects often can be ignored. At higher frequencies, however, that is clearly not the case. At frequencies where the tee dimensions are a significant fraction of the line lengths, designing an acceptable hybrid requires the use of modern discontinuity models, often ones based on a database of EM simulations. These are discussed more completely in Section 6.1.3.3. When the line lengths and impedances are properly corrected to account for the discontinuities, these hybrids can be used at surprisingly high frequencies, where the line lengths are only a few times the junction widths.

Via-hole inductance and unavoidable interconnections in the Marchand, broadband rat-race, and all the lumped designs can affect the hybrids' performance. Sometimes it is possible to compensate for them by changing line lengths or other dimensions. Often the only option is to make the vias as large as possible, minimizing their inductances, and to minimize the lengths of interconnections.

Lange hybrids require bond wires or air bridges to realize connections between the strips at the center and ends of the structure. The inductances of those connections are on the order of a few tenths of a nanohenry in hybrid circuits, somewhat less in monolithic integrated circuits. The use of multiple bond wires reduces the inductance, of course, but generally not in proportion to the number of wires, the result of magnetic coupling between closely spaced wires. Another parasitic in Lange hybrids involves the complicated metal interconnections at the ends of the coupled-line section. As with tee junctions, the kinds of things that a designer can get away with can be surprising. In any case, in all high-frequency designs, it is a good practice to EM simulate the ends of the Lange hybrid. EM simulating the ends but treating the coupled lines as in Section 2.2 is an efficient way to design a Lange hybrid. It is rarely necessary to EM simulate the whole structure.

8.2 Quadrature-Coupled Circuits

Sometimes incorrectly called balanced circuits, quadrature-coupled circuits are extremely valuable in microwave technology. Quadrature coupling is used most often in amplifiers, but it is also valuable for many other kinds of circuits, including switches, phase shifters, and mixers. We begin by examining the properties of hybrids as they are used in such circuits.

8.2.1 The Terminated Quadrature Hybrid

In this section we examine the properties of a terminated Lange hybrid. Since the S parameters of an ideal Lange hybrid are those of an ideal quadrature hybrid when its electrical length θ is π / 2, this involves no loss of generality. To determine the important properties, we shall make use of network graphs. The first task is to generate the graphs in a useful form.

The S-parameter matrix of an ideal quadrature hybrid can be written

$$S = \begin{bmatrix} 0 & 0 & S_a & S_b \\ 0 & 0 & S_b & S_a \\ S_a & S_b & 0 & 0 \\ S_b & S_a & 0 & 0 \end{bmatrix} \tag{8.19}$$

For an ideal hybrid,

$$\begin{aligned} S_a &= \frac{-j}{\sqrt{2}} \\ S_b &= \frac{1}{\sqrt{2}} \end{aligned} \tag{8.20}$$

but for an ideal Lange hybrid, from (8.3) and (8.4),

$$S_a = \frac{\sqrt{2}/2}{(\sqrt{2}/2)\cos(\theta) + j\sin(\theta)} \tag{8.21}$$

$$S_b = \frac{j(\sqrt{2}/2)\sin(\theta)}{(\sqrt{2}/2)\cos(\theta) + j\sin(\theta)} \tag{8.22}$$

where we have used the fact that c, the voltage coupling factor, is $\sqrt{2}/2$ for a hybrid. S_a is the path from the input to the –90-degree port; S_b to the 0-degree port. From (8.19) we can draw the network graph, shown in Figure 8.9.

We first wish to determine the input reflection coefficient of the hybrid terminated at two ports with loads that are not necessarily identical. The result will be a useful in the design of many kinds of high-frequency components. We choose port 1 as the excitation, terminate its isolated port, port 2, with an ideal load, and terminate ports 3 and 4, the outputs, with reflection coefficients Γ_3 and Γ_4, respectively. The hybrid and its network graph are shown in Figure 8.10.

The graph has no loops or other such complexity, so the transfer function can be written by inspection. There are only two paths from a_1 to b_1, so we write immediately

$$b_1 = a_1(S_b\Gamma_4 S_b) + a_1(S_a\Gamma_3 S_a) \tag{8.23}$$

or

$$\Gamma_{in} = \frac{b_1}{a_1} = S_b^2\Gamma_4 + S_a^2\Gamma_3 \tag{8.24}$$

For the ideal case, we substitute (8.20) and obtain

$$\Gamma_{in} = \frac{1}{2}(\Gamma_4 - \Gamma_3) \tag{8.25}$$

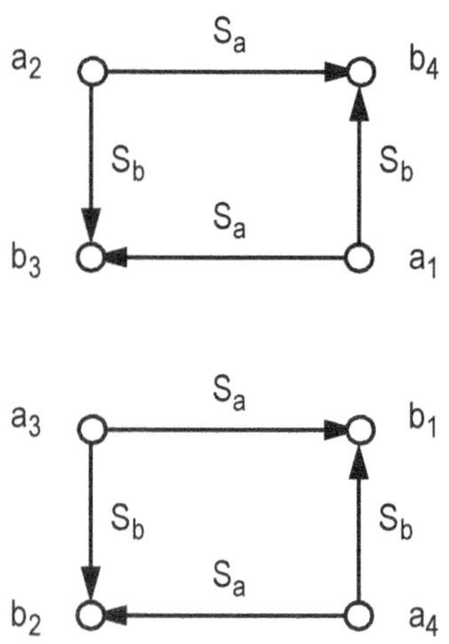

Figure 8.9 Network graph of the quadrature hybrid described by (8.19).

When $\Gamma_3 = \Gamma_4$, the input reflection coefficient is ideally zero. If, for example, the Γ_3 and Γ_4 are the input reflection coefficients of a pair of amplifiers, the input reflection coefficient of the combination is zero, as long as the amplifiers are identical. Of course, by driving them in this manner, we have lost 3 dB of gain, but that can be recovered, as we shall see, by combining them similarly at the output.

For a more general expression, we substitute (8.21) and (8.22) into (8.24). The result is

$$\Gamma_{in} = \frac{\frac{1}{2}(\Gamma_3 - \Gamma_4 \sin^2(\theta))}{[(\sqrt{2}/2)\cos(\theta) + j\sin(\theta)]^2} \tag{8.26}$$

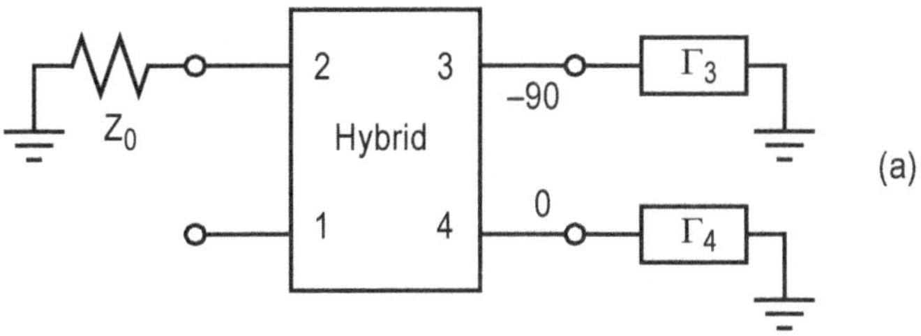

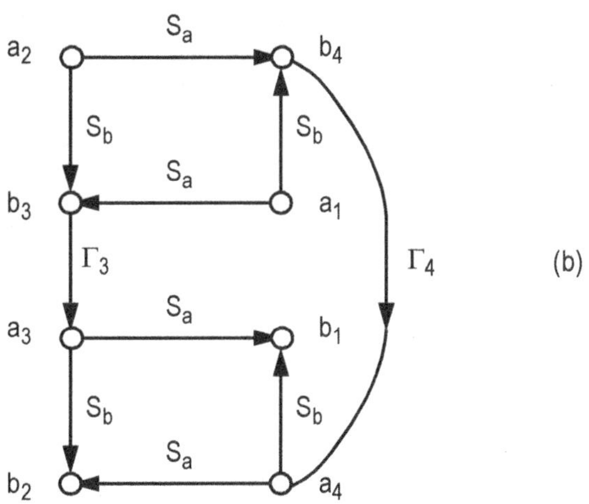

Figure 8.10 Network graph of the terminated hybrid. The hybrid is terminated at its output ports, ports 3 and 4, by Γ_3 and Γ_4, respectively.

This shows that perfect cancellation occurs only when $\theta = \pi / 2$. In practice, the input return loss is improved significantly over the entire passband of the hybrid. We can show this with the magnitude of the input reflection coefficient when $\Gamma_3 = \Gamma_4 = \Gamma$. Some simple algebra gives the result

$$|\Gamma_{in}| = \frac{|\Gamma| \cos^2(\theta)}{2 - \cos^2(\theta)} \tag{8.27}$$

At 70% of the center frequency, $\theta = 0.7\,\pi / 2$; then, we find the input reflection coefficient to be only 0.038 $|\Gamma|$, still much lower than the load's reflection coefficient.

Next, we would like to see how much power is transferred to the termination. The same approach gives us

$$b_2 = a_1(S_a \Gamma_3 S_b) + a_1(S_b \Gamma_4 S_a) \tag{8.28}$$

or

$$S_{H21} = \frac{b_2}{a_1} = S_a S_b(\Gamma_3 + \Gamma_4) \tag{8.29}$$

where S_{H21} is the transfer function of the terminated hybrid. This becomes, with the same substitutions,

$$S_{H21} = \frac{\frac{1}{2} j(\Gamma_3 + \Gamma_4)\sin(\theta)}{[(\sqrt{2}/2)\cos(\theta) + j\sin(\theta)]^2} \tag{8.30}$$

If we again assume that $\Gamma_3 = \Gamma_4 = \Gamma$, we obtain

$$|S_{H21}| = \frac{2|\Gamma| \sin(\theta)}{1 + \sin^2(\theta)} \tag{8.31}$$

which shows that the transmission path to the termination, when $\theta = \pi / 2$, is simply Γ. This illustrates more clearly how the circuit works: power applied to port 1 is split and delivered equally to ports 3 and 4; the power reflected from ports 3 and 4 is then directed to the termination, so the input reflection coefficient is zero.

When $\theta = \pi / 2$,

$$S_{H21} = \frac{-j}{2}(\Gamma_3 + \Gamma_4) \tag{8.32}$$

This implies that the terminated hybrid can work as a phase shifter; when ports 3 and 4 are open-circuited, $\Gamma_3 = \Gamma_4 = 1$ and it provides a –90-degree phase shift. When $\Gamma_3 = \Gamma_4 = -1$, short-circuited ports, the phase shift is +90 degrees. Other values of phase can be obtained by terminating the ports in a reactance, and a variable phase shift can be obtained by using a varactor as a termination.

8.2.2 Quadrature-Coupled Amplifier

Quadrature-coupled amplifiers are among the most commonly used circuit structures in microwave electronics. Such an amplifier consists of two identical, simple amplifier blocks coupled by quadrature hybrids at the input and output in the manner shown in Figure 8.11(a). The 90-degree port of hybrid A is connected to the input of one amplifier, and the 90-degree port of hybrid B is connected to the output of the other amplifier.

We make the same assumptions as in Section 8.2.1: the hybrids are identical, ideal, lossless, coupled-line structures described by (8.19) through (8.22). Invariably they are Lange hybrids. Additionally, we assume that the amplifiers, whose S matrices are designated $\mathbf{S_x}$ and $\mathbf{S_y}$, are unilateral; we could include their reverse transmission, but that would complicate things considerably and would not change the results. The normalizing impedance of all ports of both the hybrids and amplifiers is Z_0. The entire structure realizes a two-port whose input is port 1 of hybrid A and output is port 1 of hybrid B. The port numbers of that two-port are 1 and 2, respectively, shown in boxes in Figure 8.11(a). $\mathbf{S_c}$ is the S matrix of the combination.

8.2.2.1 Gain and Port Reflection Coefficients

From the previous section, it should be clear that the input and output reflection coefficients of the combination are zero when the amplifiers are identical and $\theta = \pi / 2$. We saw also that the input reflection coefficient rises only gradually as we move away from the hybrids' center frequency. From (8.26),

$$S_{c11} = \Gamma_{\text{in}} = \frac{\frac{1}{2}(S_{x11} - S_{y11}\sin^2(\theta))}{[(\sqrt{2}/2)\cos(\theta) + j\sin(\theta)]^2} \tag{8.33}$$

and from (8.27), when $S_{x11} = S_{y11} = S_{11}$,

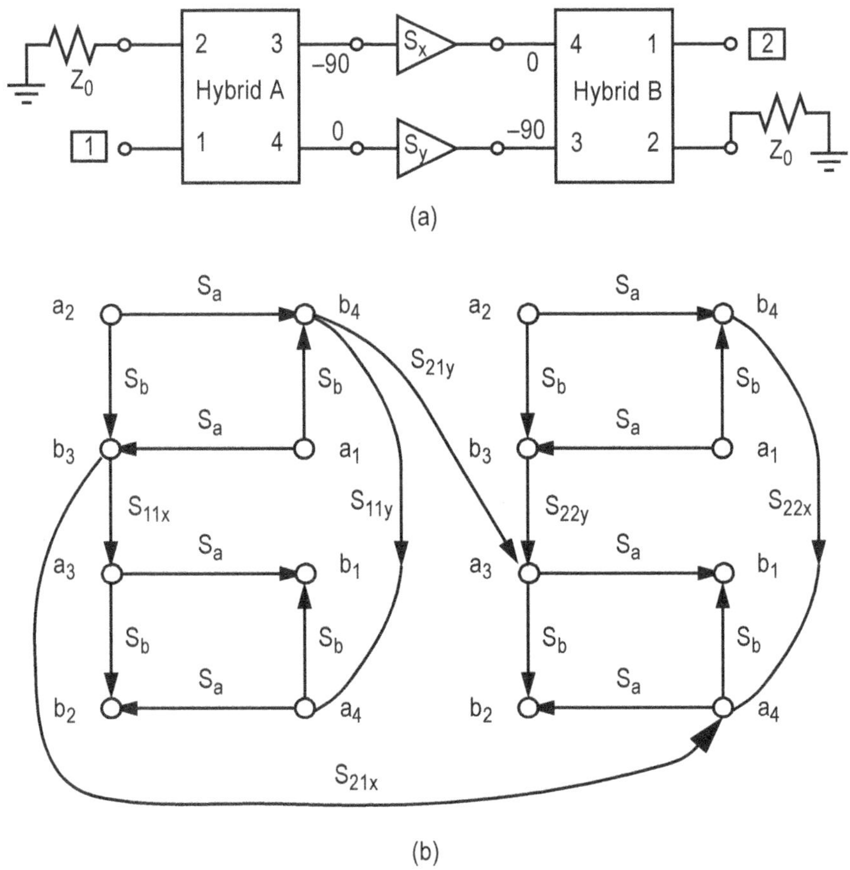

Figure 8.11 A quadrature-coupled pair of amplifiers (a) and their network graph (b). The port numbers 1 and 2 in boxes are those of the complete quadrature-coupled amplifier.

$$|S_{c11}| = \frac{|S_{11}|\cos^2(\theta)}{2 - \cos^2(\theta)} \tag{8.34}$$

Similarly,

$$S_{c22} = \Gamma_{\text{out}} = \frac{\frac{1}{2}(S_{y22} - S_{x22}\sin^2(\theta))}{[(\sqrt{2}/2)\cos(\theta) + j\sin(\theta)]^2} \tag{8.35}$$

and, with identical amplifiers, $S_{x22} = S_{y22} = S_{22}$ and

$$|S_{c22}| = \frac{|S_{22}|\cos^2(\theta)}{2 - \cos^2(\theta)} \tag{8.36}$$

Both the input and output reflection coefficients are improved considerably.

We now find the transmission through the combination. Again, the graph in Figure 8.11(b) has no loops or other such complexities, so we can find S_{c21} by inspection:

$$b_{1B} = a_{1A}(S_a S_{21x} S_b) + a_{1A}(S_b S_{21y} S_a) \tag{8.37}$$

So

$$S_{c21} = \frac{b_{1B}}{a_{1A}} = S_a S_b(S_{x21} + S_{y21}) \tag{8.38}$$

where b_{1B} is the b wave from port 1 of hybrid B, and so on. We found the quantity S_a S_b in the previous section, so we immediately can write

$$S_{c21} = \frac{\frac{1}{2}j(S_{x21} + S_{y21})\sin(\theta)}{[(\sqrt{2}/2)\cos(\theta) + j\sin(\theta)]^2} \tag{8.39}$$

If we again assume that the amplifiers are identical, such that $S_{x21} = S_{y21} = S_{21}$, we obtain

$$|S_{c21}| = \frac{2|S_{21}|\sin(\theta)}{1 + \sin^2(\theta)} \tag{8.40}$$

The transducer gain is

$$G_t = |S_{c21}|^2 = \frac{4|S_{21}|^2\sin^2(\theta)}{(1 + \sin^2(\theta))^2} \tag{8.41}$$

which, at center frequency, is simply $|S_{21}|^2$.

A little investigation will show that the gain of the quadrature-coupled amplifier is significantly greater than that of the individual hybrids. This occurs because the output hybrid compensates somewhat for the roll-off of the input one. While this is useful for achieving broad bandwidth, and, indeed, quadrature-coupled amplifiers having several octaves of bandwidth are regularly produced, the phenomenon has an important disadvantage. We shall see that the noise performance of the combination tends to increase rapidly around 50% of the hybrids' center frequency. It is likely that this problem can be reduced by overcoupling the hybrids at center frequency to reduce their imbalance at large frequency offsets, but at the cost of an increase in bandcenter noise figure.

With identical amplifiers, at the center frequency of the hybrid,

$$S_{c21} = -jS_{21} \tag{8.42}$$

The combination has the same bandcenter gain as the individual amplifiers. This occurs because the signals from the two amplifier blocks combine in phase in the output hybrid, compensating for the 3-dB loss due to the power split at the input.

8.2.2.2 Large-Signal Performance

The two amplifier blocks in Figure 8.11(a) each operate at a 3-dB lower level than they would without the hybrid coupling, so the power-handling ability of the combination is 3 dB greater than that of the individual amplifiers. Thus, both the 1-dB compression point and the third-order intercept point are increased by 3 dB; the structure acts as a power combiner.

If the hybrid had infinite bandwidth, the effect of intermodulation intercepts of all orders would be the same as for the third order, a 3-dB increase. The even-order products, however, are likely to be outside the hybrid's passband, so it is somewhat more difficult to assess them. For second-order IM products, for example, which occur close to zero frequency, one amplifier is decoupled from the output but the other is connected to the output with virtually zero loss. The power split at the input causes the second-order IM products of that connected amplifier to be 6 dB lower than it would be if the amplifier were operated by itself. From (7.41), the IP_2 for that product is 6 dB greater. This value is better than what would be obtained by simple power combining. The improvement results from eliminating the distortion output of one stage.

8.2.2.3 Noise

We now consider the effect of the combination on the noise figure of the amplifier. We must consider three phenomena:

1. The frequency-dependent coupling of the amplifiers by the hybrid might affect its noise figure.
2. The noise from the termination on port 2 of the input hybrid is applied to the inputs of the two amplifier blocks, so it is amplified and appears at the output. We need to know whether that noise increases the amplifier's noise figure.
3. Similarly, the noise from the output termination is reflected by the amplifier blocks' output ports and ends up in the hybrid-coupled amplifier's output.

To find the noise temperature of the quadrature-coupled amplifier, we must first determine the output power in terms of noise temperature and then divide by the gain of the hybrid-coupled combination. The output noise from the two amplifiers is uncorrelated, so it combines powerwise in the output. That noise power, T_L, is

$$T_L = T_{nx}|S_{x21}|^2|S_b|^2 + T_{ny}|S_{y21}|^2|S_a|^2 \tag{8.43}$$

where $T_{nx,y}$ are the noise temperatures of the amplifiers x and y. Substituting our earlier expressions for S_a and S_b gives

$$T_L = \frac{T_{nx}|S_{x21}|^2 + T_{ny}|S_{y21}|^2 \sin^2(\theta)}{1 + \sin^2(\theta)} \tag{8.44}$$

Interestingly, if we assume that the amplifiers are identical, so $S_{x21} = S_{y21} = S_{21}$ and $T_{nx} = T_{ny} = T_n$, we find that

$$T_L = |S_{21}|^2 T_n \tag{8.45}$$

where T_n is the noise temperature of either stage. The output noise level at all frequencies is the same as that of a single amplifier block.

The gain of the quadrature-coupled amplifier is found as the magnitude squared of S_{c21} from (8.39). We obtain

$$G_t = |S_{c21}|^2 = \frac{|S_{x21} + S_{y21}|^2 \sin^2(\theta)}{(1 + \sin^2(\theta))^2} \tag{8.46}$$

Dividing T_L by the gain gives the noise temperature of the quadrature-coupled amplifier, T_{nc},

$$T_{nc} = \frac{(T_{nx}|S_{x21}|^2 + T_{ny}|S_{y21}|^2 \sin^2(\theta))(1 + \sin^2(\theta))}{|S_{x21} + S_{y21}|^2 \sin^2(\theta)} \tag{8.47}$$

Again, if the amplifiers are identical, with $T_{nx} = T_{ny} = T_n$, we obtain

$$T_{nc} = T_n \frac{(1 + \sin^2(\theta))^2}{4 \sin^2(\theta)} \tag{8.48}$$

This expression shows that the noise temperature at center frequency is the same as that of the individual amplifier blocks. It indicates a modest increase in noise temperature at other frequencies; at 70% of the center frequency, the increase is only 1.3%.

To examine the second point, we must find the transfer function from port 2 of the input hybrid to port 1 of the output hybrid. Again, by inspection,

$$b_{1B} = a_{2A}(S_b S_{x21} S_b) + a_{2A}(S_a S_{y21} S_a) \tag{8.49}$$

So

$$\frac{b_{1B}}{a_{2A}} = S_b^2 S_{x21} + S_a^2 S_{y21} \tag{8.50}$$

Substituting our previous values for S_a and S_b, we obtain

$$\frac{b_{1B}}{a_{2A}} = \frac{\frac{1}{2}(S_{y21} - S_{x21}\sin^2(\theta))}{[(\sqrt{2}/2)\cos(\theta) + j\sin(\theta)]^2} \tag{8.51}$$

At center frequency, with identical amplifiers, $b_{1B}/a_{2A} = 0$.

The output noise from the termination, ΔT_{No}, in terms of noise temperature, is the transducer gain between these ports times the termination's temperature, T_t:

$$\Delta T_{No} = T_t\left|\frac{b_{1B}}{a_{2A}}\right|^2 = T_t\frac{|S_{y21} - S_{x21}\sin^2(\theta)|^2}{(1 + \sin^2(\theta))^2} \tag{8.52}$$

Dividing this by the transducer gain from (8.46) gives the increase in the equivalent input noise temperature due to this phenomenon, ΔT_{Ni}:

$$\Delta T_{Ni} = \frac{T_{No}}{G_t} = T_t\frac{|S_{y21} - S_{x21}\sin^2(\theta)|^2}{\sin^2(\theta)|S_{x21} + S_{y21}|^2} \tag{8.53}$$

If we make our customary assumption that the amplifiers are identical, we obtain

$$\Delta T_{Ni} = T_t\frac{(1 - \sin^2(\theta))^2}{4\sin^2(\theta)} \tag{8.54}$$

This phenomenon adds approximately 4K to the noise temperature at 70% of the center frequency and 36K at 50%.

If the amplifier gains are not identical, the phenomenon is exacerbated. Unfortunately, (8.53) does not provide much intuitive sense of the effect of dissimilar gains on the input noise temperature. The simplest view is probably to let $S_{y21} = k\,S_{x21}$, where k is a complex imbalance factor, which accounts for both gain and phase imbalance. Then (8.53) becomes

$$\Delta T_{Ni} = T_t \frac{|1 - k\sin^2(\theta)|^2}{|1 + k|^2 \sin^2(\theta)} \tag{8.55}$$

which should be a little easier to use. Clearly, (8.55) is identical to (8.54) when $k = 1$.

The third phenomenon is easy to deal with. Using (8.29) to (8.31), assuming that the amplifiers are identical, and recognizing that the hybrid is reciprocal, we have

$$|S_{B21}| = \left|\frac{b_{1B}}{a_{2B}}\right| = \left|\frac{b_{2B}}{a_{1B}}\right| = \frac{2|S_{22}|}{2 - \cos^2(\theta)} \tag{8.56}$$

so the output noise increase is

$$\Delta T_{No} = T_t |S_{B21}|^2 = T_t \frac{4|S_{22}|^2}{[2 - \cos^2(\theta)]^2} \tag{8.57}$$

The equivalent input noise temperature increase is

$$\Delta T_{ni} = \frac{\Delta T_{No}}{|S_{c21}|^2} = T_t \frac{|S_{22}|^2}{|S_{21}|^2} \frac{1}{\sin^2(\theta)} \tag{8.58}$$

where the transducer gain of the hybrid-coupled amplifier, $|S_{c21}|^2$ was found from (8.40) and again we have assumed the amplifiers to be identical. Equation (8.58) is one we might have guessed with fair accuracy. The transmission from the termination to the output port is clearly $|S_{22}|^2$, and we must divide the output noise temperature by the transducer gain of the amplifier, $|S_{21}|^2$, to obtain the input noise temperature. The only (mild) surprise is the sine term, which accounts for the frequency deviation from the hybrid's center.

The contribution of this termination noise to the amplifier's noise temperature can be surprisingly great. In broadband amplifiers, it is not unusual to have $|S_{22}|^2 \sim 1$ and $|S_{21}|^2 \sim 10$ over at least a part of the band. Then, at 70% of the center frequency, the load contributes 38K to the input noise temperature. In many low-noise amplifiers, this is a significant increase.

Figure 8.12 shows the gain and noise temperature of a pair of quadrature-coupled amplifiers with all phenomena included. The amplifier blocks have a 10-dB gain, a 100K noise temperature, and an output VSWR of 2.0.

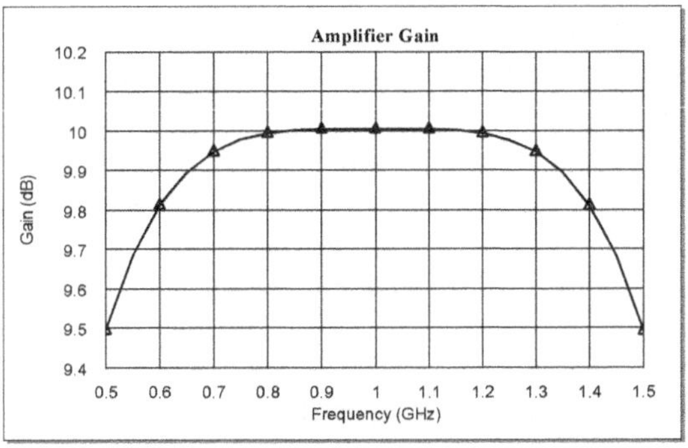

(a)

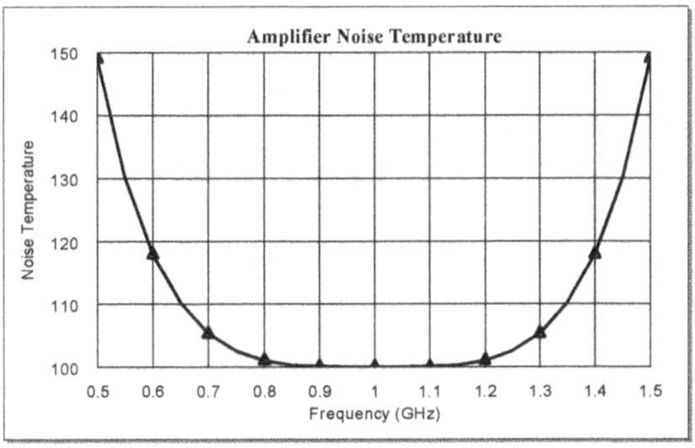

(b)

Figure 8.12 Gain (a) and noise temperature (b) of a quadrature-coupled amplifier. The individual amplifier blocks have a 10-dB gain, a 100K noise temperature, and an output VSWR of 2.0. The temperature of the hybrid's terminations is 20°C.

The temperature of both terminations is 20°C. The combined effect of these phenomena is modest as long as the amplifier bandwidth is kept below an octave. Beyond this range, the noise temperature rises rapidly, in this case increasing almost 50% at half the center frequency.

We must close this section with two points. First, we recognize that the quadrature-coupled amplifier is not completely isolated from source-impedance variations, as would be an amplifier having an input isolator. In the quadrature-coupled amplifier, the individual amplifier blocks see a source reflection coefficient that is half that at the input to the combined amplifier. This is still an improvement, of course, but a lesser one than would be achieved with an isolator. Second, the resistive losses in real hybrids affect the noise temperature of the amplifiers in the manner described in Section 7.2.3. The excess loss through each path of the isolator increases the noise temperature of the amplifier in that path and reduces the gain. If the losses are close to the same value, they can be treated as a single attenuator at the input and output of the quadrature-coupled amplifier; if they are significantly different, they can be treated as losses at the inputs and outputs of the individual amplifier blocks. In the latter case, T_{nx} and T_{ny} differ. Fortunately, modern circuit simulators easily account for this case.

8.3 Balanced Amplifiers Using Baluns and 180-Degree Hybrids

It is also possible to couple two amplifiers by power dividers, hybrids, or baluns. When a balun is used, or a hybrid in which the amplifiers are connected to the 0- and 180-degree ports, it is called a *balanced amplifier*; in other cases, it is simply a pair of amplifiers that are power combined.

In discussing quadrature-coupled amplifiers, we took advantage of the fact that a particular type of hybrid, the Lange hybrid, was used for the combination. In this case, a variety of types of hybrids or baluns may be employed, so we must treat them as ideal elements.

8.3.1 The Terminated Balun

In previous discussions of baluns, we made the point that a balun is a two-port device that provides a transition from a balanced transmission medium to an unbalanced one; in analyzing or evaluating a balun, however, it is useful to treat the two output terminals as separate ports. We take that latter view in this section, and show that it can tell us a lot about the way baluns work in balanced circuits.

8.3.1.1 Input Reflection Coefficient

In Section 3.2.4 we derived the properties of a lossless, symmetrical balun with a matched input and determined that its S matrix must have the form

$$\mathbf{S} = \begin{bmatrix} 0 & S & -S \\ S & \Gamma & I \\ -S & I & \Gamma \end{bmatrix} \tag{8.59}$$

where Γ is the reflection coefficient of ports 2 and 3, the output ports; I is the isolation term; and S is the transfer term between the input and outputs. Because of the 3-dB power division, $|S|^2 = 0.5$, and we found additionally that $\Gamma = I$ and that $|\Gamma|^2 = |I|^2 = 0.25$. Thus, $\Gamma = I = \pm 0.5$.

Figure 8.13 shows the terminated balun and its network graph. Unlike the quadrature hybrid, this graph has a number of paths and loops, so the use of graph-reduction techniques or Mason's rule is essential.

The only quantity of interest is the input reflection coefficient, b_1 / a_1. We use Mason's rule (5.18) to determine it. The loops and paths are as follows:

Paths:

$$\begin{aligned} P_1 &= (-S)\Gamma_3 I \Gamma_2 S = -S^2 I \Gamma_3 \Gamma_2 \\ P_2 &= S\Gamma_3 I \Gamma_2 (-S) = -S^2 I \Gamma_3 \Gamma_2 \\ P_3 &= S\Gamma_2 S = S^2 \Gamma_2 \\ P_4 &= (-S)\Gamma_3(-S) = S^2 \Gamma_3 \end{aligned} \tag{8.60}$$

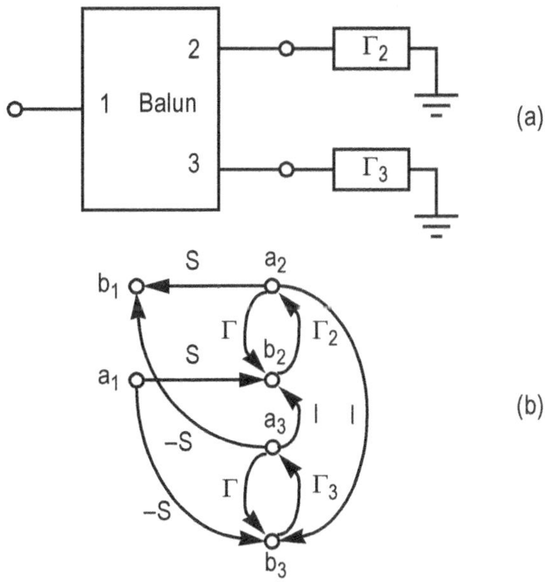

Figure 8.13 An ideal balun terminated at its output ports (a); its network graph (b).

First-Order Loops:

$$\begin{aligned} L_1 &= \Gamma\Gamma_2 \\ L_2 &= \Gamma\Gamma_3 \\ L_3 &= \Gamma_3 I \Gamma_2 I = I^2\Gamma_3\Gamma_2 \end{aligned} \tag{8.61}$$

Second-Order Loops:

$$L_4 = \Gamma\Gamma_2\Gamma\Gamma_3 = \Gamma^2\Gamma_2\Gamma_3 \tag{8.62}$$

Paths P_1 and P_2 touch all loops, P_3 touches all but L_2, and P_4 touches all but L_1.

Substituting these into (5.18) creates the usual algebraic mess, which can be reduced to

$$\Gamma_{in} = \frac{b_1}{a_1} = \frac{S^2[\Gamma_2 + \Gamma_3 - 2(I + \Gamma)\Gamma_2\Gamma_3]}{1 - \Gamma(\Gamma_2 + \Gamma_3)} \tag{8.63}$$

Substituting $|S|^2 = 0.5$ and $\Gamma = I = 1/2$, we obtain

$$\Gamma_{in} = \frac{\frac{1}{2}(\Gamma_2 + \Gamma_3) - \Gamma_2\Gamma_3}{1 - \frac{1}{2}(\Gamma_2 + \Gamma_3)} \tag{8.64}$$

If we assume the loads to be identical, so $\Gamma_2 = \Gamma_3 = \Gamma_L$, we obtain, finally,

$$\Gamma_{in} = \Gamma_L \tag{8.65}$$

This result applies only to an ideal balun. Real baluns have a phase shift, which changes the conclusions a bit. If we assume that

$$S = \frac{\sqrt{2}}{2}\exp(j\phi) \tag{8.66}$$

$$\Gamma = I = \frac{1}{2}\exp(j\theta) \tag{8.67}$$

we find that

$$\Gamma_{\text{in}} = \Gamma_L \exp(2j\phi) \tag{8.68}$$

That is, coupling two circuit elements by a balun does not change the magnitude of the input reflection coefficient, as it did with the quadrature hybrid, but it does add a phase term equal to twice the balun's phase delay.

8.3.1.2 Even- and Odd-Mode Port Reflection Coefficients

Suppose port 1 is terminated with a load, Γ_1, and we wish to determine the even- and odd-mode reflection coefficients at ports 2 and 3. To find the reflection coefficient at, say, port 2, we excite ports 2 and 3 with *a* waves in either an even or odd mode, so $a_3 = \pm a_2$, and determine b_2 / a_2. Ports 2 and 3 are terminated in Z_0, so $\Gamma_2 = \Gamma_3 = 0$. The configuration is shown in Figure 8.14(a). At port 2,

$$b_2 = \hat{S}_{22}a_2 + \hat{S}_{23}(\pm a_2) \tag{8.69}$$

where the circumflex indicates the S parameters of the balun with the Γ_1 termination on port 1. The network graph has no loops, so $\hat{S}_{22}$ and $\hat{S}_{23}$ can be found directly from the network graph.

The paths for $\hat{S}_{22}$ are

$$\begin{aligned} P_1 &= \Gamma \\ P_2 &= S\Gamma_1 S \end{aligned} \tag{8.70}$$

and for $\hat{S}_{23}$,

$$\begin{aligned} P_1 &= I \\ P_2 &= (-S)\Gamma_1 S \end{aligned} \tag{8.71}$$

Therefore,

$$\frac{b_2}{a_2} = (\Gamma + S^2\Gamma_1) \pm (I - S^2\Gamma_1) \tag{8.72}$$

where the plus sign is for the even mode and minus for the odd. We obtain

$$\left.\frac{b_2}{a_2}\right|_e = \Gamma + I = \pm 1 \tag{8.73}$$

and

$$\left.\frac{b_2}{a_2}\right|_o = 2S^2\Gamma_1 \tag{8.74}$$

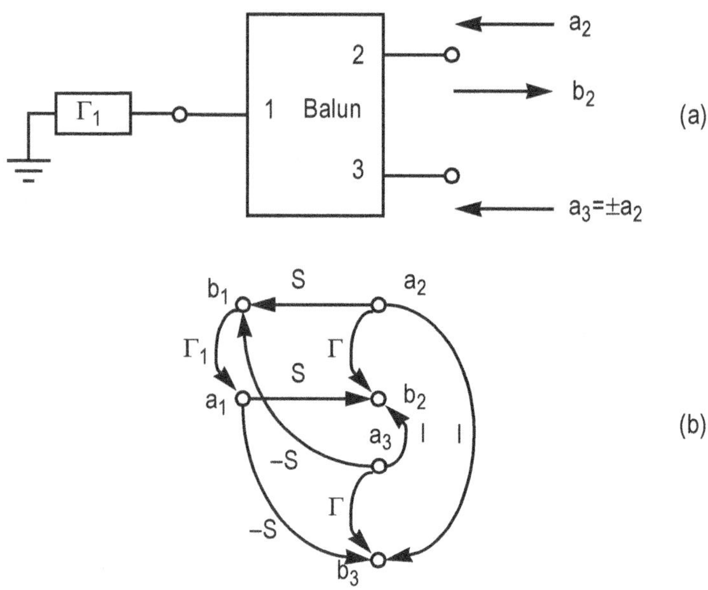

Figure 8.14 To determine the even- and odd-mode port reflection coefficients of the balun, we excite the ports as shown in (a). Ports 2 and 3 are terminated in Z_0. The network graph is shown in (b).

Recall that $|S|^2 = 0.5$ so

$$\left.\frac{b_2}{a_2}\right|_o = \Gamma_1 \angle 2\phi \tag{8.75}$$

where ϕ is the phase angle of S. In an ideal Marchand balun, for example, $\phi = \pi / 2$; in that case,

$$\left.\frac{b_2}{a_2}\right|_o = -\Gamma_1 \tag{8.76}$$

The port looks like a short or open circuit to even-mode excitation, depending on whether Γ and I are +0.5 or –0.5; that, in turn, depends on the type of balun. For most types of baluns, $\Gamma = I = +0.5$. The odd mode sees an inverted version of the terminating impedance on port 1; the Marchand balun behaves like a quarter-wave transmission line of characteristic impedance Z_0 between port 2 and port 1. In all types of baluns, however, when port 1 is terminated in Z_0, the odd-mode impedance at ports 2 and 3 is also Z_0.

We have not examined the even- and odd-mode impedances at port 3, but a similar exercise shows that they are the same as at port 2.

Since the balun is a lossless device, an odd-mode reflection coefficient equal to the load reflection coefficient implies that all the power of the excitation is directed to the load on port 1. This means that the balun, when excited in an odd mode, is a perfect power combiner. Similarly, the even-mode reflection coefficient of unity implies that the balun completely rejects even-mode excitations.

This exercise shows why baluns, in spite of their poor isolations and reflection coefficients at ports 2 and 3, are useful microwave devices. The components in which they are used are invariably ones where the ports are excited with an odd mode, so the balun combines those outputs effectively and presents a good reflection coefficient to the individual blocks. The even-mode open- or short-circuit port can also be of value in such components as balanced mixers and frequency multipliers, as it may provide an optimum termination to unwanted mixing or intermodulation products, which exit the pair of components in an even mode. More information on this can be found in [7.1].

8.3.2 Balun-Coupled Balanced Circuits

Two two-port components can be combined with baluns at the input and output. This kind of balanced configuration, shown in Figure 8.15, is used for many types of circuits, especially for power amplifiers. In amplifiers, it provides rejection of even-order mixing products, most importantly even harmonics of the input frequency.

As with the quadrature-coupled amplifier, the input signal is split between the two amplifier inputs with 3-dB loss. At the output balun, the two signals are combined in phase, so the gain of the combination is the same as that of the individual amplifier blocks. The combination also provides a power-combining function, so 1-dB compression point is increased 3 dB, and the odd-order output intercept points are 3 dB higher for products within the bandwidth of the output balun. From the results in Section 8.3.1, it should be clear that the input and output VSWRs of the balun-coupled amplifier are that of the individual amplifier blocks. This configuration provides no improvement in input or output VSWR. Its advantages are power combining and, as we shall see, rejection of even harmonics and even-order intermodulation products.

8.3.3 Even Harmonics and Even-Order Distortion

Amplifiers generate both odd- and even-order distortion products, either because of weak nonlinearities or from being driven into saturation. They also generate harmonics of both orders; these can be viewed as a kind of single-tone distortion product. In class-B and class-AB amplifiers, the device is biased at or near its gate or base turn-on voltage, so it conducts over only half the excitation cycle. In that case, the waveform in each amplifier block is strongly distorted, but, since the output amplitude is proportional

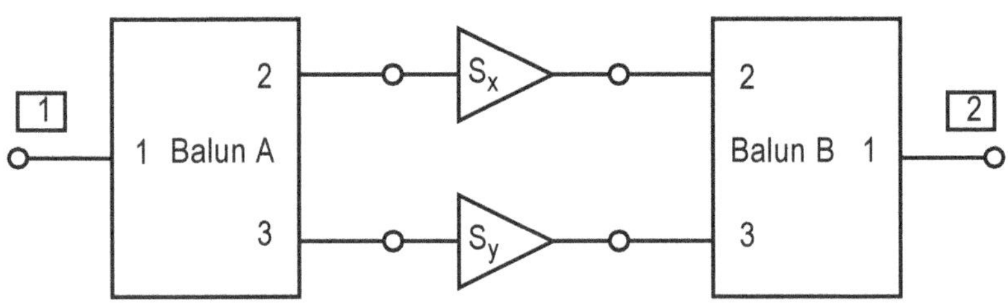

Figure 8.15 Balanced amplifier using baluns.

to the input, the amplification is still linear. It is necessary, however, to reject or filter harmonic components from the output.

It is common to interconnect Class-B/AB amplifiers with baluns as in Figure 8.15. In this case, one amplifier carries one-half of the excitation cycle, while the other carries the other half. The circuit is ideally linear, but in the practical case, a certain amount of distortion is generated, primarily by the soft turn-on of the real devices. This distortion contains significant even harmonics and, for multitone signals, even-order distortion.

From the approach described in Section 7.1.6, it is possible to show that even-order distortion products, including harmonics, are in phase at the amplifier outputs and thus excite the output balun in an even mode. As a simple example, consider a single-tone excitation of magnitude V. At the input of amplifier x we have

$$V_{xi} = \frac{\sqrt{2}}{2} V \cos(\omega t) \tag{8.77}$$

and at amplifier y,

$$V_{yi} = \frac{\sqrt{2}}{2} V \cos(\omega t + \pi) \tag{8.78}$$

where V_{xi} and V_{yi} are the input waveforms. The quadratic term in (7.38) gives, for both amplifiers, V_{xo} and V_{yo},

$$V_{xo} = V_{yo} = \frac{\sqrt{2}}{4} B V^2 (1 + \cos(2\omega t)) \tag{8.79}$$

which indicates an even-mode, second-harmonic component plus a dc term, which is of no interest. For reasons that go well beyond the scope of this book, it is usually best for the balun to present an even-mode short circuit to the amplifiers. Therefore, a wirewound transformer or some type of high-frequency equivalent is preferred for balanced amplifiers. See [7.1] for more information. From (8.79) it should be clear that effective operation requires that the balun's bandwidth include significant even harmonics and other even-order distortion products.

The even-order intercept points, as with the quadrature-coupled amplifier, can be problematical. Some products may be outside the bandwidth of the baluns, so their effects are difficult to predict. The intercept points of in-band, even-order products are ideally infinite, but in practice depend on the amount of rejection by the balun. That depends, in turn, on the output balun's even-mode performance at those frequencies. In any case, it is usually greater than that of the individual amplifiers.

8.3.4 Hybrid-Coupled Balanced Circuits

Balanced circuits can be realized by 180-degree hybrids instead of baluns. In that case, the hybrid's 0- and 180-degree ports are used in place of the baluns' output ports and the isolated port is terminated in its port-normalizing impedance. In some kinds of nonlinear circuits, the isolated port can be used. In certain kinds of mixers, for example, the isolated port is used for the IF output.

Most of the properties of hybrid-coupled circuits are the same as for balun-coupled ones. The main difference, however, is that the hybrid presents an even-mode port reflection coefficient of zero instead of one. Thus, even-mode power from the individual circuit blocks sees a matched load. This can be helpful in providing stability when the balance is imperfect, as is common in active circuits.

One consideration in the use of hybrids in balanced circuits is that most 180-degree hybrids are relatively narrowband. Most hybrids do not have enough bandwidth to include even-mode harmonic components, so, depending on the type of hybrid, significant harmonic output is possible.

Hybrids are most useful for certain kinds of balanced circuits. Figure 8.16, for example, shows a mixer using a rat-race hybrid. This type of mixer is used frequently in microwave circuits, as it provides convenient input ports for the RF and LO excitations. Since those inputs are isolated ports of the hybrid, it provides LO-to-RF isolation without requiring filters or other such circuits. Furthermore, if the Δ port of the hybrid is used as the LO, the IF output node is a virtual ground for the LO, so LO-to-IF isolation is ensured. The circuit also provides rejection of certain unwanted even-order mixing products and spurious responses. Its main disadvantage is narrow bandwidth; as with virtually all diode mixers, the bandwidth is limited by that of the hybrid. The hybrid, however, is not helpful in terminating unwanted mixing products and LO harmonics optimally. For more information on this and similar balanced nonlinear circuits, see [7.1].

One of the main reasons for the use of hybrid-coupled circuits is power combining. Power-combined amplifiers can use either the 0-degree or 180-degree ports of the hybrid; the performance, in terms of output power and input VSWR, is identical. A pair of hybrid-coupled amplifiers can be combined with another pair of hybrids, creating an amplifier having four individual blocks and 6-dB greater output power. This process can be continued, combining amplifiers numbering in any power of two, although

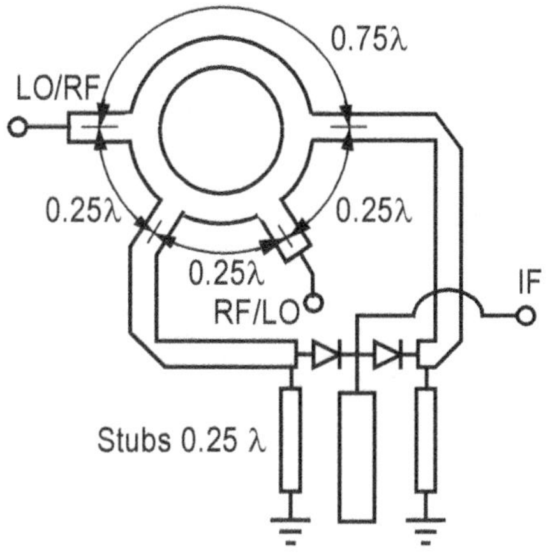

Figure 8.16 A balanced mixer using a rat-race hybrid. One pair of mutually isolated ports is used as the RF and LO, and the second pair connects to the diodes. The diodes are, in effect, individual mixer circuits. The quarter-wavelength stubs provide IF and RF/LO bypassing.

eventually the structure becomes cumbersome. Such "tree" structures require simple hybrids or power dividers; one favored type is a Wilkinson power divider [1.3], which can be viewed as a 180-degree hybrid with a terminated internal port.

References

[8.1] Ho, C., "Design of Lumped Quadrature Couplers," *Microwave J.*, Vol. 22, Sept., 1979, p. 67.

[8.2] Hogheiden, J., M. Ciminera, and G. Jue, "Improved Planar Spiral Transformer Theory Applied to a Miniature Lumped Element Quadrature Hybrid," *IEEE Trans. Microwave Theory Tech.*, Vol. MTT-45, 1997, p. 543.

[8.3] Maas, S., "Broadband VHF Lumped-Element Couplers Using Imperfect Transformers," *Applied Microwaves and Wireless*, November 1999, p. 70.

[8.4] March, S., "A Wideband Stripline Ring Hybrid," *IEEE Trans. Microwave Theory Tech.*, Vol. MTT-16, 1968, p. 361.

[8.5] Parisi, S. J., A Lumped-Element Rat-Race Coupler," *Applied Microwaves*, Aug., 1989, p. 84.

About the Author

Stephen Maas received BSEE and MSEE degrees in electrical engineering from the University of Pennsylvania in 1971 and 1972, respectively, and a Ph.D. in electrical engineering from UCLA in 1984. He joined the National Radio Astronomy Observatory in 1974, where he designed the low-noise receivers for the Very Large Array radio telescope. Subsequently, at Hughes Aircraft Company and TRW, he developed low-noise microwave and millimeter-wave systems and components—primarily FET amplifiers and diode and FET mixers—for space communication. He has also been employed as a research scientist at the Aerospace Corporation, where he worked on the optimization of nonlinear microwave circuits and the development of circuit-design software based on harmonic-balance, Volterra-series, and time-domain methods. He joined the UCLA Electrical Engineering Faculty in 1990 and left it in 1992. Since then, he has worked as an independent consultant. He is also the Chief Scientist of AWR Corporation.

Dr. Maas is the author of several other books, *Microwave Mixers* (1986 and 1992), *Nonlinear Microwave Circuits* (1988 and 2003), *The RF and Microwave Circuit Design Cookbook* (1998), and *Noise in Linear and Nonlinear Circuits* (2005), all published by Artech House. From 1990 until 1992, he was the editor of the *IEEE Transactions on Microwave Theory and Techniques*, and from 1990 to 1993 he was a member of the MTT Administrative Committee and publications chairman of the IEEE MTT Society. He received the MTT Society's Microwave Prize in 1989 for his work on distortion in diode mixers and its Application Award in 2002 for his invention of the FET resistive mixer. He is a fellow of the IEEE.

Index

Recent Titles in the Artech House Microwave Library

Chipless RFID Reader Architecture, Nemai Chandra Karmakar, Prasanna Kalansuriya, Randika Koswatta, and Rubayet E-Azim

Design of Linear RF Outphasing Power Amplifiers, Xuejun Zhang, Lawrence E. Larson, and Peter M. Asbeck

Design Methodology for RF CMOS Phase Locked Loops, Carlos Quemada, Guillermo Bistué, and Iñigo Adin

Design of CMOS Operational Amplifiers, Rasoul Dehghani

Design of RF and Microwave Amplifiers and Oscillators, Second Edition, Pieter L. D. Abrie

Digital Filter Design Solutions, Jolyon M. De Freitas

Discrete Oscillator Design Linear, Nonlinear, Transient, and Noise Domains, Randall W. Rhea

Distortion in RF Power Amplifiers, Joel Vuolevi and Timo Rahkonen

EMPLAN: Electromagnetic Analysis of Printed Structures in Planarly Layered Media, Software and User's Manual, Noyan Kinayman and M. I. Aksun

An Engineer's Guide to Automated Testing of High-Speed Interfaces, José Moreira and Hubert Werkmann

Essentials of RF and Microwave Grounding, Eric Holzman

FAST: Fast Amplifier Synthesis Tool—Software and User's Guide, Dale D. Henkes

Feedforward Linear Power Amplifiers, Nick Pothecary

Filter Synthesis Using Genesys S/Filter, Randall W. Rhea

Foundations of Oscillator Circuit Design, Guillermo Gonzalez

Frequency Synthesizers: Concept to Product, Alexander Chenakin

Fundamentals of Nonlinear Behavioral Modeling for RF and Microwave Design, John Wood and David E. Root, editors

Generalized Filter Design by Computer Optimization, Djuradj Budimir

Handbook of Dielectric and Thermal Properties of Materials at Microwave Frequencies, Vyacheslav V. Komarov

Handbook of RF, Microwave, and Millimeter-Wave Components, Leonid A. Belov, Sergey M. Smolskiy, and Victor N. Kochemasov

High-Linearity RF Amplifier Design, Peter B. Kenington

High-Speed Circuit Board Signal Integrity, Stephen C. Thierauf

Integrated Microwave Front-Ends with Avionics Applications, Leo G. Maloratsky

Intermodulation Distortion in Microwave and Wireless Circuits, José Carlos Pedro and Nuno Borges Carvalho

Introduction to Modeling HBTs, Matthias Rudolph

Introduction to RF Design Using EM Simulators, Hiroaki Kogure, Yoshie Kogure, and James C. Rautio

Klystrons, Traveling Wave Tubes, Magnetrons, Crossed-Field Amplifiers, and Gyrotrons, A. S. Gilmour, Jr.

Lumped Elements for RF and Microwave Circuits, Inder Bahl

Lumped Element Quadrature Hybrids, David Andrews

Microstrip Lines and Slotlines, Third Edition, Ramesh Garg, Inder Bahl, and Maurizio Bozzi

Microwave Circuit Modeling Using Electromagnetic Field Simulation, Daniel G. Swanson, Jr. and Wolfgang J. R. Hoefer

Microwave Component Mechanics, Harri Eskelinen and Pekka Eskelinen

Microwave Differential Circuit Design Using Mixed-Mode S-Parameters, William R. Eisenstadt, Robert Stengel, and Bruce M. Thompson

Microwave Engineers' Handbook, Two Volumes, Theodore Saad, editor

Microwave Filters, Impedance-Matching Networks, and Coupling Structures, George L. Matthaei, Leo Young, and E. M. T. Jones

Microwave Materials and Fabrication Techniques, Second Edition, Thomas S. Laverghetta

Microwave Materials for Wireless Applications, David B. Cruickshank

Microwave Mixer Technology and Applications, Bert Henderson and Edmar Camargo

Microwave Mixers, Second Edition, Stephen A. Maas

Microwave Network Design Using the Scattering Matrix, Janusz A. Dobrowolski

Microwave Radio Transmission Design Guide, Second Edition, Trevor Manning

Microwave Transmission Line Circuits, William T. Joines, W. Devereux Palmer, and Jennifer T. Bernhard

Microwaves and Wireless Simplified, Third Edition, Thomas S. Laverghetta

Modern Microwave Circuits, Noyan Kinayman and M. I. Aksun

Modern Microwave Measurements and Techniques, Second Edition, Thomas S. Laverghetta

Neural Networks for RF and Microwave Design, Q. J. Zhang and K. C. Gupta

Noise in Linear and Nonlinear Circuits, Stephen A. Maas

Nonlinear Microwave and RF Circuits, Second Edition, Stephen A. Maas

Q Factor Measurements Using MATLAB®, Darko Kajfez

QMATCH: Lumped-Element Impedance Matching, Software and User's Guide, Pieter L. D. Abrie

Passive RF Component Technology: Materials, Techniques, and Applications, Guoan Wang and Bo Pan, editors

Practical Analog and Digital Filter Design, Les Thede

Practical Microstrip Design and Applications, Günter Kompa

Practical Microwave Circuits, Stephen Maas

Practical RF Circuit Design for Modern Wireless Systems, Volume I: Passive Circuits and Systems, Les Besser and Rowan Gilmore

Practical RF Circuit Design for Modern Wireless Systems, Volume II: Active Circuits and Systems, Rowan Gilmore and Les Besser

Production Testing of RF and System-on-a-Chip Devices for Wireless Communications, Keith B. Schaub and Joe Kelly

Radio Frequency Integrated Circuit Design, Second Edition, John W. M. Rogers and Calvin Plett

RF Bulk Acoustic Wave Filters for Communications, Ken-ya Hashimoto

RF Design Guide: Systems, Circuits, and Equations, Peter Vizmuller

RF Linear Accelerators for Medical and Industrial Applications, Samy Hanna

RF Measurements of Die and Packages, Scott A. Wartenberg

The RF and Microwave Circuit Design Handbook, Stephen A. Maas

RF and Microwave Coupled-Line Circuits, Rajesh Mongia, Inder Bahl, and Prakash Bhartia

RF and Microwave Oscillator Design, Michal Odyniec, editor

RF Power Amplifiers for Wireless Communications, Second Edition, Steve C. Cripps

RF Systems, Components, and Circuits Handbook, Ferril A. Losee

The Six-Port Technique with Microwave and Wireless Applications, Fadhel M. Ghannouchi and Abbas Mohammadi

Solid-State Microwave High-Power Amplifiers, Franco Sechi and Marina Bujatti

Stability Analysis of Nonlinear Microwave Circuits, Almudena Suárez and Raymond Quéré

Substrate Noise Coupling in Analog/RF Circuits, Stephane Bronckers, Geert Van der Plas, Gerd Vandersteen, and Yves Rolain

System-in-Package RF Design and Applications, Michael P. Gaynor

TRAVIS 2.0: Transmission Line Visualization Software and User's Guide, Version 2.0, Robert G. Kaires and Barton T. Hickman

Understanding Microwave Heating Cavities, Tse V. Chow Ting Chan and Howard C. Reader

Understanding Quartz Crystals and Oscillators, Ramòn M. Cerda

For further information on these and other Artech House titles, including previously considered out-of-print books now available through our In-Print-Forever® (IPF®) program, contact:

Artech House Publishers
685 Canton Street
Norwood, MA 02062
Phone: 781-769-9750
Fax: 781-769-6334
e-mail: artech@artechhouse.com

Artech House Books
16 Sussex Street
London SW1V 4RW UK
Phone: +44 (0)20 7596 8750
Fax: +44 (0)20 7630 0166
e-mail: artech-uk@artechhouse.com

Find us on the World Wide Web at: www.artechhouse.com